Protocols in Molecular Neurobiology

Methods in Molecular Biology

John M. Walker, SERIES EDITOR

1. **Proteins,** edited by *John M. Walker, 1984*
2. **Nucleic Acids,** edited by *John M. Walker, 1984*
3. **New Protein Techniques,** edited by
   John M. Walker, 1988
4. **New Nucleic Acid Techniques,** edited by
   John M. Walker, 1988
5. **Animal Cell Culture,** edited by *Jeffrey W. Pollard
   and John M. Walker, 1990*
6. **Plant Cell and Tissue Culture,** edited by
   Jeffrey W. Pollard and John M. Walker, 1990
7. **Gene Transfer and Expression Protocols,**
   edited by *E. J. Murray, 1991*
8. **Practical Molecular Virology,** edited by
   Mary K. L. Collins, 1991
9. **Protocols in Human Molecular Genetics,**
   edited by *Christopher G. Mathew, 1991*
10. **Immunochemical Protocols ,** edited by
   Margaret M. Manson, 1992
11. **Practical Protein Chromatography,** edited by
   Andrew Kenney and Susan Fowell, 1992
12. **Pulsed-Field Gel Electrophoresis:**
   Protocols, Methods, and Theories, edited by
   Margit Burmeister and Levy Ulanovsky, 1992
13. **Protocols in Molecular Neurobiology,** edited by
   Alan Longstaff and Patricia Revest, 1992

Methods in Molecular Biology • 13

Protocols in Molecular Neurobiology

Edited by

Alan Longstaff
University of Hertfordshire, Hatfield, UK

Patricia Revest
King's College, University of London, UK

Humana Press ✳ Totowa, New Jersey

Library of Congress Cataloging in Publication Data
Main entry under title:

Methods in molecular biology.

Protocols in molecular neurobiology / edited by Alan Longstaff ,
   Patricia Revest
      p. cm. — (Methods in molecular biology ; 13)
      Includes index.
      ISBN 0-89603-199-3
      1. Molecular neurobiology—Methodology. I. Longstaff, Alan.
   II. Revest, Patricia. III. Series: Methods in molecular biology
   (Totowa, N.J.) ; 13.
   QP356.2.P76 1992
   599'.0188—dc20   92-30701
                                                         CIP

Preface

Neurobiologists are bound to differ in their perceptions of what the discipline of molecular neurobiology should encompass. We have taken the view that molecular neurobiology should cover any aspect of brain science that uses the techniques of modern molecular biology, though we accept the fact that classification of a technique as a "biochemical" or "molecular biological" one is in itself somewhat arbitrary.

Each chapter of this volume sets out to identify a clear problem in neurobiology and to place it in its context within the literature—i.e., indicating how the solution of the problem will advance knowledge in the field. The core of the chapter then details the approaches taken to solve the problem, in sufficient detail that the reader can appreciate exactly why a specific strategy was adopted and how it was executed. Each chapter also includes detailed protocols providing all the information necessary to reproduce the technique and its results in any appropriately equipped laboratory.

Moving in the general direction of the central dogma, the volume starts with chapters on manipulation of DNA. It seems likely that many genomic regions of special interest to neuroscientists will turn out to be quite large, as, for example, the dystrophin gene. Large DNA molecules pose particular problems for the experimenter, and Shaw in the first chapter of this volume gives an account of how DNA fragments greater than 30 kilobases, which cannot be successfully resolved using conventional agarose gel electrophoresis, can be separated by using pulsed-field gel electrophoresis. With its precisely timed inversions of the electric field direction, the technique separates smaller from larger molecules, larger ones take longer to reorient within the electric field vector than do smaller molecules, and thus migrate more slowly.

A standard and widely used strategy of molecular biologists interested in isolating and sequencing DNA coding for molecules of interest is to isolate total mRNA from the appropriate tissue and from this to generate complementary DNA (cDNA). Vreugdenhil and Darlison point out that this approach may cause difficulties for mRNAs of low abundance (frequently the case for molecules of interest to neurobiologists, such as receptors) and that the alternative use of genomic libraries is particularly applicable for invertebrates with—when compared to vertebrates—relatively small genomes. Although preparation of vertebrate DNA is simple and widely described, isolation of genomic DNA from invertebrates can be problematic. The continuing importance of invertebrate neurobiology guarantees the wide utility of this procedure.

Often the amount of DNA available for study is limited. This is the case for mitochondrial DNA (mtDNA), mutations of which are associated with various neurological disorders, including Parkinson's disease. The characterization of small amounts of DNA is made possible by iterative procedures that successively increase the amount of specified DNA fragments—the polymerase chain reaction (PCR). PCR allows millions of copies of a specific DNA fragment to be synthesized in vitro from, theoretically, only a single copy of the original sequence. Two oligonucleotide primers complementary to regions flanking the DNA fragment of interest are chosen, one for each strand of the double-stranded DNA. Repeated cycles of denaturation of the double-stranded DNA, annealing of the primers to their complementary DNA, and replication of the complementary strand by DNA polymerase allow rapid production of copies of the region between the primers. Since the DNA produced in each cycle acts as the template for subsequent cycles, the growth in the number of copies is exponential. The uses of different varieties of PCR, in relation to mtDNA, is described in detail by Tanaka and Ozawa.

A key approach to isolating a specific fragment of DNA from heterogeneous fragments distributed among a large number of recombinant DNA clones requires the synthesis of oligonucleotide probes, which are predicted to be complementary to, and thus will hybridize with, only the DNA of interest. Two chapters, one from Bateson and Darlison and another from Webb and Bateson, deal with the use of these probes. The first chapter provides crucial theoretical insights into the design of suitable oligonucleotides—including PCR primers—using many examples from neurobiology as illustration. The second chapter gives a detailed example, describing precisely how the design of degenerate oligonucleotide primers allows the isolation of G-protein-coupled receptors using PCR amplification.

There are a large number of proteins present in cells that are thought to be involved in specialized functions and a first step in their identification is the isolation of tissue-specific mRNA. The use of labeled cDNA to probe a retinal cDNA library is described by Kuo, who provides a detailed account of differential colony hybridization in the analysis of photoreceptor-specific molecules implicated in visual transduction. The generation of the retinal-specific probes is accomplished by the hybridization of retinal and brain cDNA with the subsequent removal of both the hybridized double-stranded DNA and the brain-specific cDNA, leaving only the retinal-specific cDNA.

There are now many documented instances of molecules that occur in a variety of isoforms arising either by differential mutation of copies of an ancestral gene or by alternative splicing of heterologous RNA transcripts. These isoforms, though displaying high homology with each other, often exhibit significant differences in cellular compartmentation, tissue by localization, developmental expression, and pharmacological properties, all of which strongly implies functional differences. Clearly, antibodies that are specific for particular isoforms could prove invaluable in investigating the roles of these isoforms. How can specific antibodies be generated if the proteins are highly homologous or when sufficiently pure preparations of the protein are unavailable? This is the question tackled by Shyjan and Levenson in their study of isoforms of Na/K-ATPase in brain. The strategy they report is to ligate cDNA fragments derived from the distinct isoforms to an *E. coli* reporter gene using a plasmid vector. The *E. coli* clones generate fusion proteins that are then used as antigens to produce polyclonal antisera. The fusion proteins that are monospecific for a particular isoform can then be purified by immunoabsorption.

The expression of brain-specific proteins is regulated at least in part by modulation of the mechanisms responsible for initiating transcription of RNA by RNA polymerase II. The study of these processes is obviously advanced by a knowledge of initiation sites, which can be deduced by use of either the S_1 nuclease digestion assay or primer extension. This is discussed by Weisinger, DeCristofaro, and La Gamma. The essence of the S_1 assay is that a labeled oligonucleotide probe is constructed that will hybridize to the putative 5' end of the mRNA under investigation. This 5' end is now protected—being double-stranded—from digestion by S_1 nuclease, which removes DNA probe material upstream of the RNA start site and downstream RNA not recognized by the probe. The labeled fragments may now be fractionated by gel electrophoresis and visualized by autoradiography. Primer extension is useful if a labeled deoxyribonucleotide probe can be constructed that is complementary to a sequence of RNA close to the start site. The probe is hybridized to the RNA and reverse transcriptase catalyzes the extension of the probe up to the start of the transcript.

Modulation of transcription by nuclear transcription factors that bind to DNA is of considerable physiological importance. Weisinger and La Gamma provide detailed protocols for making nuclear extracts and for methylation interference footprinting, which enables the identification of the DNA binding sites that recognize and bind transcription factors. In the technique of methylation interference assays, the end-labeled DNA fragment of interest is randomly methylated and then incubated with the putative DNA binding factor(s). The regulatory proteins can only bind to nonmethylated DNA and both free and bound fractions can be separated by gel-retardation assays. The different fractions can then be analyzed in sequencing gels. The DNA is cleaved at the methylated residues by piperidine and the different sized fragments separated on a gel. The band corresponding to the region of protein binding will be missing in the bound fraction because of the interference of protein binding by the methylation of the binding site. This can be seen as a gap or footprint in the gel that corresponds to a band in the free fraction gel from which the sequence of the protein binding region can be deduced.

Low provides a detailed account of how transgenic mice can be produced by microinjection of cloned DNA into fertilized oocytes. The motivation for this procedure is the study of how neuropeptide genes are regulated. The central idea is the construction of reporter fusion genes, made from a reporter gene—the expression of which can be readily distinguished—linked to the upstream regulatory elements of interest. It is these fusion genes that are used to generate the transgenic animals. Thus, the regulation of the neuropeptide gene can be monitored in the transgenic animals by the concomitant expression of the reporter gene.

The key technique of *in situ* hybridization histochemistry, directed at particular mRNAs, is discussed in two chapters. Pasinetti describes the use of [^{35}S]-labeled antisense complementary RNA to localize RNAs in either paraffin or frozen brain sections by hybridization and autoradiography. These protocols have been devised with a view to executing combined *in situ* hybridization and immunocytochemistry, an approach that may become increasingly important in determining the efficiency with which specific mRNA molecules are translated; a low transcript abundance does not necessarily imply low levels of expression of the coded protein. Subsequently, Kiyama, Emson, and Tohyama consider *in situ* hybridization using probes coupled to a reporter molecule—specifically alkaline phosphatase—rather than a radiolabel. This strategy is speedier than autoradiography and can be particularly useful when hunting for mRNA molecules present only in low concentrations.

Once a particular DNA molecule has been cloned, it is necessary to confirm whether it codes for the specific protein of interest. The transcription of a cloned DNA molecule and the subsequent translation of the RNA transcript in a suitable expression vector followed by its identification by an appropriate assay, pharmacological response, or electrophysiological signature is the ultimate test of what is coded by the DNA. Alternatively, faced with mRNA extracted from brain, the RNA can be fractionated by size and each fraction examined using the expression vector to determine which fraction contains the transcript of interest. An iterative procedure using cycles of RNA fractionation may permit identification of the required mRNA. Indeed, this was exactly the method used in the recent cloning of the NMDA glutamate receptor subtype. The most widely used expression vector is the *Xenopus* oocyte, which will translate many (but not all) mRNA molecules injected into it. The chapter by Dascal and Lotan provides protocols for several total RNA extraction methods, affinity chromatography to purify mRNA and for preparation, injection, and care of the oocytes. Additionally, these authors show how the use of antisense oligodeoxynucleotides directed at a specific mRNA, in this case coding for a voltage-dependent calcium channel, can selectively inhibit its expression in the oocyte.

At this point the emphasis of the volume changes from techniques applicable to nucleic acid research to those applicable to proteins. Indeed the next five chapters are effectively case studies illustrating how particular techniques have been used to reveal the molecular biology of proteins with importance for neuroscientists. Although the methodologies were designed for specific experimental models, all are widely applicable. Adamo and his colleagues begin the section by providing a battery of protocols they use for the study of insulin and insulin-like growth factor I receptors. There are methods for the preparation and primary culture of neurons and astrocytes, receptor binding studies, receptor solubilization, and partial purification using wheat germ agglutinin. There is also a description of receptor phosphorylation methods for both cell-free and intact-cell systems, and of the methods by which phosphorylation sites may be determined. Nakata, using a selective antagonist as an immobilizing ligand, describes the purification by affinity chromatography of an adenosine receptor. The strategy involves two affinity chromatographic separations with an intervening hydroxyapatite column chromatography to achieve a 50,000-fold purification of the receptor.

The regulation of calcium fluxes into the cytosol is of considerable importance to nervous system function and two chapters examine how molecular biology has revealed the structure of two of the channels involved. Schneider et al. label the voltage-dependent, L-type calcium channel with a dihydropyridine having a high binding affinity for the channel, so that they can track it throughout a purification procedure that uses affinity chromatography (with wheat germ agglutinin), ion exchange chromatography, and sucrose density gradients. Lai and Meissner turn their attention to the ryanodine receptor, an ion channel in the sarcoplasmic reticulum of vertebrate skeletal and heart muscle through which calcium ions flood into the cytosol on depolarization. Interestingly, ryanodine receptors are present in brain and they are homologous with the inositol trisphosphate-sensitive calcium channels in the endoplasmic reticulum. These authors provide protocols for the purification and characterization of ryanodine receptors using sucrose density gradients and sodium dodecyl sulfate polyacrylamide gels. They also describe how the molecules can be reconstituted into planar lipid bilayers for further functional studies.

The nicotinic acetylcholine receptor (nAChR) was the first receptor to be completely sequenced and as such has been the most extensively studied. One important approach in identifying the functional sites on the receptor is photoaffinity labeling, in which a radiolabeled ligand with an affinity for specific regions of a receptor or ion channel— an agonist binding site, for example—can be covalently bonded to the receptor by a brief pulse of ultraviolet light. The complex between the ligand and the protein is now stable, and standard protein chemistry techniques can be used to identify the ligand binding site. Hucho reports the use of a channel blocker to elucidate those portions of the nAChR that form the ion channel wall.

An introduction to the seminal role being played by patch clamping needs no apology since it involves measuring current flow through single ion channels and may thus be regarded as molecular electrophysiology. Patch clamping may be used as an "assay" that allows characterization of the exact nature of a molecule coded for by cloned DNA or its mRNA transcript when expressed in an appropriate vector.

It is no exaggeration to claim that the techniques of patch clamping have revolutionized our understanding of cell membrane function. Indeed, in recognition of this, Sakmann and Neher, who invented the original methods, received the Nobel prize for physiology or medicine in 1991. Standen and Stanfield provide a theoretical background that illustrates how these powerful techniques can be appllied to the analysis of channel function. Using methods applicable to mammalian cell lines, *Xenopus* oocytes that have been injected with mRNA or cell lines that can be transfected with DNA, they describe how both macroscopic (whole-cell) and microscopic (single-channel) currents can be analyzed to derive models of the kinetic behavior of channels. Scott and Dolphin explain their use of whole-cell patch clamping to study the regulation of voltage-dependent calcium channels by G-proteins. They use a variety of GTP/GDP analogs, and of particular note is their use of caged compounds. These may be esters of substrates of interest—for example, GTP or inositol trisphosphate, which can be rapidly cleaved by brief illumination, thus enabling the time course of processes involving the substrates to be studied accurately without problems of diffusion, or substrate instability. They also indicate some of the problems involved in single channel recording of calcium channels caused by channel rundown.

Although it has to be accepted that there is no better way of learning a new technique than working in a laboratory where it is used on a regular basis, we believe this volume provides a viable alternative. The format of detailed protocols, coupled with the "insider's" tips on crucial points, will enable a greater understanding of the experiments than is possible with the level of detail available from a standard research paper. We hope that this book will give some ideas both to newcomers to the diverse methods of molecular neurobiology and to those who are already familiar with some of the techniques detailed here.

Alan Longstaff
Patricia Revest

Contents

Preface .... *v*

Contributors .... *xv*

CH. 1. Separation of Large DNA Molecules by Pulsed-Field
Gel Electrophoresis,
Duncan J. Shaw *1*

CH. 2. The Isolation of Genomic DNA from Invertebrates,
Erno Vreugdenhil and Mark G. Darlison *15*

CH. 3. Analysis of Mitochondrial DNA Mutations,
Masashi Tanaka and Takayuki Ozawa *25*

CH. 4. The Design and Use of Oligonucleotides,
Alan N. Bateson and Mark G. Darlison *55*

CH. 5. The Use of Degenerate Oligonucleotides for Polymerase Chain-
Reaction-Based Isolation of Related DNA Sequences,
Tania E. Webb and Alan N. Bateson *67*

CH. 6. Isolation of Photoreceptor Cell-Specific MEKA cDNA
by Differential Hybridization,
Che-Hui Kuo *79*

CH. 7. Generation of Isoform-Specific Antisera from Cloned cDNAs:
*Application to Multiple Forms of the Na/K-ATPase Expressed
in Rat Brain,*
Andrew W. Shyjan and Robert Levenson *93*

CH. 8. Determination of Transcriptional Initiation Sites and Their
Usage in the Nervous System,
*Gary Weisinger, Joseph D. DeCristofaro,
and Edmund F. La Gamma* *115*

CH. 9. Mapping of *Trans*-Acting Factor Binding in the Nervous System,
Gary Weisinger and Edmund F. La Gamma *139*

CH. 10. *In Situ* Hybridization to Brain Tissue Sections Using Labeled
Single-Strand Complementary RNA Probes,
Giulio M. Pasinetti *155*

CH. 11. *In Situ* Hybridization Histochemistry Using Alkaline
Phosphatase-Labeled Oligodeoxynucleotide Probe,
Hiroshi Kiyama, Piers C. Emson, and Masaya Tohyama *167*

xiii

CH. 12. The Identification of Neuropeptide Gene Regulatory Elements in Transgenic Mice,
Malcolm J. Low 181

CH. 13. Expression of Exogenous Ion Channels and Neurotransmitter Receptors in RNA-Injected *Xenopus* Oocytes,
Nathan Dascal and Ilana Lotan 205

CH. 14. Analysis of Insulin and Insulin-Like Growth Factor-I Receptors in Neural Tissues,
Martin L. Adamo, Mohan K. Raizada, Joshua Shemer, Akira Ota, Colin Sumners, John Olson, and Derek LeRoith 227

CH. 15. Use of Affinity Chromatography in Purification of A_1 Adenosine Receptors from Rat Brain Membranes,
Hiroyasu Nakata 261

CH. 16. Purification and Structure of L-Type Calcium Channels,
Toni Schneider, Stefan Regulla, Wolfgang Nastainczyk, and Franz Hofmann 273

CH. 17. Purification and Reconstitution of the Ryanodine-Sensitive Ca^{2+} Release Channel Complex from Muscle Sarcoplasmic Reticulum,
F. Anthony Lai and Gerhard Meissner 287

CH. 18. Identification of a Ligand-Gated Ion Channel by Photoaffinity Labeling and Microsequencing,
Ferdinand Hucho 307

CH. 19. Voltage-Gated Ion Channels: *Electrophysiological Approaches,*
Nicholas B. Standen and Peter R. Stanfield 325

CH. 20. An Electrophysiological Approach to the Regulation of Neuronal Voltage-Activated Calcium Channels by Guanine Nucleotide Binding Proteins,
Roderick H. Scott and Annette C. Dolphin 347

Index 367

Contributors

Martin L. Adamo • *Diabetes Branch, National Institutes of Diabetes, and Digestive and Kidney Diseases, National Institutes of Health, Bethesda, MD*

Alan N. Bateson • *MRC Molecular Biology Unit, Cambridge, UK (Present address: Department of Pharmacology, University of Alberta, Edmonton, Alberta, Canada)*

Mark G. Darlison • *Institut fur Zellbiochemie und Klinische Neurobiologie, Universitäts-Krankenhaus Eppendorf, Hamburg, Germany*

Nathan Dascal • *Sackler School of Medicine, Division of Physiology and Pharmacology, Tel Aviv University, Tel Aviv, Israel*

Joseph D. DeCristofaro • *Department of Pediatrics and Neurobiology, State University of New York at Stony Brook, NY*

Annette C. Dolphin • *Department of Pharmacology, Royal Free Hospital School of Medicine, London, UK*

Piers C. Emson • *MRC Group, Institute of Animal Physiology and Genetics Research, Babraham, Cambridge, UK*

Franz Hofmann • *Institut fur Pharmakologie und Toxikologie der Technische Universitat Munchen, Munchen, Germany*

Ferdinand Hucho • *Fachbereich Chemie, Institut fur Biochemie, Freie Universitat Berlin, Berlin, Germany*

Hiroshi Kiyama • *Department of Anatomy and Neuroscience, Osaka University Medical School, Osaka, Japan*

Che-Hui Kuo • *Department of Pharmacology, Cancer Research Institute, Kanazawa University, Kanazawa, Japan*

Edmund F. La Gamma • *Department of Pediatrics and Neurobiology, State University of New York at Stony Brook, NY*

F. Anthony Lai • *Division of Physical Biochemistry, National Institute for Medical Research, London, UK*

Derek LeRoith • *Diabetes Branch, National Institutes of Diabetes, and Digestive and Kidney Diseases, National Institutes of Health, Bethesda, MD*

Robert Levenson • *Department of Cell Biology, Yale University School of Medicine, New Haven, CT*

MALCOLM J. LOW • *Vollum Institute for Advanced Biomedical Research, Oregon Health Sciences University, Portland, OR*

ILANA LOTAN • *Sackler School of Medicine, Division of Physiology and Pharmacology, Tel Aviv University, Tel Aviv, Israel*

GERHARD MEISSNER • *Department of Biochemistry, University of North Carolina, Chapel Hill, NC*

HIROYASU NAKATA • *Department of Neurochemistry, Tokyo Metropolitan Institute for Neuroscience, Tokyo, Japan*

WOLFGANG NASTAINCZYK • *Medizinische Biochemie, Medizinische Fakultat der Universitat des Saarlandes, Homber/Saar, Germany*

JOHN OLSON • *Department of Physiology, University of Florida, College of Medicine, Gainesville, FL*

AKIRA OTA • *Diabetes Branch, National Institutes of Diabetes, and Digestive and Kidney Diseases, National Institutes of Health, Bethesda, MD*

TAKAYUKI OZAWA • *Department of Biomedical Chemistry, Faculty of Medicine, University of Nagoya, Nagoya, Japan*

GIULIO M. PASINETTI • *Ethel Percy Andrus Gerontology Center, University of Southern California, Los Angeles, CA*

MOHAN K. RAIZADA • *Department of Physiology, University of Florida, College of Medicine, Gainesville, FL*

STEFAN REGULLA • *Institut fur Pharmakologie und Toxikologie der Technischen Universitat Munchen, Munchen, Germany*

TONI SCHNEIDER • *Institut fur Pharmakologie und Toxikologie der Technischen Universitat Munchen, Munchen, Germany*

RODERICK H. SCOTT • *Department of Physiology, St. George's Hospital Medical School, London, UK*

DUNCAN J. SHAW • *Department of Medical Genetics, University of Wales College of Medicine, Cardiff, UK*

JOSHUA SHEMER • *Diabetes Branch, National Institutes of Diabetes, and Digestive and Kidney Diseases, National Institutes of Health, Bethesda, MD*

ANDREW W. SHYJAN • *Department of Cell Biology, Yale University School of Medicine, New Haven, CT*

NICHOLAS B. STANDEN • *Department of Physiology, University of Leicester, UK*

PETER R. STANFIELD • *Department of Physiology, University of Leicester, UK*

COLIN SUMNERS • *Department of Physiology, University of Florida, College of Medicine, Gainesville, FL*

MASASHI TANAKA • *Department of Biomedical Chemistry, Faculty of Medicine, University of Nagoya, Japan*

MASAYA TOHYAMA • *Department of Anatomy and Neuroscience, Osaka University Medical School, Osaka, Japan*
ERNO VREUGDENHIL • *Biochemisch Laboratorium, Vrije Universiteit, Amsterdam, The Netherlands*
TANIA E. WEBB • *MRC Molecular Neurobiology Unit, Cambridge, UK*
GARY WEISINGER • *Department of Pediatrics and Neurobiology, State University of New York at Stony Brook, NY*

CHAPTER 1

Separation of Large DNA Molecules by Pulsed-Field Gel Electrophoresis

Duncan J. Shaw

1. Introduction

Conventional agarose gel electrophoresis is a widely used technique for the analysis of many kinds of biological molecules, including fragments of DNA. It has one major limitation, namely its inability to resolve DNA fragments of greater than approx 30 kb in length. In human genetics, the enormous length of the genome (3 million kb) makes it necessary to have techniques capable of analyzing much larger DNA molecules, and it was in response to this that the methods of pulsed field-gel electrophoresis (PFGE) were developed. The procedure was first described by Schwartz and Cantor *(1)*, and since then numerous variations and modifications have been published *(2–4)*. These methods all have in common the use of two alternately switched (pulsed) electric fields, arranged at an angle of between 90 and 180 degrees. Conventional gel electrophoresis uses a single, continuous electric field. In PFGE, the molecules are forced to change direction each time the field is switched, and the time taken for a large DNA molecule to reorient itself in response to the change in field is a direct function of its size. Hence, at each pulse, longer molecules become retarded relative to shorter ones because of their longer reorientation time, and over the course of the electrophoretic run, a separation is achieved. The theory of the method was described in detail by Southern et al. *(5)*.

From: *Methods in Molecular Biology, Vol. 13: Protocols in Molecular Neurobiology*
Edited by: A. Longstaff and P. Revest Copyright © 1992 The Humana Press, Totowa, NJ

1

1.1. Applications

There does not appear to be an upper limit to the size of molecule that can be separated by PFGE. The initial descriptions of the method showed separations of the chromosomes of the yeast *Saccharomyces cerevisiae*, which range in size from 250–2500 kb. Subsequently, other organisms have since been analyzed, for example, the yeasts *Candida albicans (6)* and *Schizosaccharomyces pombe*, whose largest chromosome is at least 5 million bases (5 Mb) *(7)*. Antigenic variation in the parasite *Trypanosoma brucei* was shown to be associated with chromosomes of altered length *(8)*. As well as providing electrophoretic karyotypes of lower organisms, PFGE has been widely used in mapping large parts of mammalian (including human) chromosomes. It is the latter application that is the major emphasis of this chapter. PFGE has provided a means of bridging the gap that has existed between the molecular biological techniques, such as conventional electrophoresis and DNA cloning, and traditional human genetic methods, such as linkage analysis and cytogenetics.

The methods described here relate to the use of PFGE to construct maps of regions of human chromosomes, which one would want for a specific reason, such as the presence of a disease-associated gene. For an example of a published PFGE map of such a region, the reader is referred to the papers describing the isolation of the cystic fibrosis gene *(9)* and the mapping of the chromosomal region containing the Huntington's disease gene *(10)*. This form of restriction mapping makes use of enzymes that cut mammalian DNA infrequently, because their recognition sequences are either very long (8 bases as opposed to the usual 4 or 6) and/or contain one or more occurrences of the DNA sequence CG, which is not only underrepresented in mammalian DNA, but is also often methylated, which makes it resistant to restriction enzyme cleavage. A list of such enzymes is shown in Table 1, and a survey of the distribution in human DNA of sites for a range of restriction enzymes was published in reference *11*.

Following electrophoresis, the gel is stained to visualize the DNA (Fig. 1), and can then be treated by normal Southern blotting and hybridization protocols. In the case of simple organisms, such as yeast, the chromosomes are visible as discrete bands after staining the gel, and it is possible to carry out experiments, such as the localization of a new gene probe to a specific chromosome, by blotting the gel and hybridizing the blot with the probe. For the analysis of human-size

Table 1
Restriction Enzymes Suitable for PFGE Mapping

Enzyme	Recognition sequence
*BssH*II	G'CGCGC
*Cla*I	AT'CGAT
*Eag*I	C'GGCCG
*Mlu*I	A'CGCGT
*Nae*I	GCC'GGC
*Nar*I	GG'CGCC
*Not*I	GC'GGCCGC
*Nru*I	TCG'CGA
*Pac*I	TTAAT'TAA
*Pvu*I	CGAT'CG
*Rsr*II	CG'G (A/T) CCG
*Sac*II	CCGC'GG
*Sal*I	G'TCGAC
*Sfi*I	GGCCNNNN'NGGCC
*Sma*I	CCC'GGG
*Spl*I	C'GTACG
*Xho*I	C'TCGAG
*Xma*I	C'CCGGG

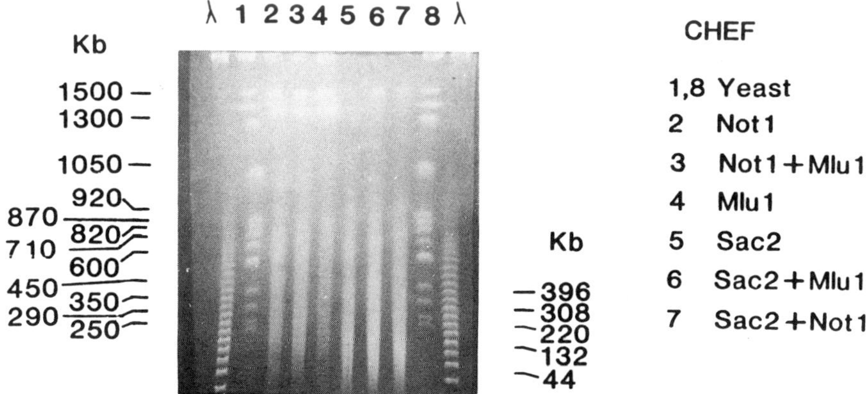

Fig. 1. Ethidium bromide-stained PFGE gel, run using a Bio-Rad Chef apparatus. Tracks 1 and 8, *Saccharomyces cerevisiae* chromosomes; tracks λ, oligomers of phage λ; tracks 2–7, human DNA digested with various rare-cutter restriction enzymes.

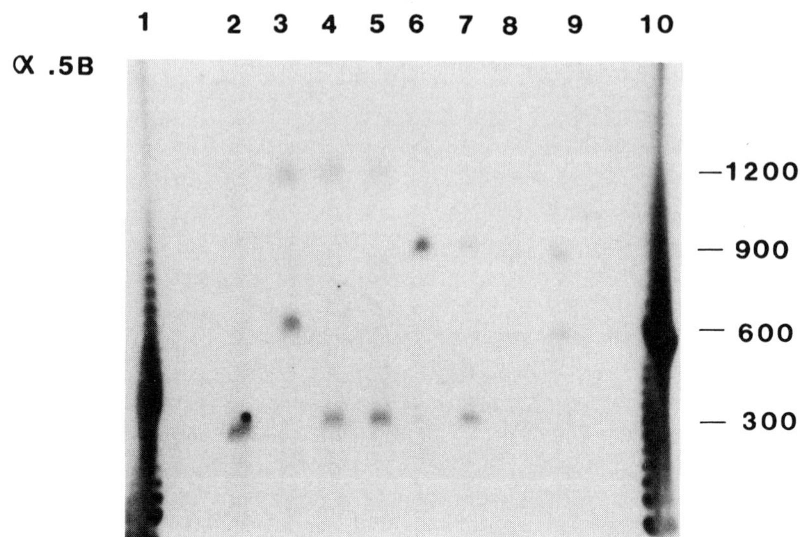

Fig. 2. Southern blot of a PFGE gel hybridized with a DNA probe derived from human chromosome 19 (probe name α.5B). Tracks 1 and 10, phage λ oligomers; tracks 2–9, human DNA digested with *Mlu*I; scale on right, sizes of DNA fragments in kilobases.

genomes, blotting and hybridization allow gene probes and other DNA sequences to be assigned to specific large DNA restriction fragments, which may or may not be part of an existing map (Fig. 2).

2. Materials

2.1. Apparatus

The prospective user of PFGE has a choice between commercially available apparatus or construction "in house." The author has experience with both and is currently using Bio-Rad (Richmond, CA) "Chef" systems, which offer excellent performance. Similar equipment, such as the Pharmacia-LKB (Uppsala, Sweden) "Pulsaphor" system, has been recommended by colleagues.

The Bio-Rad system is based on the PFGE variant known as CHEF (contour-clamped homogeneous electric field) electrophoresis (4), and it is possible to construct a homemade system from the description in the paper (4). This consists of a shallow hexagonal perspex tank, 36 cm in width, with 24 evenly spaced platinum electrodes around the

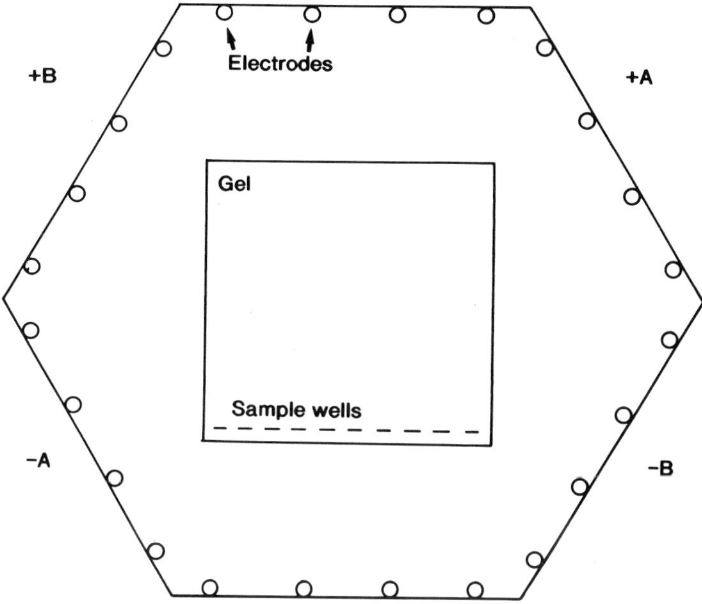

Fig. 3. Chef-type apparatus.

periphery. Each electrode is connected to its neighbors by resistors, and the active electrodes are fed with current via diodes and a switching box (*see* Fig. 3). A requirement common to all forms of PFGE is a device for switching the electric current between the two pairs of electrodes; this can be constructed in a workshop, based either on an electronic timer or alternatively, by using relays controlled from the output port of a personal computer. The latter allows sophisticated switching regimes to be programmed. Commercial units are also available. Normal laboratory electrophoresis power supplies are generally suitable, since PFGE does not demand very high currents or voltages.

Because PFGE is usually carried out under controlled temperature conditions, some form of recirculating thermostatic waterbath will be needed. Suitable units can be obtained from a number of laboratory equipment suppliers.

The other specialized equipment that will be needed is a mold in which to cast the agarose blocks containing the DNA samples. In the author's laboratory a homemade system is used, consisting of strips of perspex 160 × 8 × 6 mm, arranged in parallel, separated at their ends by small perspex spacers (20 × 8 × 2 mm) and secured by

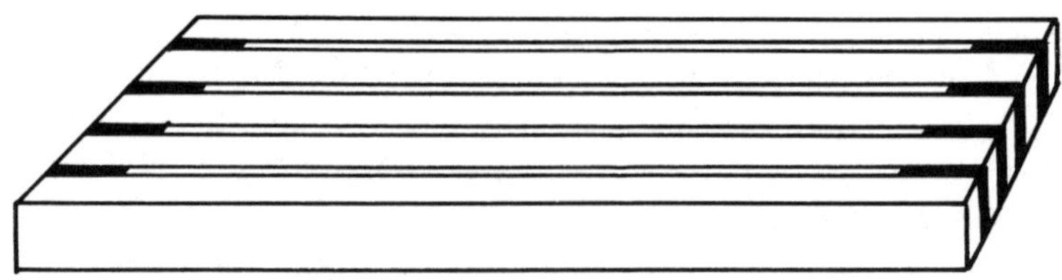

Fig. 4. Mold for casting agarose sample blocks.

adhesive tape so as to form channels 120 mm long, 8 mm deep, and 2 mm wide (Fig. 4).

2.2. Solutions

Unless stated otherwise, reagents are stored at room temperature.

1. YPD broth (per 100 mL): 2 g peptone, 1 g yeast extract, and 2 g D-glucose. Autoclave.
2. SE buffer: 75 mM NaCl, and 25 mM EDTA, pH 8. Autoclave.
3. Lyticase (yeast cell wall removing enzyme; Sigma Chemicals, St. Louis, MO): 300 U/mL in SE. Store at –20°C.
4. 0.5M DTT solution: 0.077 g of dithiothreitol in 1 mL water. Store at –20°C.
5. Proteinase K: Dissolve in SE at 40 mg/mL and store in 1-mL aliquots at –20°C.
6. Lysis buffer for mammalian cells: 0.15M NH$_4$Cl, 0.01M KHCO$_3$, 0.1 mM EDTA. Per 500 mL: 4 g NH$_4$Cl, 0.5 g KHCO$_3$, and 0.1 mL 0.5M EDTA, pH 8.
7. 0.5M EDTA solution for mammalian or yeast cells (per 500 mL): 95 g ethylene diamine tetra-acetic acid (anhydrous); add NaOH pellets until pH is 9.5. Filter sterilize or autoclave.
8. 10% Sarcosyl solution (per 100 mL): 10 g sodium lauryl sarcosinate, water to 100 mL. Filter sterilize.
9. Phosphate-buffered saline (PBS): Obtain from any supplier of tissue culture products.
10. Phenyl methyl sulfonyl fluoride (PMSF): Dissolve in 100% ethanol at 10 mg/mL and discard after use. (**Warning:** This compound is very toxic, *see* Note 1.)
11. Low-melting-point agarose (recommended supplier Gibco-BRL Ltd, Gaithersburg, MD): Dissolve in SE or PBS at 1 g/100 mL, by heating in a microwave or waterbath. Before use, remelt at 70°C and cool to 37°C.
12. TE buffer: 10 mM Tris-HCl, 1 mM EDTA, pH 8.0. Autoclave.

13. Trypsin/EDTA solution: Obtain from any tissue culture supplier.
14. 0.5X TBE buffer. 45 mM Tris base, 45 mM boric acid, and 1 mM EDTA: 2 L contain 10.8 g Tris, 5.5 g boric acid, and 0.74 g EDTA.
15. Gel denaturing solution: 0.5M NaOH, 1.5M NaCl; 500 mL contain 10 g NaOH and 44 g NaCl.
16. Gel-neutralizing and blotting solution is 1M ammonium acetate, 0.02M NaOH; 1500 mL contain 115 g ammonium acetate and 1.2 g NaOH.
17. Ethidium bromide solution: Dissolve ethidium bromide at 10 mg/mL in water. Store in a darkened bottle. Always wear gloves when handling this solution, since it is carcinogenic (*see* Note 1).
18. 2X SSC: 0.3M sodium chloride, 0.03M sodium citrate.

3. Methods

3.1. Sample Preparation

Conventional methods for preparation of DNA in solution result in a degree of shearing of the molecules such that the average fragment length is approx 100–200 kb. This is obviously unsatisfactory for analysis by PFGE, so methods have been devised for the preparation of intact, unsheared DNA. These involve encapsulation of the cells, prior to lysis, in an agarose matrix that is permeable to small molecules, but traps the DNA. In this section, simple procedures for the isolation of yeast chromosomes suitable for use as size markers and DNA from human cells that can be digested with restriction enzymes are described.

3.1.1. Yeast Chromosomes

1. Grow a 5-mL overnight culture of *Saccharomyces cerevisiae* in YPD broth at 30°C or 37°C. Use this to inoculate a 100-mL culture (YPD broth in a 1-L flask), and grow overnight with shaking.
2. Harvest cells by centrifugation (1000g, 10 min, room temperature). Wash by gentle resuspension in 20 mL of SE buffer, and harvest the cells again.
3. Resuspend the cells in 3 mL of SE buffer, and pour into a Universal bottle (30-mL plastic screw-top bottle) at 37°C. Add 6.5 mL of 1% low-melting-point agarose in SE buffer (melted at 70°C and cooled to 37°C), 0.25 mL of lyticase enzyme solution, and 0.5 mL of DTT solution. Mix well. Working reasonably quickly so that the agarose remains liquid, pipet the mixture into the perspex mold. This should have been assembled to provide five channels, with a total vol of approx 10 mL and one side sealed with adhesive tape.

4. When the agarose is set (this can be hastened by placing the mold on ice), the strips of agarose containing the yeast cells are cut into blocks and removed from the mould. This is done by removing the end perspex strip, thus exposing the strip of agarose, and cutting it into 20 equal portions with a scalpel. The little agarose blocks are then gently pushed into a Universal containing lysis solution (10 mL of SE buffer, 0.25 mL lyticase, and 0.5 mL DTT). This procedure is repeated for each strip of agarose, so that eventually 100 blocks have been made and placed in the bottle of lysis solution. Incubate at 37°C for at least 2 h or overnight.

5. Pour off the lysis solution, taking care not to lose any blocks, and replace with proteinase K solution (10 mL of 0.5M EDTA, pH 9.5, 1 mL of 40 mg/mL proteinase K and 2 mL of 10% sarcosyl). Incubate at 50°C for 48 h. Do not allow the temperature to go any higher, because there is a danger of the blocks melting.

6. The blocks may now be stored indefinitely at 4°C. Before use, wash the blocks three times for 1 h at 37°C or 50°C, in 10 vol of 50 mM EDTA, pH 8.0.

3.1.2. Preparation of DNA from Human Cells

The following procedure is used routinely with fresh whole blood or cultured mammalian cell lines. It should be possible to adapt it for use with any tissue that can be reduced to a suspension of single cells. Great care should be taken when handling human materials because of the potential for infection. All vessels used and all waste generated should be sterilized before cleaning or disposal, and gloves should be worn at all times.

1. If the material is cultured cells, then harvest by centrifugation (1000g, 5 min, room temperature). Cells that grow attached to the flask should first be released by treatment with trypsin/EDTA solution according to standard tissue culture procedures. Proceed to step 7. If the material is blood, proceed to step 2.

2. Mix 20 mL of whole blood with 60 mL of lysis buffer in a centrifuge bottle. Place on ice for 15 min (*see* Note 2).

3. Harvest the white cells by centrifugation in a swing-out rotor (1000g, 10 min, 4°C). Carefully pour off the supernatant into a bucket of bleach. The pellet of cells will be red at this stage.

4. Resuspend the pellet in 8 mL PBS by gently pipeting up and down, and transfer to a Universal bottle.

5. Add 16 mL lysis buffer, mix, and leave on ice for 5 min.

6. Harvest the white cells by centrifugation as in step 3. The pellet of cells, which will be slightly pink and may be covered by a layer of red debris, should be visible.
7. Resuspend the cells in 10 mL PBS. Take out 0.1 mL of the suspension, and dilute to 1.0 mL in PBS.
8. Count the cells using a hemocytometer (a counting chamber viewed with a microscope). Be careful not to confuse genuine white cells, which are the larger, rounded objects, with bits of debris. Calculate the cell density in your undiluted sample, according to the design of the counting chamber you are using. Next, calculate the number of blocks you will make. With mammalian cells, it is usual to use 1 million cells/block. Therefore, if your 10-mL sample has a density of 10^7 cells/mL, 100 blocks can be made.
9. Spin down the cells again as above. For 100 blocks, resuspend the cells in 3.5 mL PBS, warm to 37°C, add 6.5 mL of 1% low-melting-point agarose in PBS at 37°C, and mix. For amounts other than 100 blocks, scale the volumes up or down accordingly.
10. Working reasonably quickly so that the agarose does not set, pipet the agarose/cells mixture into the mold. When using the apparatus described above, this should set up to provide the correct capacity for the number of blocks to be made. With this design, one strip has a vol of 2 mL, corresponding to 20 blocks. Allow the agarose to set by placing it on ice for 15 min.
11. As described in Section 3.1.1., step 4, cut up the agarose strips into blocks and put into a Universal bottle containing proteinase K solution (for 100 blocks, the quantities are 10 mL 0.5M EDTA, pH 9.5, 2 mL 10% sarcosyl, and 1 mL 40 mg/mL proteinase K). Incubate at 50°C for 48 h. At this stage, the samples may be stored indefinitely at 4°C.

3.2. Restriction Enzyme Digestion

DNA samples prepared as agarose blocks are usually as susceptible to restriction enzyme digestion as DNA made in the conventional way. Any problems encountered are usually because of the reagents used, and the recommended suppliers for some products are given in Section 2.2. However, other makes of reagent may be equally satisfactory.

1. To prepare blocks for digestion, put a block into 1 mL of TE, add 4 µL PMSF solution, and incubate at 50°C for 30 min. Repeat this treatment once, and then repeat twice more, but without PMSF. Pipet off the TE.
2. Blocks are digested in a total vol of 0.2 mL (the block is itself 0.1 mL), and the enzyme manufacturer's recommendation or standard labo-

ratory protocols should be followed with regard to buffer components. A typical digest would contain 1 block, 0.08 mL distilled water, 0.02 mL 10X buffer, and 20 U of restriction enzyme. Most digests are complete after 2–4 h incubation at the recommended temperature.

3.3. Running Gels

The exact conditions under which gels are run will depend on the size range of the molecules under study and on the particular apparatus used. There are three important parameters to consider: the voltage, the switching interval, and the buffer temperature. The concentration of agarose used is usually 1%, and varying this has a less profound effect on the results.

In general, increasing either the temperature, the switching interval, or the voltage will enable larger molecules to be fractionated and will also increase the absolute mobility of all the molecules. All of these factors, together with the length of time for which the gel is run and the design of the apparatus, are interdependent. It is therefore not possible to give a generally applicable protocol (*see* Note 3). Instead, a method suitable for the Bio-Rad "Chef" apparatus will be described. This would be easily adaptable to other apparatus of similar geometry, including homemade equipment, and in any case when using a commercial design, the manufacturer's instructions should be used as a starting point for experimentation. A detailed analysis of the effects of altering the various parameters in PFGE was presented by Birren et al. *(12)*.

1. Prepare 2 L of 0.5X TBE buffer. Use a portion of this to make the gel. This is usually 1.0% agarose, but can be reduced to 0.7% for fractionation of extremely large fragments (>2000 kb). The volume of agarose required will depend on the size of the gel; for a 15-cm square gel, use 200 mL. Melt the agarose in a microwave oven and cool to 50°C before pouring the gel. Use a sample comb whose teeth match the dimensions of the agarose sample blocks.
2. Put the remaining buffer into the gel tank, and switch on the cooling device. This allows the buffer to reach working temperature before the run is started. A temperature of 15°C measured in the tank during the run is the object; this corresponds to a setting of about 5°C on the cooler thermostat, but the user will have to determine the setting for his/her own apparatus.
3. When the gel is set, remove the comb, and place the gel in the tank. Leave for 1 h or until the buffer temperature has reached 15°C.
4. Remove the gel, and load the sample blocks into the wells using a

sterilized microspatula. Try to get them in without breaking them (this takes practice). It may be best to avoid using the outside lanes on the gel, since these sometimes do not run straight. Normally, yeast chromosomes are used as markers, and a block of these is placed in each of the end lanes. For calibration down to approx 50 kb, intact bacteriophage λ DNA can also be used (approx 1–2 µg/lane). λ DNA spontaneously forms oligomers on storage, which will be found to be useful markers since their sizes are exact multiples of the monomer length (49 kb for wild-type λ). Replace the gel in the tank with the row of samples between the two negative electrodes.

5. Set the voltage and pulse time, and switch on. The exact settings depend on what size range is to be separated, but a good starting point is to use a 90-s switch time, 200 V, and run the gel for 24 h. This separates DNA molecules in the range 50–1100 kb. The Bio-Rad apparatus and some other designs allow the switching time to be continuously increased during the course of a run. This is known as a "ramp," and gives enhanced separation of a wider range of molecules. A ramp starting and finishing with 50- and 250-s switching times, respectively, is recommended for separation in the range 50–2500 kb.

6. After the run, take out the gel, and expose it to medium-wavelength UV light on a transilluminator. This nicks the very high-mol-wt DNA and allows it to be Southern blotted efficiently. The optimum UV exposure time has to be determined for a particular transilluminator, but is likely to be in the range of 30 s to 2 min. Calibration is best achieved by running a gel with a series of identical DNA samples, exposing different parts of the gel to UV for a range of times, and blotting and hybridizing with a suitable probe to determine which conditions result in the best signal.

7. Remove the buffer from the tank, and add half of it to the gel together with 80 µL of 10 mg/mL ethidium bromide solution (carcinogenic) to stain the DNA. Incubate the gel at room temperature with gentle shaking for 30 min, then replace the staining solution with the remainder of the buffer, and incubate for a further 30 min.

8. Photograph the gel.

9. At this stage, the user may wish to employ his/her standard gel-blotting protocol. In the author's laboratory, Hybond N membranes (Amersham International PLC, Aylesbury, UK) are used, in conjunction with the following procedure (all steps at room temperature).

10. Denature the DNA in the gel by 30 min gentle shaking in 500 mL of gel denaturing solution.

11. Neutralize for 30 min in 500 mL of neutralizing solution. Repeat this step.

12. Blot the gel on to a Hybond N membrane overnight, using 500 mL of neutralizing solution.
13. Remove the blot from the gel, rinse it briefly in 2X SSC solution, allow it to dry (at room temperature or in an 80°C oven), and irradiate with medium-wavelength UV to fix the DNA to the filter. The optimum exposure time must be determined for the UV source in use. With a normal transilluminator, it is likely to be in the range 5–30 s. Calibration may be carried out in a similar manner to that described earlier in step 6.

4. Notes

1. Two of the chemicals used in these procedures are potentially harmful: phenyl-methyl-sulphonyl-fluoride (PMSF), which is highly toxic, and ethidium bromide, which is carcinogenic. Appropriate precautions should be taken when handling these compounds or their solutions, including the use of gloves and masks, and disposal should be carried out according to the local safety regulations.
2. Human blood samples may be contaminated with hepatitis or HIV viruses. It is essential to obtain guidance from the local safety committee concerning the handling of such samples and the disposal of contaminated waste.
3. The purchaser of a commercial PFGE apparatus will be able to obtain practical guidance, supplementary to this chapter, from the manufacturer.

Acknowledgments

I would like to thank the numerous colleagues who helped and advised me in the early days of PFGE.

References

1. Schwartz, D. C. and Cantor, C. R. (1984) Separation of yeast chromosome-sized DNAs by pulsed field gradient gel electrophoresis. *Cell* **37,** 67–75.
2. Carle, G. F. and Olson, M. V. (1984) Separation of chromosomal DNA molecules from yeast by orthogonal-field-alternation gel electrophoresis. *Nucleic Acids Res.* **12,** 5647–5664.
3. Carle, G. F., Frank, M., and Olson, M. V. (1986) Electrophoretic separations of large DNA molecules by periodic inversion of the electric field. *Science* **232,** 64–68.
4. Chu, G., Vollrath, D., and Davis, R. W. (1986) Separation of large DNA molecules by contour-clamped homogeneous electric fields. *Science* **234,** 1582–1585.

5. Southern, E. M., Anand, R., Brown, W. R. A., and Fletcher, D. S. (1987) A model for the separation of large DNA molecules by crossed field gel electrophoresis. *Nucleic Acids Res.* **15,** 5925–5943.
6. Snell, R. G. and Wilkins, R. J. (1986) Separation of chromosomal DNA molecules from C. albicans by pulsed field gel electrophoresis. *Nucleic Acids Res.* **14,** 4401–4406.
7. Vollrath, D. and Davis, R. W. (1987) Resolution of DNA molecules greater than 5 megabases by contour-clamped homogeneous electric fields. *Nucleic Acids Res.* **15,** 7865–7876.
8. Van der Ploeg, L. H. T., Schwartz, D. C., Cantor, C. R., and Borst, P. (1984) Antigenic variation in Trypanosoma brucei analysed by electrophoretic separation of chromosome-sized DNA molecules. *Cell* **37,** 77–84.
9. Rommens, J. M., Iannuzzi, M. C., Kerem, B., Drumm, M. L., Melmer, G., Dean, M., Rozmahel, R., Cole, J. L., Kennedy, D., Hidaka, N., Zsiga, M., Buchwald, M., Riordan, J. R., Tsui, L. C., and Collins, F. S. (1989) Identification of the cystic fibrosis gene: chromosome walking and jumping. *Science* **245,** 1059–1065.
10. Bucan, M., Zimmer, M., Whaley, W. L., Poustka, A., Youngman, S., Allitto, B. A., Ormondroyd, E., Smith, B., Pohl, T. M., MacDonald, M., Bates, G. P., Richards, J., Volinia, S., Gilliam, T. C., Sedlacek, Z., Collins, F. S., Wasmuth, J. J., Shaw, D. J., Gusella, J. F., Frischauf, A. M., and Lehrach H. (1990) Physical maps of 4p16. 3, the area expected to contain the Huntington disease mutation. *Genomics* **6,** 1–15.
11. Drmanac, R., Petrovic, N., Glisin, V., and Crkvenjakov, R. (1986) A calculation of fragment lengths obtainable from human DNA with 78 restriction enzymes: an aid for cloning and mapping. *Nucleic Acids Res.* **14,** 4691–4692.
12. Birren, B. W., Lai, E., Clark, S. M., Hood, L., and Simon, M. I. (1988) Optimised conditions for pulsed field gel electrophoretic separations of DNA. *Nucleic Acids Res.* **16,** 7563–7582.

CHAPTER 2

The Isolation of Genomic DNA from Invertebrates

Erno Vreugdenhil and Mark G. Darlison

1. Introduction

Generally, genomic DNA is used either for the construction of genomic libraries or for Southern blot analysis. For several reasons, it is frequently preferable in the field of invertebrate neurobiology to screen, at least initially, genomic rather than complementary DNA (cDNA) libraries for sequences of interest (for example, *see* refs. *1–3*). First, low-abundance mRNAs (e.g., those encoding neuroreceptors; *see* refs. *2–4*) have only a very small probability of being present in cDNA libraries. Second, the expression of genes that encode neuronal proteins may be subject to developmental regulation, and the corresponding mRNAs may be present in the organism only at certain stages *(4)*. Third, it is well established, at least in vertebrates, that certain neuron-specific proteins (e.g., ligand-gated ion channels, voltage-gated ion channels, and guanine nucleotide-binding protein-coupled receptors) are encoded by gene families (*see* refs. *5–7*). For the complete characterization of such families, therefore, the screening of genomic libraries may be preferable, since each gene should ideally be present in the library at an equivalent level. In contrast, sequences encoding individual members of gene families are normally present in cDNA libraries at frequencies that are dependent on the abundance of their corresponding mRNAs. Fourth, the genome sizes of some well studied invertebrate species are quite small when compared with those of vertebrates: For example, the haploid genome sizes of the nematode

From: *Methods in Molecular Biology, Vol. 13: Protocols in Molecular Neurobiology*
Edited by: A. Longstaff and P. Revest Copyright © 1992 The Humana Press, Totowa, NJ

Caenorhabditis elegans and of the fruit fly *Drosophila melanogaster* are approx 8.0×10^7 and 1.4×10^8 bp, respectively *(8)*. In contrast, the haploid genome size of humans is 2.8×10^9 bp *(8)*. Therefore, when using representative genomic libraries from such invertebrates, it is possible to know whether the entire genome has, indeed, been screened. After DNA sequence analysis has confirmed the identity of a particular genomic clone, subfragments can be used to determine the abundance of the corresponding neuronal mRNA by Northern blot analysis and to investigate the genomic complexity by Southern blot analysis.

The isolation of genomic DNA from vertebrate tissues is both straightforward and well documented (*see* ref. *9*). In contrast, the preparation of high-mol-wt genomic DNA from invertebrate sources is frequently more troublesome. This difference is probably owing to the specific structure and biochemical content of certain invertebrate tissues. For example, the presence of chitin in insects may prevent the complete disruption of tissues with which it is associated. Also, in our experience, deoxyribonuclease (DNase) activity that is present in invertebrate digestive systems is often difficult to block fully with classical DNase inhibitors. We therefore describe here two methods that have been used in our laboratory to isolate genomic DNA routinely from a variety of invertebrate sources, including molluscan ganglia, molluscan glands, and insect muscles. The first method involves the isolation of nuclei and yields very pure, high-mol-wt DNA (>100 kb in size) that is especially suitable for the construction of representative genomic libraries. The second method is much quicker and results in larger yields of DNA; however, this is usually of a poorer quality and is suitable only for genomic Southern blot analysis.

2. Materials

2.1. Method A:
Genomic DNA from Isolated Nuclei

1. Homogenization buffer: 60 m*M* KCl, 15 m*M* NaCl, 0.15 m*M* spermine, 0.5 m*M* spermidine, 15 m*M* *N*-(2-hydroxyethyl)-piperazine-*N'*-ethane-sulfonic acid (HEPES), 2 m*M* ethylenediaminetetra-acetic acid (EDTA), 0.5 m*M* ethylene glycol-*bis*-(β-aminoethyl ether) *N,N,N',N'*-tetra-acetic acid (EGTA), 14 m*M* β-mercaptoethanol, 0.3*M* sucrose, pH 7.5.
2. TE buffer: 10 m*M* Tris-HCl, 1 m*M* EDTA, pH 7.5.
3. TNE buffer: 10 m*M* Tris-HCl, 0.1*M* NaCl, 5 m*M* EDTA, pH 7.5.

4. Proteinase K: 10 mg/mL in TE buffer.
5. 10% (w/v) sodium dodecyl sulfate (SDS).
6. Recrystallized phenol (or molecular biology grade, ultra-pure phenol) equilibrated with $0.1 M$ Tris-HCl, pH 8.0. Equilibrate by mixing vigorously with an equal vol of $0.1 M$ Tris-HCl, pH 8.0. Allow the phases to separate, remove the upper aqueous layer, and repeat the equilibration of the lower phenol layer until the pH is 8.0; check this with pH paper. Gloves should be worn at all times when handling phenol, since it is both toxic and caustic. In addition, it may be readily absorbed through the skin.
7. Phenol:chloroform: a 1:1 (v/v) mixture of equilibrated phenol and chloroform/isoamylalcohol (the latter in a v/v ratio of 24:1).
8. Chloroform.
9. 0.1X SSC: 15 mM NaCl, 1.5 mM sodium citrate, pH 7.0.
10. Layers of gauze (mesh size approx 0.5 mm): Boil for 10 min in TE buffer prior to use. Store at 4°C.
11. Dialysis tubing. Preparation: Cut the dialysis tubing (2–2.5 cm wide) into pieces of convenient length, i.e., 10–20 cm. Boil for 10 min in 1–2 L of 2% (w/v) sodium bicarbonate containing 1 mM EDTA, pH 8.0. Allow to cool, and then rinse the tubing thoroughly in distilled water. Boil again for 10 min in 1 mM EDTA, pH 8.0, allow to cool, and store (in EDTA) at 4°C. Always handle the dialysis tubing with gloved hands to avoid DNase contamination, and rinse the tubing both inside and out with sterile distilled water before use.
12. Sterile Potter homogenizer and Teflon™ pestle with a clearance of 0.1 mm—the homogenizer should be able to hold a vol of 20 mL.
13. Autoclaved 30-mL Corex centrifuge tubes.
14. Sterile 50-mL Falcon tubes.
15. Sterile Pasteur pipets: The narrow end of these is removed using a glass cutter to leave a bore of approx 5 mm. The edges of the glass are smoothed by heating over a Bunsen burner.
16. Glass funnel (approx 10 cm wide at its widest point).
17. A sterile Teflon™ pestle of approx 1.9 cm diameter.

2.2. Method B:
Hot Phenol Extraction

1. Extraction buffer: 300 mM sodium acetate, 10 mM Tris-HCl, 15 mM EDTA, pH 7.4.
2. Sterile double-distilled water.
3. $3M$ sodium acetate adjusted to pH 5.2 with glacial acetic acid.
4. Equilibrated phenol and phenol:chloroform (*see* Section 2.1., steps 6 and 7).

5. Chloroform.
6. 96% (v/v) ethanol.
7. 70% (v/v) ethanol.
8. Liquid nitrogen: This can cause serious burns; therefore, insulated gloves and a face visor should be worn.
9. Mortar (approx 15 cm diameter) and pestle.
10. Clean razor blades.
11. Sterile 50-mL Falcon tubes.
12. Sterile 1.5-mL microcentrifuge tubes.
13. Sterile Pasteur pipet with a sealed, hooked end (produced by heating the pipet over a Bunsen burner).

3. Methods

3.1. Method A: Genomic DNA from Isolated Nuclei

1. Add ice-cold homogenization buffer (10 mL/g of tissue) to a Potter homogenizer, and leave on ice to chill thoroughly (*see* Note 1).
2. Quickly cut a maximum of 1 g of freshly dissected tissue into small pieces (<0.5 cm cubes), preferably on ice (*see* Note 2).
3. Transfer this to the Potter tube containing the homogenization buffer.
4. Homogenize the tissue using up to five strokes of the pestle, by hand (*see* Note 3).
5. Remove lumps of tissue by filtering the homogenate through two layers of gauze. This is done by putting the layers of gauze over a prechilled glass funnel, which is placed over a 30-mL Corex tube on ice. Wash the Potter homogenizer with approx 2 mL of cold homogenization buffer, and filter this through the same gauze.
6. Centrifuge the Corex tube (1500*g*, 10 min, 4°C) in a precooled rotor.
7. Decant off the supernatant, and gently rehomogenize the pellet using three hand-strokes of a Teflon™ pestle with a diameter just less than that of the Corex tube. Recentrifuge as just described in the same tube (*see* Note 4).
8. Decant off the supernatant, and gently resuspend the pellet in 3 mL of TNE buffer/g tissue using the Teflon™ pestle, and transfer the suspension to a 50-mL Falcon tube. Wash the Corex tube with a further 2 mL TNE buffer, and transfer this to the same Falcon tube.
9. Add 250 µL of 10% (w/v) SDS (to give a final concentration of 0.5% [w/v]).
10. Add 25 µL of Proteinase K (10 mg/mL), which has been preincubated for 1 h at 37°C, to give a final concentration of 50 µg/mL. Mix very gently (*see* Note 5).

11. Incubate the sample for 16 h at 37°C without agitation.
12. Extract the mixture by adding an equal vol (5 mL) of equilibrated phenol and gently rotating the Falcon tube in a horizontal position (e.g., on a rotary shaker) for 10 min.
13. Separate the phases by centrifugation (8000*g*, 5 min, 4°C).
14. Remove the aqueous (top) phase, which contains the DNA, using a wide-bore Pasteur pipet. Transfer to a fresh 50-mL Falcon tube, and repeat this phenol extraction procedure until no interface is visible (normally two to three times).
15. Remove the aqueous phase, and extract once with an equal vol (5 mL) of phenol:chloroform. Remove the upper aqueous phase, and extract with an equal vol of chloroform (*see* Note 6).
16. Carefully transfer the aqueous phase (upper) to dialysis tubing using a wide-bore sterile Pasteur pipet and dialyze against 2 L of 0.1X SSC for 2 h at 4°C.
17. Repeat the dialysis twice, changing the buffer each time.
18. Transfer the contents of the dialysis tubing to a 50-mL Falcon tube, and store at 4°C. Do not store frozen (*see* Note 7).

3.2. Method B:
Hot Phenol Extraction

1. Take 1 g of dissected tissue, cut this into small pieces (<0.5 cm cubes) using a clean razor blade, and grind it to a fine powder, under liquid nitrogen, in a mortar and pestle. Ensure that all of the liquid nitrogen does not evaporate (*see* Note 8).
2. Carefully pour 10 mL of extraction buffer and 10 mL of equilibrated phenol, in small aliquots, into the mortar, and grind this to a fine paste under liquid nitrogen.
3. Transfer the paste to a sterile 50-mL Falcon tube on dry ice. After evaporation of the liquid nitrogen, place the tube on wet ice to allow temperature equilibration.
4. Cap and transfer the tube to a 65°C waterbath, and incubate for 10 min. Shake the tube vigorously every 2 min to homogenize the mixture. Ensure that all traces of liquid nitrogen have evaporated before putting the tube at 65°C, and do not screw the cap of the tube down too tightly (*see* Note 12).
5. Separate the aqueous and phenolic phases by centrifugation (8000*g*, 10 min, room temperature). Remove the aqueous, DNA-containing (upper) phase to a fresh 50-mL Falcon tube.
6. Add 10 mL of equilibrated phenol, and extract at 65°C, as described earlier. Repeat the extraction procedure until no interface is visible (normally four to six times).

7. Remove the aqueous phase, and extract once with an equal vol (10 mL) of phenol:chloroform by gently rotating (e.g., on a rotary shaker) for 10 min. Remove the upper aqueous layer, and extract with an equal vol (10 mL) of chloroform, as described earlier for the phenol:chloroform extraction.

8. Remove the upper aqueous layer, and add 1/10 vol (1 mL) of 3M sodium acetate, pH 5.2, and 2.2 vol (22 mL) of 96% (v/v) ethanol. Mix gently. The genomic DNA should spool, or a cloudy suspension should form. Either collect the spooled DNA using a Pasteur pipet with a hooked end and proceed to step 9, or centrifuge immediately at 3000g for 5 min at room temperature and continue with step 10.

9. Transfer the DNA, on the end of the pipet, to a 1.5-mL microcentrifuge tube containing 1 mL of 70% (v/v) ethanol. Leave for 30 s, and then transfer the pipet and DNA to a microcentrifuge tube containing 1 mL of 96% (v/v) ethanol. Leave for 30 s, air-dry for a few minutes, and finally allow the DNA to dissolve, overnight, in an appropriate vol (e.g., 1–2 mL) of sterile distilled water at 4°C. Do not store frozen.

10. If the genomic DNA is centrifuged, then wash the pellet (this should have a white color) by adding 1 mL 70% (v/v) ethanol. Gently flick the tube, leave at room temperature for 1 min, and then decant off the ethanol. Place the tube in an inverted position on a paper towel to allow all of the fluid to drain away from the pellet. Air-dry the genomic DNA for 10 min, add 2 mL sterile, double-distilled water, and leave to resuspend, overnight, at 4°C. Do not allow the DNA pellet to dry completely, since it will then be very difficult to resuspend. Do not store frozen (*see* Notes 9–11).

3.3. Estimation of the Quality and Yield of DNA

1. To determine the size of the genomic DNA, electrophorese an aliquot in a 0.4% (w/v) agarose gel (for details, *see* ref. *9*) in parallel with suitable size markers, such as intact λ DNA (approx 50 kb in size) and intact bacteriophage T4 DNA (approx 165 kb in size).

2. Determine the DNA concentration by measuring the absorbance of a suitable dilution (in TE buffer or sterile water) at 260 nm. 1 OD 260 = 50 µg/mL. Alternatively, electrophorese dilutions of the genomic DNA in parallel with known concentrations of λ DNA in a 0.8% (w/v) agarose gel. The concentration is then estimated by comparison of the ethidium bromide fluorescence of the sample and visualized under UV irradiation, with that of known standards (*9*).

3. The genomic DNA from both preparations is usually concentrated enough for immediate use. If it is not, it may be concentrated by successive extractions with an equal vol of 2-butanol by gently inverting the Falcon tube. The phases will separate after standing the extraction vertically. Centrifugation is not necessary. Note that the aqueous layer, containing the DNA, will be at the bottom. After concentration, dialyze the sample extensively against 0.1X SSC to remove any traces of 2-butanol.

4. Notes

4.1. Notes Relating to Method A

1. Successful isolation of high-mol-wt DNA, using the first protocol, depends largely on the isolation of intact nuclei. Because of this, the first seven steps are critical; therefore, great care should be taken here.
2. An important point to consider is the choice of invertebrate tissue. If at all possible, dissect soft tissues (e.g., glands and muscles), and try to prevent contamination with parts of the digestive system, which is a rich source of DNase, and skeletal material, which often impedes the complete disruption of tissues. Cell lines, if they are available, are a preferred starting material; these exist for certain insects, such as *Drosophila (10)* and *Periplaneta americana (11)*.
3. It is important to homogenize the tissue gently; vigorous homogenization will lead to damaged nuclei and, therefore, both a reduced yield and shearing of genomic DNA. It is unnecessary to achieve a homogeneous suspension by using many repetitive strokes of the homogenizer (*see* Section 3.1., step 4); the presence of small lumps of tissue after the first homogenization are not a problem, since these will be removed by filtration through the layers of gauze.
4. Another important parameter is the osmolarity of the sucrose buffer. After resuspending the nuclear pellet, which normally has a white color, the solution should be milky and nonviscous. Agglutination of the nuclei results in a rather "sticky" solution at this stage and suggests that the osmotic strength of the sucrose buffer is incorrect. In such a circumstance, it is better to prepare a fresh sucrose solution carefully and start again, rather than proceeding with the DNA isolation. It is also advisable to use high-quality sucrose, since lower grades often contain many contaminants, especially metal ions.
5. After adding SDS and Proteinase K (Section 3.1., steps 9 and 10), the suspension should become viscous. This can be checked by gently tilting the tube first one way and then the other. After incubating

overnight at 37°C, the solution should be clear. If it is not, then add a further 25 µL of preincubated Proteinase K (10 mg/mL), and incubate for a further 2 h at 37°C.

6. Care should also be taken during the sequential extractions with phenol, phenol:chloroform, and with chloroform, since high-mol-wt DNA is easily degraded by mechanical shearing (e.g., by mixing too vigorously or by using small-bore pipets).

7. The overall yield of DNA using Method A is, in our hands, generally around 400 µg/g of starting material. The molecular size of this is typically >100 kb. We have used this DNA for the construction of representative λ genomic libraries (e.g., in λEMBL3), and the isolation therefrom of molluscan GABA$_A$ receptor and acetylcholine receptor subunit genes *(3,12,13)*.

4.2. Notes Relating to Method B

8. The second protocol is much more rapid than the first; genomic DNA can be obtained in 6–8 h. Furthermore, this method usually yields a large amount of genomic DNA (2–3 mg/g of starting material).

9. The main disadvantage of Method B, however, is that the DNA may be slightly degraded (only 20–100 kb in size) and contain some contamination by protein. Therefore, genomic DNA isolated by this method is more suitable for Southern blot analysis than for the construction of genomic libraries, which requires high-quality nucleic acid (e.g., that prepared using Method A). We have used such DNA to investigate the complexity of neurotransmitter receptor/ion channel subunit sequences in the genome of the fresh-water snail *Lymnaea stagnalis* (E. Vreugdenhil, R. J. Harvey, and M. G. Darlison, unpublished data).

10. If necessary, the DNA prepared by Method B can be further purified (e.g., by adjusting the concentration of NaCl to 0.1 *M* and following the instructions from step 9 of Method A onwards). This will, however, result in the loss of DNA. Such treatment will not improve the quality of the DNA to that required for the construction of genomic libraries.

11. Restriction endonuclease digestion of DNA, isolated by Method B, may be found to be incomplete. In our experience, the three following treatments can improve digestion. First, heating the DNA for 10 min at 65°C prior to digestion may cause dissociation of DNA-bound proteins and, therefore, render the DNA more accessible to restriction enzymes. Second, the DNA preparation may contain contaminants that inhibit the action of restriction enzymes. This problem

may be overcome, at least partially, by performing digestions in a relatively large vol (e.g., 0.5 mL rather than the more usual vol of 20 µL or so). Third, endonuclease digestion in the presence of RNase A (at a final concentration of 0.05 µg/µL) may also improve the extent of cleavage.

12. Note that the presence of any liquid nitrogen at step 4 (Section 3.2.) may cause the Falcon tube to explode. It is therefore essential to ensure that all traces have evaporated before transferring the tube to a 65°C waterbath.

References

1. Bossy, B., Ballivet, M., and Spierer, P. (1988) Conservation of neural nicotinic acetylcholine receptors from *Drosophila* to vertebrate central nervous systems. *EMBO J.* **7,** 611–618.

2. Marshall, J., Darlison, M. G., Lunt, G. G., and Barnard, E. A. (1988) Cloning of putative nicotinic acetylcholine receptor genes from the locust. *Biochem. Soc. Trans.* **16,** 463–465.

3. Harvey, R. J., Vreugdenhil, E., Barnard, E. A., and Darlison, M. G. (1990) Cloning of genomic and cDNA sequences encoding an invertebrate γ-aminobutyric acid$_A$ receptor subunit. *Biochem. Soc. Trans.* **18,** 438–439.

4. Hermans-Borgmeyer, I., Zopf, D., Ryseck, R.-P., Hovemann, B., Betz, H., and Gundelfinger, E. D. (1986) Primary structure of a developmentally regulated nicotinic acetylcholine receptor protein from *Drosophila. EMBO J.* **5,** 1503–1508.

5. Barnard, E. A., Darlison, M. G., and Seeburg, P. (1987) Molecular biology of the GABA$_A$ receptor: the receptor/channel superfamily. *Trends Neurosci.* **10,** 502–509.

6. Stühmer, W., Ruppersberg, J. P., Schröter, K. H., Sakmann, B., Stocker, M., Giese, K. P., Perschke, A., Baumann, A., and Pongs, O. (1989) Molecular basis of functional diversity of voltage-gated potassium channels in mammalian brain. *EMBO J.* **8,** 3235–3244.

7. Schofield, P. R., Shivers, B. D., and Seeburg, P. H. (1990) The role of receptor subtype diversity in the CNS. *Trends Neurosci.* **13,** 8–11.

8. Lewin, B., ed. (1980) *Gene Expression,* vol. 2: *Eucaryotic Chromosomes,* 2nd ed., Wiley, New York.

9. Sambrook, J., Fritsch, E. F., and Maniatis, T. (1989) *Molecular Cloning, A Laboratory Manual,* 2nd ed., Cold Spring Harbor Laboratory, Cold Spring Harbor, NY.

10. Debec, A. (1978) Haploid cell cultures of *Drosophila melanogaster. Nature* **274,** 255,256.

11. Philippe, C. and Landureau, J. C. (1975) Culture de cellules embryonnaires et d'hemocytes de blatte d'origine parthenogenetique. *Exp. Cell Res.* **96,** 287–296.

12. Vreugdenhil, E., Harvey, R. J., van Marle, A., Barnard, E. A., and Darlison, M. G. (1991) Molecular biological characterisation of ligand-gated ion channel/receptors in *Lymnaea*, in *Molluscan Neurobiology* (Kits, K. S., Boer, H. H., and Joosse, J., eds.), North Holland, Amsterdam, pp. 353–358.
13. Harvey, R. J., Vreugdenhil, E., Zaman, S. H., Bhandal, N. S., Usherwood, P. N. R., Barnard, E. A., and Darlison, M. G. (1991) Sequence of a functional invertebrate GABA$_A$ receptor subunit which can form a chimeric receptor with a vertebrate α subunit. *EMBO J.* **10**, 3239–3245.

Analysis of Mitochondrial DNA Mutations

Masashi Tanaka and Takayuki Ozawa

1. Introduction

Human mitochondrial DNA (mtDNA) is a closed circular genome of 16,569 bp *(1)*, encoding 13 subunits of four enzyme complexes (Complexes I, III, IV, and V) in the oxidative phosphorylation system *(2)*. Mutations of mtDNA have been demonstrated to be associated with various neuromuscular diseases. Point mutations of mtDNA are reported in MELAS syndrome (mitochondrial myopathy, encephalopathy, lactic acidosis, and stroke-like episodes) *(3,4)*, MERRF syndrome (myoclonus epilepsy associated with ragged-red fibers) *(5,6)*, Leber's disease (hereditary optic neuropathy) *(7)*, a type of encephalomyopathy *(8)*, and fatal infantile cardiomyopathy *(9)*. Deletions of mtDNA are observed in Kearns–Sayre syndrome and chronic progressive external ophthalmoplegia (CPEO) *(10)*. It is also proposed that accumulation of mtDNA mutations is an important contributor to several degenerative diseases and the aging process *(11)*. Evidence supporting this hypothesis has been presented in Parkinson's disease *(12,13)*, cardiomyopathy *(14,15)*, and presbycardia *(16)*. Therefore, the analysis of mtDNA mutations seem to be increasing their importance both in clinical neurology and in the basic neurological science.

This chapter covers the molecular biological approaches to mtDNA mutations. We present various gene amplification techniques for analyzing mutations of mitochondrial genome without using radioisotopes.

From: *Methods in Molecular Biology, Vol. 13: Protocols in Molecular Neurobiology*
Edited by: A. Longstaff and P. Revest Copyright © 1992 The Humana Press, Totowa, NJ

The first section of this chapter describes the methods for characterizing point mutations by direct DNA sequencing of amplified DNA by the polymerase chain reaction (PCR). The following sections describe the Southern blot method, and the PCR-S_1 method for identification and localization of mtDNA deletion. In both methods, mtDNA fragments amplified by PCR from genomic DNA are used as the probes for hybridization or S_1 nuclease digestion. In order to amplify and to analyze any target region of mtDNA, we present a complete list of primers both for amplification and sequencing in the Appendix.

2. Materials

2.1. Fluorescence-Based Direct Sequencing

2.1.1. Reagents

1. 10X PCR buffer: 100 mM Tris-HCl, pH 8.3, 500 mM KCl, 15 mM MgCl$_2$, 0.1% gelatin. Store at –20°C in 300-µL aliquots. Keep stable at –20°C for 3 mo. Freeze-thaw cycles less than three times.
2. 1.25 mM dNTP mix: Mix 12.5 µL each of 0.1M solutions of deoxynucleotide-5'-triphosphates (dATP, dCTP, dGTP, dTTP) from Boehringer Mannheim, and sterile H$_2$O to 1 mL. Store at –20°C in 300-µL aliquots. Stable at –20°C for 3 mo. Freeze-thaw cycles less than three times.
3. Oligonucleotide primers: Primers were synthesized using an Applied Biosystems model 391 DNA synthesizer and then purified with Oligonucleotide Purification Cartridges from Applied Biosystems (Foster City, CA) according to the manufacturer's instruction. The sequences of primers L and H are listed in the Appendix, and Tables 1 and 2.
4. *Taq* DNA polymerases: Ampli*Taq* DNA polymerase (Cetus) and *Taq* DNA polymerase (Promega). Store at –20°C.
5. Mineral oil: Light white oil. Store at room temperature.
6. Agarose gel electrophoresis buffer (TAE): 10 mM Tris-acetate, pH 8.0, 2 mM EDTA. Store at room temperature. Add ethidium bromide to 5 µg/mL before use.
7. 1% Agarose gel: 1% agarose gel in TE containing 5 µg/mL ethidium bromide. Stable for 1 wk at 4°C. Handle with gloves because ethidium bromide is a strong mutagen.
8. 3M sodium acetate (pH 7.4): Sterilize by autoclaving. Store at room temperature.
9. Ethanol: Keep at –20°C.
10. 70% Ethanol: Keep at –20°C
11. *Taq* sequencing kit: Applied Biosystems.

Table 1
List of L Primers for Amplification of the Light Strand of mtDNA

Primer	Position From	To	Sequence (5'→3')
L4	41	60	CTCCATGCATTTGGTATTTT
L32	321	340	TGGCCACAGCACTTAAACAC
L87	871	890	AGGGTTGGTCAATTTCGTGC
L116	1161	1180	AACTCAAAGGACCTGGCGGT
L146	1461	1480	AGGGCCCTGAAGCGCGTACA
L173	1731	1750	ACCCAAATAAAGTATAGGCG
L201	2011	2030	GATAGCTGGTTGTCCAAGAT
L231	2311	2330	TAACATGAAAACATTCTCCT
L260	2601	2620	ATCACTTGTTCCTTAAATAG
L288	2881	2900	CTACTATACTCAATTGATCC
L317	3171	3190	CCGTAAATGATATCATCTCA
L344	3441	3460	ACTACAACCCTTCGCTGACG
L372	3721	3740	ACAATCTCATATGAAGTCAC
L403	4031	4050	TCCTAGGAACAACATATGAC
L434	4341	4360	GAATCGAACCCATCCCTGAG
L462	4621	4640	GTTCCACAGAAGCTGCCATC
L488	4881	4900	CCCATCTCAATCATATACCA
L515	5151	5170	CTACTACTATCTCGCACCTG
L540	5401	5420	TAAAAATAAAATGACAGTTT
L568	5681	5700	CAAACACTTAGTTAACAGCT
L596	5961	5980	CTATTATTCGGCGCATGAGC
L625	6251	6270	TATAGTGGAGGCCGGAGCAG
L651	6511	6530	CTGCTGGCATCACTATACTA
L680	6801	6820	GACACACGAGCATATTTCAC
L704	7041	7060	GTCCTATCAATAGGAGCTGT
L731	7311	7330	TTCATGATTTGAGAAGCCTT
L761	7611	7630	TACAAGACGCTACTTCCCCT
L790	7901	7920	TGAACCTACGAGTACACCGA
L820	8201	8220	TTCATGCCCATCGTCCTAGA
L853	8531	8550	ACGAAAATCTGTTCGCTTCA
L881	8811	8830	CACCCAACTATCTATAAACC
L909	9091	9110	ACACTTATCATCTTCACAAT
L932	9321	9340	TCCATAACGCTCCTCATACT
L962	9621	9640	GCATCAGGAGTATCAATCAC
L989	9891	9910	TCCAAACATCACTTTGGCTT
L1019	10191	10210	TCCCCCGCCCGCGTCCCTTT
L1047	10471	10490	TGCCCCTCATTTACATAAAT
L1076	10761	10780	TGCTAAAACTAATCGTCCCA
L1108	11081	11100	ATAACATTCACAGCCACAGA
L1138	11381	11400	CCTCTTTACGGACTCCACTT
L1167	11671	11690	AACCCCCTGAAGCTTCACCG

(continued)

Table 1 *(Continued)*
List of L Primers for Amplification of the Light Strand of mtDNA

| Primer | Position | | Sequence (5'→3') |
	From	To	
L1192	11921	11940	TGATCAAATATCACTCTCCT
L1222	12221	12240	GAACTGCTAACTCATGCCCC
L1250	12501	12520	GTGCCTAGACCAAGAAGTTA
L1280	12801	12820	CAGTTGATGATACGCCCGAG
L1310	13101	13120	AGGAATCTTCTTACTCATCC
L1341	13411	13430	ATAGGAGGACTACTCAAAAC
L1371	13711	13730	GCCGGAAGCCTATTCGCAGG
L1396	13961	13980	TTCTTACGAGCCAAAACCTG
L1429	14291	14310	TCATAAATTATTCAGCTTCC
L1451	14511	14530	CTATTAAACCCATATAACCT
L1482	14821	14840	CAACATCTCCGCATGATGAA
L1512	15121	15140	AACAGCCTTCATAGGCTATG
L1543	15431	15450	GCCCTCGGCTTACTTCTCTT
L1569	15691	15710	AATATTTCGCCCACTAAGCC
L1594	15941	15960	TTTTCCAAGGACAAATCAGA
L1619	16191	16210	CCCATGCTTACAAGCAAGTA
L1641	16411	16430	CGTGAAATCAATATCCCGCA

Table 2
List of H Primers for Amplification of the Heavy Strand of mtDNA

| Primer | Position | | Sequence (5'→3') |
	From	To	
H12	140	121	GAATCAAAGACAGATACTGC
H38	400	381	AAATTTGAAATCTGGTTAGG
H60	620	601	AAACATTTTCAGTGTATTGC
H82	840	821	TTATTGCTAAAGGTTAATCA
H107	1090	1071	TAGTGGGGTATCTAATCCCA
H166	1680	1661	TAGGTTTAGCTCAGAGCGGT
H196	1980	1961	TATAAATCTTCCCACTATTT
H223	2250	2231	TCAGTTATATGTTTGGGATT
H253	2550	2531	TGGGCAGGCGGTGCCTCTAA
H280	2820	2801	TCGCCCCAACCGAAATTTTT
H318	3200	3181	AATACTAAGTTGAGATGATA
H366	3680	3661	GAGTTTGATGCTCACCCTGA
H394	3960	3941	GCCTGCGGCGTATTCGATGT
H426	4280	4261	TTTATCAGACATATTTCTTA

(continued)

Table 2 *(Continued)*
List of H Primers for Amplification of the Heavy Strand of mtDNA

Primer	Position From	To	Sequence (5'→3')
H483	4850	4831	GCCGGATGTCAGAGGGGTGC
H509	5110	5091	GTAGTAGTTAGGATAATATA
H536	5380	5361	AGTGTGATTGAGGTGGAGTA
H580	5820	5801	GAGGTGATTTTCATATTGAA
H594	5960	5941	GTATAGTGTTCCAATGTCTT
H617	6190	6171	CGGGGAAACGCCATATCGGG
H650	6520	6501	ATGCCAGCAGCTAGGACTGG
H674	6760	6741	ATAAACCCTAGGAAGCCAAT
H726	7280	7261	GAATGAGCCTACAGATGATA
H753	7550	7531	GACAAAGTTATGAAATGGTT
H784	7860	7841	TCGTTGACCTCGTCTGTTAT
H815	8170	8151	TTGACCGTAGTATACCCCCG
H832	8340	8321	CTTAATCTTTAACTTAAAAG
H854	8560	8541	GGGCAATGAATGAAGCGAAC
H884	8860	8841	TGCCCGCTCATAAGGGGATG
H902	9040	9021	GCATGAGTAGGTGGCCTGCA
H929	9310	9291	AATCACATGGCTAGGCCGGA
H954	9560	9541	TCGGGGCCAGTGCCCTCCTA
H982	9840	9821	AGCCAATAATGACGTGAAGT
H1014	10160	10141	GGATTTTTCTATGTAGCCGT
H1043	10450	10431	TAATTTAATGAGTCGAAATC
H1072	10740	10721	GTACGTAGTCTAGGCCATAT
H1103	11050	11031	AGAGAGGTAGAGTTTTTTTC
H1136	11380	11361	TATCTTTACTATAAAAGCTA
H1189	11910	11891	GTTACTAGCACAGAGAGTTC
H1219	12210	12191	TCTCGGTAAATAAGGGGTCG
H1243	12450	12431	GGATTTTACATAATGGGGGT
H1275	12770	12751	TCTCAGCCGATGAACAGTTG
H1308	13100	13081	GCTACAACTATAGTGCTTGA
H1338	13400	13381	TCTTGTTCATTGTTAAGGTT
H1363	13650	13631	GGGGAAGCGAGGTTGACCTG
H1393	13950	13931	GGGGATTGTGCGGTGTGTGA
H1420	14220	14201	TTAGTAGTAGTTACTGGTTG
H1450	14520	14501	GGTTTAATAGTTTTTTTAAT
H1479	14810	14791	GGAGGTCGATGAATGAGTGG
H1506	15080	15061	TTTCTGAGTAGAGAAATGAT
H1560	15620	15601	GGACGCCTCCTAGTTTGTTA
H1578	15800	15781	GTCCAATGATGGTAAAAGGG
H1599	16010	15991	AAATTAGAATCTTAGCTTTG
H1619	16210	16191	TACTTGCTTGTAAGCATGGG
H1643	16450	16431	CGAGGAGAGTAGCACTCTTG

12. 5X *Taq* sequencing buffer: 250 mM NaCl, 50 mM MgCl$_2$, 50 mM Tris-HCl, pH 8.5. Sterilize by autoclaving. Store at –20°C.
13. A termination mix: 1.5 mM ddATP, 62.5 μM dATP, 250 μM dCTP, 375 μM dGTP, 250 μM dTTP.
14. C termination mix: 0.75 mM ddCTP, 250 μM dATP, 62.5 μM dCTP, 375 μM dGTP, 250 μM dTTP.
15. G termination mix: 0.125 mM ddGTP, 250 μM dATP, 250 μM dCTP, 94 μM 7-deaza-dGTP, 250 μM dTTP.
16. T termination mix: 1.25 mM ddTTP, 250 μM dATP, 250 μM dCTP, 375 μM dGTP, 62.5 μM dTTP. Store these termination mixes at –20°C.
17. TE buffer: 10 mM Tris-HCl, pH 8.0, and 1 mM EDTA. Store at room temperature.
18. Dye universal primers: Dissolve 80 pmol of each dye primer (M13mp18) in 200 μL of TE buffer. Complete dissolution will require vigorous voltexing. Store in the dark at –20°C .
19. 3M sodium acetate (pH 5.2): Sterilize by autoclaving. Store at room temperature.
20. Formamide (deionized): Mix 10 mL of formamide and 0.5 g of mixed-bed, ion-exchange resin (e.g., Bio-Rad, Richmond, CA, AG501-X8, 20–50 mesh). Stir for 30 min at room temperature. Filter twice through Whatman No. 1 filter paper. Dispense into 0.5-mL aliquots, and store at –20°C.
21. 50 mM EDTA (pH 8.0): Sterilize by autoclaving. Store at room temperature.

2.1.2. Specialist Equipment

1. DNA thermal cycler (Perkin-Elmer/Cetus).
2. Agarose gel electrophoresis apparatus.
3. DNA sequencing system: Model 373A (Applied Biosystems, Foster City, CA).

2.2. Nonisotopic Southern Blot Analysis

2.2.1. Reagents

1. 10X *Pvu*I buffer: 100 mM Tris-HCl, pH 7.5, 70 mM MgCl$_2$, 0.6M NaCl, and 70 mM 2-mercaptoethanol.
2. 10X *Bam*HI buffer: 100 mM Tris-HCl, pH 8.0, 70 mM MgCl$_2$, 1M NaCl, 20 mM 2-mercaptoethanol, 0.1% bovine serum albumin.
3. 1X TAE: 40 mM Tris-acetate, pH 8.0, 1 mM EDTA.
4. DNA size markers: λ phage DNA digested with *Hind*III (Takara Shuzo, Kyoto, Japan) and phage X174 DNA digested with *Hae*III (Takara Shuzo).

5. ECL gene detection system: The system (DNA labeling reagent, glutaraldehyde solution, blocking agent, hybridization buffer, and detection reagents 1 and 2) is obtained from Amersham International, Amersham, UK. Store at –20°C.
6. Hybridization buffer: Heat the ECL hybridization buffer at 65°C for 10 min, and add NaCl to $0.5M$ and the ECL blocking agent to 5% w/v. Mix for 30 min on a magnetic stirrer. The temperature should be below 42°C. This can be stored in aliquots (50 mL) at –20°C for up to 3 mo.
7. Primary wash buffer: $6M$ urea, 0.4% SDS, 75 mM NaCl, 7.5 mM sodium citrate (pH 7.0).
8. Secondary wash buffer 2X SSC: $0.3M$ NaCl, 30 mM sodium citrate, pH 7.0.

2.3. PCR-S_1 Method

2.3.1. Reagents

1. Heteroduplex formation buffer: 10 mM Tris-HCl (pH 7.5), 7 mM MgCl$_2$, 60 mM NaCl.
2. S_1 nuclease reaction mixture: 50 mM sodium acetate buffer, pH 4.6, 1 mM zinc acetate, 250 mM NaCl, 50 µg/mL bovine serum albumin. Store at –20°C.
3. S_1 nuclease (Takara Shuzo, Kyoto, Japan).

2.3.2. Specialist Equipment

Mini-agarose gel electrophoresis apparatus: Mupid-2, Cosmo-Bio (Tokyo, Japan) or equivalent.

3. Methods

3.1. Fluorescence-Based Direct Sequencing of Mitochondrial DNA

Point mutations of mtDNA have been demonstrated to be associated with various neuromuscular diseases: MELAS syndrome (A to G transition at nucleotide 3243) *(3,4)*, MERRF syndrome (A to G transition at 8344) *(5,6)*, Leber's disease (G to A transition at 11,778) *(7)*, and a type of encephalomyopathy (T to G transversion at 8993) *(8)*. Analysis of the total sequence of mtDNA is essential for identifying a pathogenic mutation in a new disease entity. Sequencing of a part of mtDNA segment is also required for genetic diagnosis of a known disease when no restriction enzymes that detect the specific point mutation are available. Therefore, nucleotide sequencing of mtDNA is an important approach in molecular neurology.

Mutations of mtDNA were previously analyzed by cloning of PCR-amplified fragments. Recently, the primer end-labeling method *(17)* and the primer extension-labeling method *(18)* have been employed for direct sequencing of PCR-amplified DNA. The former method is not suited for sequencing of a large region of mtDNA, because each sequencing primer must be radiolabeled separately. For the latter method, it is essential to remove previously used primers completely prior to sequencing either by polyethylene glycol precipitation *(19)*, gel filtration, or ultrafiltration *(20)*. These traditional methods are fairly time-consuming, and require expensive disposable columns for gel filtration or centrifugal ultrafiltration.

This section describes a method for direct sequencing of mitochondrial DNA fragments amplified from genomic DNA without using cloning procedures, radioisotopes, or columns for template purification *(3)*. This method permits the rapid, accurate, and easy determination of sequences in specific regions of mitochondrial DNA, being well suited for large-scale sequencing projects where a number of separate regions of mtDNA from different patients are to be sequenced.

The principle of fluorescence-based direct sequencing is schematically presented in Fig. 1. First, the target sequence is amplified as a double-stranded DNA fragment by "symmetric PCR," where primers L_1 and H_1 are present in equal amounts. Primers L_n and H_n have the sequences identical to the light (L) and heavy (H) strands of mtDNA, respectively. Then, an excess of the H strand of the target region is generated by "asymmetric PCR," where primer H_2 is present in a large excess over primer FL_2. Since the primer FL_n consists of both the M13 universal sequence and the L strand sequence, the H strand synthesis starting from primer H_2 extends over primer FL_2; therefore, the resulting single-stranded DNA incorporates the sequence complementary to the M13 universal sequence at its 3' end. Finally, the single-stranded DNA serves as the template for the automated sequencing system using commercially available M13 universal primers labeled with four different fluorescent dyes. In this system, extended fragments from nonlabeled primers will not interfere with the fluorescence-based sequencing, and tedious hands on work for removal of previous primers is unnecessary.

3.1.1. First PCR Reaction (Symmetric)

1. Prepare a mix containing the following: 5 µL 10X PCR buffer, 8 µL 1.25 mM dNTP mix, 5 µL 10 µM primer L_1, 5 µL 10 µM primer H_1, 0.5 µL 10 ng/µL genomic DNA, 0.25 µL 5 U/µL *Taq* DNA polymerase, and H_2O to 50 µL.

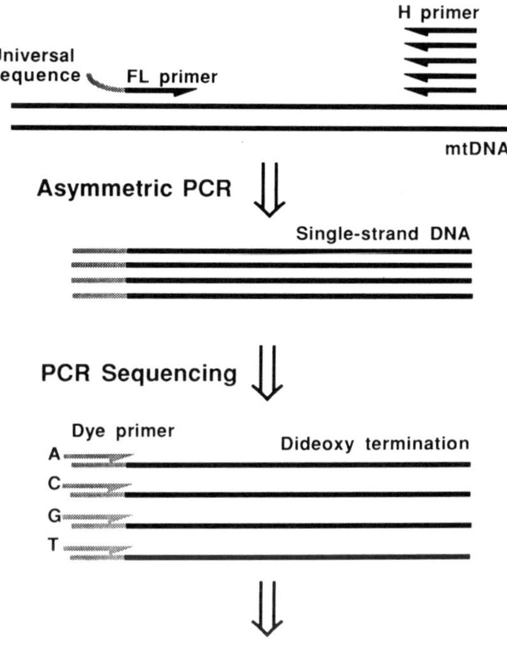

Fig. 1. Principle of fluorescence-based direct sequencing of PCR-amplified mtDNA.

2. Overlay the reaction mixture with 50 µL mineral oil. Carry out PCR in a DNA Thermal Cycler using the following conditions for 30 cycles: denaturation at 94°C for 15 s, annealing at 55°C for 15 s, and extension at 72°C for 40 s.
3. Take 5 µL of the first PCR product, separate by a 1% agarose gel, and visualize the product by fluorography after staining with ethidium bromide. The first PCR should yield a single band. When the PCR yielded several fragments, e.g., a fragment amplified from normal mtDNA and fragments derived from deleted mtDNAs, the PCR products are separated on a 1% agarose gel (ultrapure grade), stained with ethidium bromide, and cut out from the gel. The agarose gel block containing the band is frozen at −20°C for 1 h, thawed at 37°C for 5 min, and then centrifuged in an Ultrafree C3-HV column (Millipore, Yonezawa, Japan) (5000g, 10 min, 4°C). The gel is resuspended in 200 µL of TE and centrifuged again in the same column. The resultant filtrate is extracted once with phenol:chloroform: isoamyl alcohol (25:25:1), concentrated by ethanol precipitation, and then used as the template for the second PCR reaction.

4. Take out the first PCR product with a disposable tip through the layer of mineral oil. Wipe off mineral oil with tissue paper (Kimwipe™), and transfer the PCR product to another microcentrifuge tube.

5. Add the product with 5 μL 3*M* sodium acetate (pH 7.4) and 100 μL ethanol, keep the mixture at –20°C for 10 min, and then centrifuge (13,000*g*, 10 min, 4°C). The precipitate was rinsed with 150 μL of 70% ethanol and then centrifuge (13,000*g*, 5 min, 4°C), dried in a vacuum chamber for 10 min, and then resuspended in 50 μL H_2O.

3.1.2. Second PCR Reaction (Asymmetric)

1. Prepare a mix containing the following: 5 μL 10X PCR buffer, 0.8 μL 1.25 m*M* dNTP mix, 1 μL 0.5 μ*M* primer FL_2, 5 μL 10 μ*M* primer H_2, 0.5 μL the first PCR product, 0.25 μL 5 U/μL *Taq* DNA polymerase, and H_2O to 50 μL. The dNTP concentration in this step is one-tenth of that in the first PCR (*see* Note 1).

2. Overlay with 50 μL mineral oil. Carry out PCR in a Thermal Cycler using the following conditions for 30 cycles: denaturation at 94°C for 15 s, annealing at 55°C for 15 s, and extension at 72°C for 40 s.

3. Examine the second PCR product after agarose electrophoresis and ethidium bromide staining. The second PCR usually yields four bands, as shown in Fig. 2. The upper band, a, is single-stranded DNA formed between primers L_1 and H_2. The next band, b, is double-stranded DNA formed between these primers. The lower faint band, c, and the lowest sharp band, d, are single-stranded DNA and double-stranded DNA, respectively, formed between primers FL_1 and H_2. The latter two bands serve as the template for sequencing.

4. Concentrate the second PCR product by ethanol precipitation with 5 μL of 3*M* sodium acetate (pH 7.4) and 100 μL ethanol, rinse with 70% ethanol, dried in a vacuum chamber, and resuspend in 8 μL of H_2O.

3.1.3. Sequencing Reaction

1. Prepare four microcentrifuge tubes for PCR each containing the following: 1.0 μL 5X *Taq* sequencing buffer, 1.0 μL either A, C, G, or T termination mix, 1 μL 0.4 μ*M* dye primer, 1 μL the second PCR product, 0.25 μL 5 U/μL *Taq* DNA polymerase, and H_2O to 7.75 μL. The amounts of reagents are for the Joe dye primer (for A) and the Fam dye primer (for C). For the sequencing reactions with the Tamra dye primer (for G) and the Rox dye primer (for T), the amounts are doubled to a total vol of 15.5 μL (*see* Note 2).

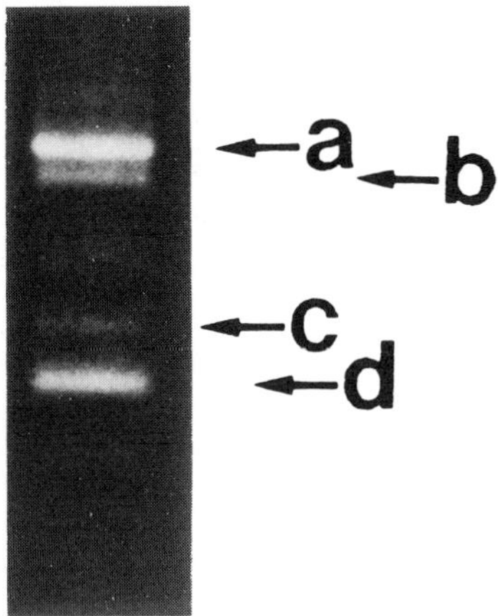

Fig. 2. Agarose gel electrophoresis of products from the asymmetric PCR. The first symmetric PCR was carried out using primers L680 and H929. The product was subjected to the second asymmetric PCR using primers FL820 and H884. The second PCR products were separated on a 1% agarose gel. The upper band, **a,** is single-stranded DNA formed between primers L680 and H884. The next band, **b,** is 2060-bp double-stranded DNA formed between these primers. The lower faint band, **c,** and the lowest sharp band, **d,** are single-stranded DNA and 678-bp double-stranded DNA, respectively, formed between primers FL820 and H884.

2. Carry out PCR in a Thermal Cycler using the following conditions for 10 cycles: denaturation at 90°C for 15 s, annealing and extension at 70°C for 60 s, but annealing at 55°C is not necessary.
3. Combine the contents of the four tubes into a tube containing a mixture of 5 µL of 3*M* sodium acetate (pH 5.2) and 120 µL ethanol. Place the tube at –20°C for 10 min, and then centrifuge at 13,000*g* for 10 min. Rinse the pellet with 120 µL of ethanol, and then centrifuge at 13,000*g* for 5 min. Dry the pellet in a vacuum chamber for 10 min, and store at –20°C in the dark until electrophoresis.
4. Just prior to loading onto a sequencing gel, resuspend the pellet in 5 µL of deionized formamide:50 m*M* EDTA (5:1 by vol).
5. Heat the sample at 90°C for 2 min to denature, immediately cool on ice, and then load the total reaction onto a 6% polyacrylamide gel that has been preelectrophoresed for 2 h.
6. Run the gel for 10 h and analyze the data. An example of the analyzed data is depicted in Fig. 3 (*see* Notes 3 and 4).

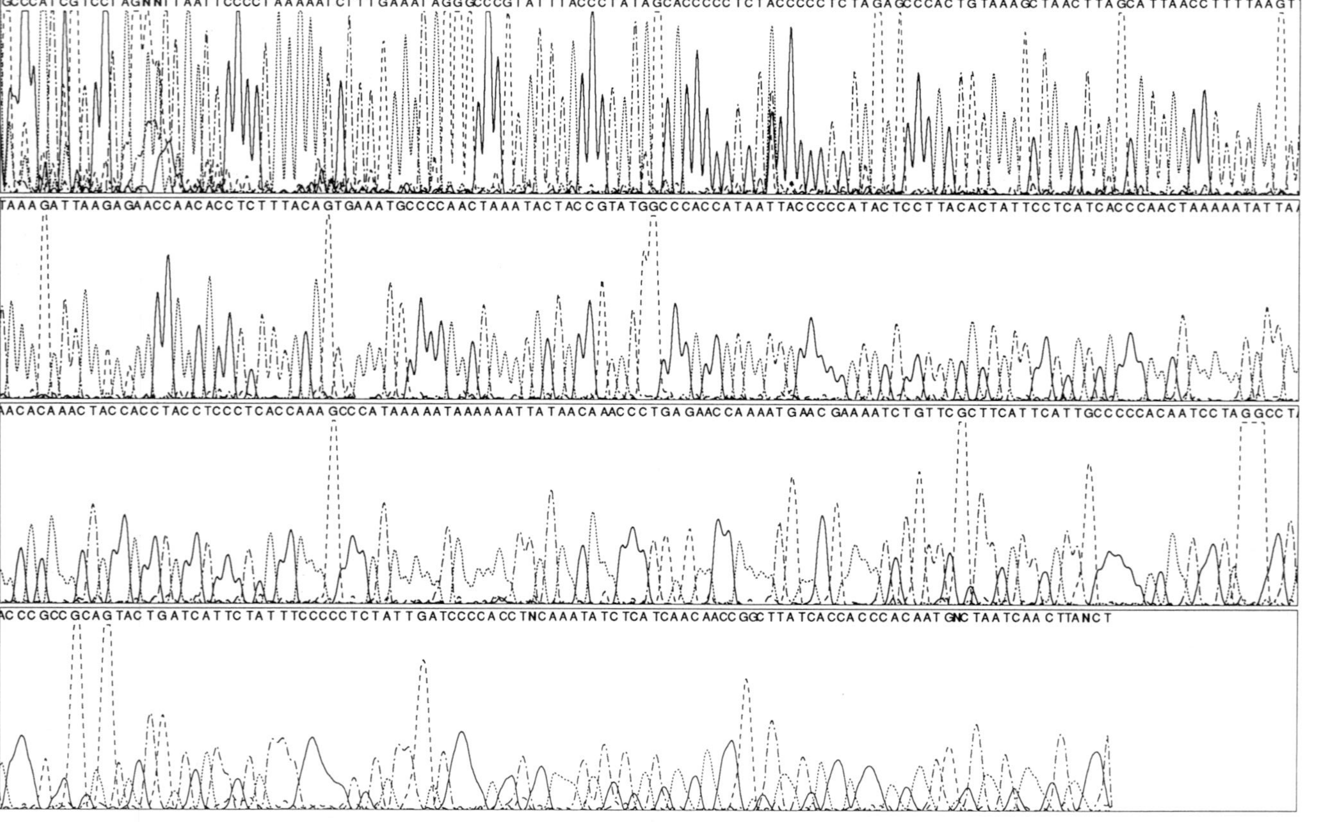

Fig. 3. Fluorescence-based direct sequencing of PCR-amplified mtDNA. The template shown in Fig. 2 was subjected to fluorescence-based direct sequencing. Dotted line (·····), adenine; solid line (—), cytosine; dashed line (- - - - -), guanine; and dot-dash line (–·—·—·—), thymine.

3.2. Nonisotopic Southern Blot Analysis of Deleted mtDNA Using PCR-Amplified Probes

Deletion of mtDNA has been observed in the tissues of patients with Kearns–Sayre syndrome and chronic progressive external ophthalmoplegia. Genetic diagnosis of these syndromes has been carried out by the Southern blot method using either purified total mtDNA or cloned fragments as the probes *(10)*. Instead of purifying or cloning mtDNA, we can easily prepare probes for Southern blot analysis by PCR *(21)*.

In this section, we describe the methods for amplifying DNA probes by using PCR, for labeling the probes with peroxidase, and for detecting by enhanced chemiluminescence (ECL). This technique does not require any cloning procedures or radioisotopes, and can be performed quickly even in ordinary clinical laboratories. In addition, we can prepare desired probes for specific regions of mtDNA and can determine the locus of deletion precisely *(14)*.

For detection of deleted mtDNA and for estimation of the ratio of deleted mtDNA to normal mtDNA, we should select restriction enzymes that cleave a closed circular mtDNA at one site yielding one linear fragment of 16.6 kb in normal subjects. This restriction site must be outside the deletion so that deleted mtDNA is cleaved into one linear fragment shorter than normal. For this purpose, we usually use *Pvu*II (restriction site 2650) because this site is rarely deleted in mutant mtDNA (Fig. 4). *Bam*HI (restriction site 14,258) can be also used, but this restriction site is sometimes lost in deleted mtDNA. Therefore, we recommend a combination of these two restriction enzymes for Southern blot analysis of deleted mtDNA.

The ECL system involves directly labeling probe DNA with horseradish peroxidase. First, the probe is completely denatured to singlestranded DNA by heating and rapid cooling. Peroxidase that has been complexed with a positively charged polymer binds to negatively charged DNA; then peroxidase and DNA probe are covalently linked with glutaraldehyde. Then, the probe is used in hybridization with target DNA immobilized on a membrane. Hybridization is carried out under the condition that denaturation of peroxidase is minimized during this step.

The ECL detection system utilizes two detection reagents. Detection reagent 1 decays to produce hydrogen peroxide. Peroxidase reduces hydrogen peroxide to superoxide radical. Detection reagent

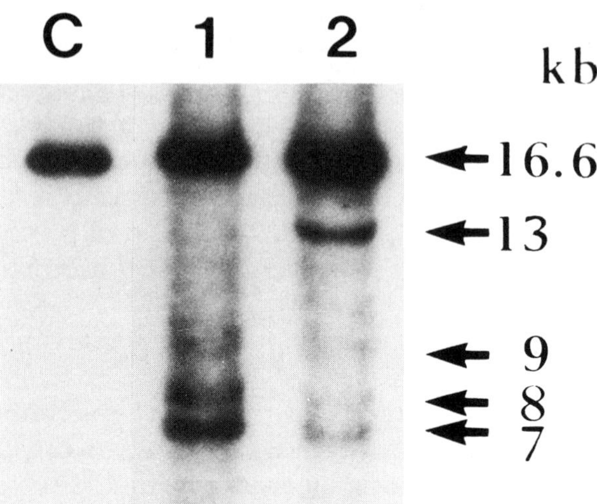

Fig. 4. Southern blot analysis of mtDNA deletion. *Pvu*II-digested total muscle DNA of two patients with exertional myoglobinuria was probed with a mixture of PCR-amplified fragments. A 16.6-kb band is derived from normal mtDNA, and multiple bands are derived from deleted mtDNAs. Although the blot indicated several common bands of 9, 8, and 7 kb between the patients, a 7-kb band indicating a deletion of 9.6 kb was dominant in Patient 1, and a 13-kb band indicating a deletion of 3.6 kb was dominant in Patient 2.

2 contains both luminol, which produces blue light on oxidation with superoxide radical, and enhancer, which increases and prolongs the light output.

3.2.1. Digestion with Restriction Enzymes

1. Prepare a microcentrifuge tube containing the following: 1 μL 10X *Pvu*I I or *Bam*HI buffer, 50 ng total muscle DNA, 5 U restriction enzyme (*Pvu*I I or *Bam*HI), and H$_2$O to 10 μL.
2. Incubate at 37°C for 1 h.

3.2.2. Gel Electrophoresis

1. Prepare 0.6% agarose solution in 1X TAE. Dissolve agarose by autoclaving.
2. Cool agarose solution to 60°C in a waterbath. Mix thoroughly, and pour into a gel former (15 × 15 cm).
3. Separate the digested DNA on the gel along with DNA size markers by electrophoresis at 100 V for 2 h or at 20 V for 10 h for a 15 cm long gel.

Table 3
Primer Pairs for Amplification of Southern Probes

Primer L	Primer H	Fragment size, bp
L596	H854	2600
L820	H1136	3180
L1108	H1338	2320
L1310	H1619	3110
L1619	H60	999
L32	H318	2880
L288	H580	2940

4. Stain the gel with 300 mL of 5 µg/mL ethidium bromide for 30 min.
5. Destain the gel with water for 10 min.
6. Photograph the gel with UV illumination.

3.2.3. Capillary Blotting

1. Under UV illumination, trim the gel leaving the lanes containing the sample.
2. Blot onto a Hybond-N⁺ membrane using 300 mL of 0.4*N* NaOH for 2 h according to the standard Southern blotting method.
3. Put the membrane between two sheets of 3MM chromatography paper (Whatman, Maidstone, UK), and dry in a clean bench.

3.2.4. Labeling of Probe with Peroxidase

1. PCR amplification of probe DNA fragment (>300 bp) in a vol of 100 µL. A mixture of six fragments (2.0–3.0 kb) covering the whole length of mtDNA was amplified from control mtDNA using the primer pairs shown in Table 3.
2. Ethanol precipitation: It is essential to remove the salt from probe DNA fragment for the subsequent labeling, because salts inhibits electrostatic association of positively charged peroxidase with negatively charged probe DNA. Therefore, aspirate ethanol completely, and wipe off the wall of microcentrifuge tube with sterile cotton swab.
3. Resuspend the DNA fragment in 30 µL of water. The concentration should be higher than 10 ng/µL.
4. Boil the DNA fragment at 100°C for 5 min for denaturation.
5. Immediately cool on ice for 5 min. Spin the tube briefly to settle down the content.
6. Add 30 µL of the ECL DNA-labeling reagent (containing horseradish peroxidase). Mix thoroughly.

7. Add 30 µL of the ECL glutaraldehyde solution. Mix thoroughly (vortexing should be shorter than 1 s).
8. Incubate at 37°C for 10 min.
9. Labeled probe may be kept on ice for <15 min. Add 90 µL of glycerol, and store at –20°C for up to 6 mo.

3.2.5. Hybridization

1. Redissolve the hybridization buffer at 42°C.
2. Place the blot membrane in a plastic container, add 25 mL of the hybridization buffer (for 100 cm² blot membrane), and incubate at 42°C for 10 min in a shaking waterbath. The face of the membrane that has been in contact with the gel should be facing upwards.
3. Add the labeled probe sufficient for a final concentration of 20 ng/mL or more. Do not pour the probe directly onto the blot membrane.
4. Incubate with agitation (60 strokes/min) at 42°C overnight. Place a plastic container on a rotary shaker in a circulating air incubator, or immerse it into a shaking waterbath.

3.2.6. Washing and Detection

1. Wash the filter twice with 200 mL of the primary wash buffer (2 mL/cm²) at 42°C for 20 min.
2. Wash the filter twice with 200 mL of the secondary wash buffer (2 mL/cm²) at room temperature for 5 min. The blot can be left in the secondary wash buffer for up to 30 min before detection.
3. Take the box containing the membrane to a dark room. Mix equal vol (12.5 mL) of the detection reagents 1 and 2—12.5 mL for 100 cm² of the blot.
4. Add the detection buffer to blot. Incubate for precisely 1 min at room temperature.
5. Wrap the blot in SaranWrap.
6. Put the wrapped membrane on a 3MM paper in a film cassette. Overlay an X-ray film, and expose for 1 min. Develop the film quickly. Estimate the appropriate exposure time (up to 60 min), and expose the second or the third film. The side of the membrane that has been facing the gel must face the X-ray film.

3.3. Primer-Shift PCR Method
for Localizing mtDNA Deletions

Some patients with Kearns–Sayre syndrome and CPEO possess such small populations of deleted mtDNAs that the deletions are undetectable by the conventional Southern blot method. In order to diagnose these disorders precisely, we have developed a method for detecting small populations of deleted mtDNA by using PCR with the combinations of primers shifting in different positions around the deletion (the primer-shift PCR method) *(22)*.

Recent studies have demonstrated that deleted mtDNA are accumulated not only in the muscle of patients with mitochondrial myopathy, but also in the hearts of patients with cardiomyopathy and in the striatum of patients with Parkinson's disease *(12,13)*. Age-dependent accumulation of multiple deletions of mtDNA is also observed in the heart *(16)*. Since mtDNA is located adjacent to the mitochondrial inner membrane, mtDNA is continuously exposed to oxidative damage by the activated oxygen leaked from the respiratory chain. The deleted mtDNA might result either directly from oxidative damage or indirectly from point mutations. Small populations of multiple-point mutations cannot be easily analyzed, but multiple deletions can be easily detected by this primer-shift PCR method even when they are present in small populations. Therefore, we can estimate the degree of damage to mtDNA by measuring the ratio of deleted mtDNA to normal mtDNA. In this section, we present the appropriate pairs of primers for detection of deleted mtDNA.

The primer-shift PCR method *(22)* is a gene-amplification method specially designed for detecting small populations of deleted mtDNA. In PCR, misannealing of primers to unexpected sites of mtDNA sometimes results in amplification of abnormal fragments. In order to ascertain that an amplified fragment is not the result of misannealing of primers, we identify the deletion of mtDNA by the primer-shift PCR method, the principle of which is as follows (Fig. 5): In an initial experiment, a fragment is amplified from the deleted mtDNA by using a pair of L and H primers surrounding the deletion. The size of the deletion can be obtained by subtracting the size of the amplified fragment from the distance between the primers. In a second experiment, another fragment is amplified from the deleted mtDNA by using a pair of L' and H primers. Then in a third experiment, a third fragment is amplified from the deleted mtDNA by using a pair of L and H' primers. The shift in the sizes of the three amplified fragments should parallel the shift in the positions of the primers from L to L', and from H to H' (Table 4). Thus, we can conclude that the amplified fragments are not owing to misannealing of the primers, but to the presence of the deleted mtDNA (*see* Note 5).

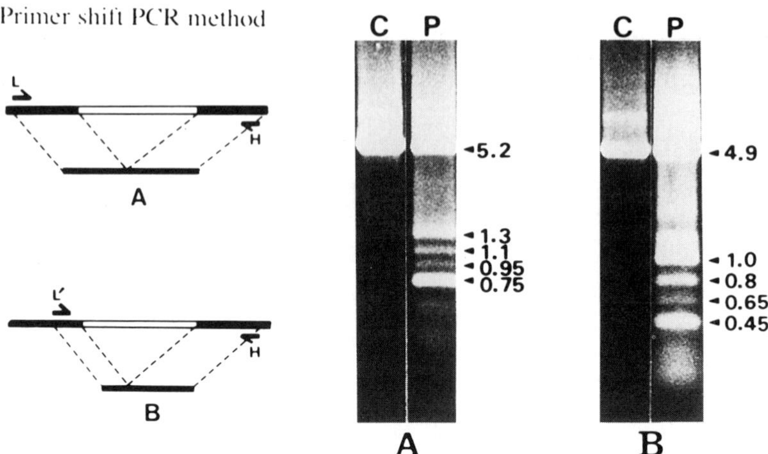

Fig. 5. Primer-shift PCR method. In experiment A (primers L820 and H1338), only a 5.2-kb fragment derived from normal mtDNA was amplified from the muscle DNA from a control (lane C). Abnormal fragments with the sizes of 1.3, 1.1, 0.95, and 0.75 kb were amplified from the muscle DNA from a patient with familial external ophthalmoplegia (lane P). In experiment B (primers L853 and H1338), only a 4.9-kb fragment derived from normal mtDNA was amplified in the control (lane C). In the patient (lane P), in addition to the 4.9-kb fragment, abnormal fragments with the sizes of 1.0, 0.8, 0.65, and 0.45 kb were amplified. Both results demonstrate that at least four populations of mutant mtDNA with deletions of 3.9, 4.1, 4.25, and 4.45 kb are present in the patient.

1. Prepare a mix containing the following: 5 μL 10X PCR buffer, 8 μL 1.25 mM dNTP mix (final concentration, 200 μM), 5 μL 10 μM primer L, 5 μL 10 μM primer H, 0.5 μL 10 ng/μL genomic DNA, 0.25 μL 5 U/μL Ampli*Taq* DNA polymerase, and H_2O to 50 μL. Primers used for PCR are listed in Tables 1 and 2, and materials as in Section 2.1.

2. Overlay the reaction mixture with one droplet of mineral oil. Carry out PCR in a DNA Thermal Cycler using the following conditions for 30 cycles: denaturation at 94°C for 15 s, annealing at 45°C or 55°C for 15 s, and extension at 72°C for 60 s.

3. Take 5 μL of the first PCR product, separate by a 1% agarose gel, and visualize the product by fluorography after staining with ethidium bromide. If the fragments are too faint, concentrate the fragments by ethanol precipitation and reanalyze by electrophoresis (*see* Notes 6–10).

Table 4
Size of the Fragments Amplified with Combinations of Two Primers

Combination of primers	Distance between two primers, kb	Size of amplified fragments, kb	Calculated size of deletion, kb
L820 + H1338	5.2	5.2	—
		1.3	3.9
		1.1	4.1
		0.95	4.25
		0.75	4.45
L853 + H1338	4.9	4.9	—
		1.0	3.9
		0.8	4.1
		0.65	4.25
		0.45	4.45
L790 + H1363	5.75	5.75	—
		2.3	3.45
		1.9	3.9
		1.5	4.25
		1.3	4.45
		0.8	4.95
		0.65	5.1

3.4. PCR-S$_1$ Method for Analysis of Deleted mtDNA

Deleted regions of mtDNA have been determined by the Southern blot method. The primer-shift PCR method (described in the previous section) and the nesting primer PCR method are useful for rapid analysis of deleted regions, but the accuracy of these two methods depends on the intervals (approx 300 bp) of available primers and, therefore, a number of primers are required for precise localization of the deletion. In contrast, the PCR-S$_1$ method *(23)* requires only up to 20 primers and can determine the localization of deletion at the accuracy of approx ±60 bp (*see* Note 11). The PCR-S$_1$ method requires no radioisotopes and can be performed in ordinary clinical laboratories. This method is so sensitive that total DNA extracted from only 5 mg of muscle tissue is sufficient for the determination of the deleted region of mtDNA. Therefore, this method is of value, especially when only a small amount of clinically biopsied sample is available.

The PCR-S$_1$ method is a combination of PCR amplification and S$_1$ nuclear analysis useful for accurate determination of deleted regions of mtDNA, the principle of which is schematically shown in Fig. 6.

PCR plus S₁ method

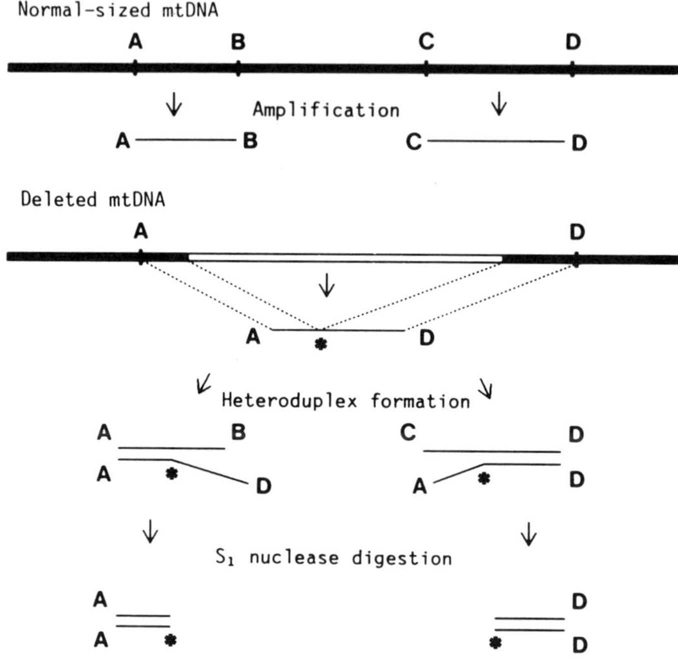

Fig. 6. Schematic presentation of the PCR-S$_1$ method. Open box indicates the region of the deletion in mutant DNA. Three fragments A–B, C–D, and A–D, are amplified with each pair of the appropriate primers. The asterisk indicates the crossover point of the deletion. The fragments A–B and A–D are mixed, and a heteroduplex is formed. Only the part from point A to the asterisk is complementary and is protected against S$_1$ nuclease digestion. The fragments C–D and A–D are subjected to the same procedure for defining the other end of the deletion.

Fragment A–B, which includes the starting point of the deletion, is amplified from the normal-sized mtDNA and fragment A–D, which includes both ends of the deletion is amplified from the mutant mtDNA. These fragments are mixed and subjected to heteroduplex formation. The complementary region of the heteroduplex formed between fragments A–D and A–B as well as the homoduplexes formed from the self-reannealing of each fragment are protected against S$_1$ nuclease. The size of the complementary region of the heteroduplex indicates the distance between point A and the starting point of the deletion (asterisks in the figure) in the mutant mtDNA. Similarly, the end point of the deletion can be determined by S_1 nuclease analysis of the heteroduplex formed from fragments C–D and A–D.

3.4.1. Amplification of Fragment from Deleted mtDNA

1. Prepare a mix containing the following: 5 μL 10X PCR buffer, 16 μL 1.25 mM dNTP mix, 10 μL 10 μM primer L_1, 10 μL 10 μM primer H_1, 1 μL 10 ng/μL genomic DNA, 0.5 μL 5 U/μL Ampli*Taq* DNA polymerase, and H_2O to 100 μL. Materials used as in Section 2.1. and 2.3.
2. Overlay the reaction mixture with 50 μL mineral oil. Carry out PCR in a Thermal Cycler using the following conditions for 30 cycles: denaturation at 94°C for 15 s, annealing at 55°C or 45°C for 15 s, extension at 72°C for 60 s.

3.4.2. Amplification of Probe Fragments from Normal mtDNA

1. For determination of the starting point of the deletion, amplify Probe A from normal mtDNA using 100 pmol each of primers L_1 and H_2.
2. For determination of the end point of the deletion, amplify Probe B from normal mtDNA using 100 pmol each of primers L_2 and H_1 (*see* Note 12).

3.4.3. Heteroduplex Formation

1. Mix two mtDNA fragments (500 ng each) amplified with PCR, and precipitate with ethanol.
2. Dissolve the pellet in 100 μL of Heteroduplex formation buffer.
3. Denature the DNA sample at 95°C for 10 min in an aluminum heating block.
4. Take the aluminum block out of the heating part, and leave the block at room temperature to cool down to about 37°C (*see* Note 13). Then transfer the microcentrifuge tube onto ice.

3.4.4. S$_1$ Nuclease Digestion

1. Precipitate the mixture with 2 vol of cold ethanol in the presence of 0.3M sodium acetate (pH 7.5).
2. Resuspend the pellet in 20 μL of the S_1 nuclease reaction mixture (*see* Note 14).
3. Add 5 U of S_1 nuclease and incubate the mixture at 37°C for 20 min.
4. Add 1 μL 100 mM EDTA (final concentration of 5 mM) to stop the reaction.
5. Separate the digested fragments by electrophoresis on 2% agarose gels, and detect fluographically after staining with ethidium bromide.

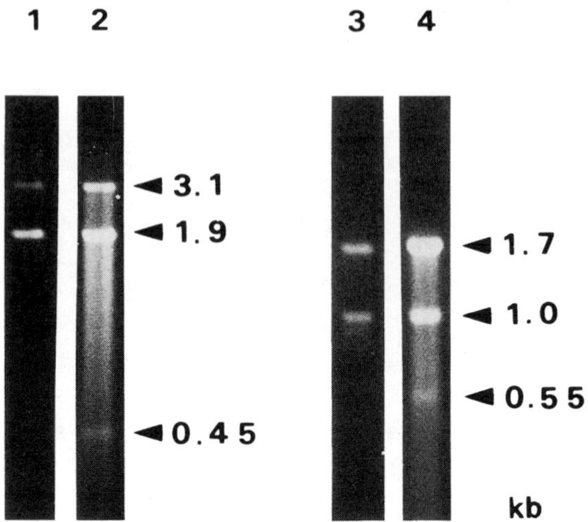

Fig. 7. Electrophoretic patterns of PCR-amplified mtDNA fragments after heteroduplex formation and S_1 nuclease digestion. Two bands in lane 1 indicate fragments A–B (3.1 kb) and A–D (1.9 kb), amplified using the pairs of primers L820 and H1136 and of primers L820 and H60, respectively. A band with the size of 0.45 kb appeared after S_1 nuclease digestion. The size of the band indicates the distance from the position of primer L820 to the beginning point of the deletion. Fragments A–B (1.7 kb) and C–D (1.0 kb) in lane 3 are amplified using the pairs of primers L1451 and H1619 and of primers L820 and H1619, respectively. The band with the size of 0.55 kb is shown in lane 4. The size indicates the distance from the end point of the deletion to the position of primer H1619.

3.4.5. Analysis of Data from PCR-S$_1$ Analysis

An example of PCR-S$_1$ analysis is shown in Fig. 7. For determination of the deletion start point, Probe A was mixed with the 1.9-kb fragment amplified from the deleted mtDNA (lane 1). After the S_1 nuclease digestion, a fragment of 0.45 ± 0.05 kb appeared (lane 2), assuming the error in estimation of the fragment size to be <10%. The start point of the deletion should be located at position 8650 ± 50 bp within the ATP6 gene.

To determine the deletion end point, Probe B was mixed with the 1.0-kb fragment amplified from the deleted mtDNA using the primers of L820 and H1619 (Fig. 7, lane 3). A fragment of 0.55 ± 0.06 kb appeared after the S_1 nuclease digestion (lane 4). Since the primer H1619 starts at position 16,209, the end point of the deletion should be located at position 15,660 ± 60 within the cytochrome *b* gene.

4. Notes

4.1. Notes on Fluorescence-Based Direct Sequencing

1. If peaks in the first part of analyzed sequence data are low and peaks in the latter part are high, the ratio of dNTP:ddNTP ratio in the sequencing reaction is too high. In ethanol precipitation, dNTP is coprecipitated with DNA. When too much dNTP is present in the template for the sequencing step, chain termination with ddNTP is inhibited, resulting in low peaks in the first part of the sequencing data and high peaks in the latter part. Therefore, in the second PCR, we use dNTP at one-tenth concentration of the first PCR. Amplification in the second PCR is barely hindered by this low concentration of dNTP. When high concentration of dNTP is used, precipitate the second PCR product with 1 vol of 4M ammonium acetate and 2 vol of ethanol in order to remove excess dNTP.

2. Dye primers are light sensitive. Keep the reaction microtubes away from light by covering with aluminum foil during PCR and vacuum drying. Protection against room light, however, is not always necessary during dispensing, when the windows of the laboratory are curtained completely against sunlight.

3. Large peaks of fluorescent primers at the beginning of sequence data and subsequent low and noisy peaks indicate either a too small amount of template DNA or inhibition of *Taq* DNA polymerase. Better sequence results are obtained by increasing the amount of template for the first PCR than for the second PCR.

4. When the nonlimiting primer H_2 is too close to the limiting primer FL_2 in the asymmetric PCR, a high peak, corresponding to the end of the template, appears in the last part of the sequence data, and neighboring peaks become low. Therefore, the distance between primers FL_2 and H_2 should be about 600 bp.

4.2. Notes on Primer-Shift PCR Method

5. The primer-shift PCR method is useful for detecting small populations of deleted mtDNA that are below the threshold for detection by the Southern blotting. There are three regions that are prone to deletion. We have selected three groups of primer pairs (Table 5). Each group of primer pairs has special purpose.

Table 5
Groups of Primer Pairs for Detection of Deleted mtDNA

Deletion start	Deletion end	Deletion size, bp	Directly repeated sequence	Primer L	Primer H	Fragment size, bp	Comment
8483	13,459	4977	ACCTCCC TCACCA	L820	H1363	0.77	Parkinson's disease
8624	15,662	7039	ATCCCCA	L820	H1619	1.00	CPEO
9192	12,908	3717	CCT	L820	H1338	1.50	CPEO
8649	16,084	7436	CATCAA CAACCG	L853	H38	1.00	Cardio-myopathy
8992	16,072	7079	None	L853	H38	1.40	Cardio-myopathy

a. Primer pairs for detecting common mtDNA deletion. The 4977-bp deletion (from 8483 to 13,459) involving a 13-bp directly repeated sequence (5'-ACCTCCCTCACCA-3') is the most common deletion among patients with Kearns–Sayre syndrome and CPEO. If PCR amplification is carried out with primers L820 and H1363 for 30 cycles, this deletion can be easily detected as a 0.77-kb fragment even in the muscle DNA from aged individuals. By using less PCR cycles (20 cycles), we have reported that accumulation of this deletion is accelerated in Parkinson's disease *(12,13)*.

b. Primer pairs for detecting age-dependent mtDNA deletion. The 7436-bp deletion from the ATP6 gene (position 8649) to the D-loop region (16,084)) involving a 12-bp directly repeated sequence (5'-CATCAACAACCG-3') is originally found in patients with primary cardiomyopathy *(14)*. This deletion can be detected with primers L853 and H38 as a 1.0-kb fragment. We have demonstrated that population of mutant mtDNA with this deletion increases with age *(16)*.

c. Primer pairs for detecting multiple mtDNA deletions. The third mtDNA region that is predisposed to deletion is between 8200 and 13,400. The multiple deletions with the sizes of 3.9, 4.1, 4.25, and 4.45 kb are detectable with primers L820 and H1338 as 1.3-, 1.1-, 0.95-, and 0.75-kb fragments. A 3460-bp deletion from the ND1 gene (position 9593) to the ND5 gene (13,052) involving a 6-bp directly repeated sequence (5'-TAGAAG-3') can be detected with primers L853 and H1338 as a 1.4-kb fragment. These deletions were originally found in patients with familial external ophthalmoplegia *(22)*. We have demonstrated that populations of mutant mtDNAs with these deletions were increased in the skeletal muscle of patients with myotonic dystrophy.

6. When no bands are seen after PCR, the following possibilities should be considered:

a. Some reagent is missing. Use positive controls.

b. *Taq* DNA polymerase is inhibited by residual phenol in the template DNA. Repeat extraction with ether and precipitation with ethanol.

c. dNTP is degraded (hydrolyzed to dNDP and Pi). Prepare fresh dNTP.

7. When only a band of about 40 bp (primer dimer) is formed:

a. A pair of the primers is not appropriate, 3' ends of the primers are complementary to each other. Use another primer.

b. The amount of template is too small. Increasing the amount of template DNA will improve the efficiency of PCR. If the DNA template is dissolved in TE buffer, increasing the volume of template means increasing EDTA, which inhibits *Taq* DNA polymerase. Therefore, DNA template should be concentrated either by ethanol precipitation or by ultrafiltration under centrifugal force using a Centricon 30 or 100.

   c. The template is missing.

8. When only a faint band of desired size is formed:

   a. The amount of template is too small. Increase the amount of template and PCR cycles.

   b. The amount of *Taq* DNA polymerase is too small. Increasing the amount of *Taq* DNA polymerase usually improves the efficiency of PCR to some extent, but this can result in greater production of nonspecific PCR products and reduced yield of the desired target. Ampli*Taq* DNA polymerase gives better results than the ordinal preparations of DNA polymerase isolated directly from thermophilic bacteria, because Ampli*Taq* that is produced in *E. coli* by cloning technology is free from thermostable DNases or proteases derived from thermophilic bacteria.

9. When multiple fragments of unexpected sizes are amplified, primer sequence is not appropriate. Check again whether the sequences of the primers are correct, because misoriented primers sometimes give multiple fragments.

10. When both the target fragment of expected size and multiple extraneous fragments are amplified:

   a. Annealing temperature is too low. Choose a higher annealing temperature. Although the higher annealing temperature increases the specificity of PCR, it decreases the efficiency of PCR amplification.

   b. Deleted mtDNA are amplified. Deleted mtDNA can be detected even in normal individuals. Choose less PCR cycles, and use template DNA isolated from young individuals.

4.3. Notes on PCR-S$_1$ Analysis

11. The accuracy in determination of the deleted regions depends on the experimental error in estimating the sizes of electrophoresed fragments. Therefore, we can improve the accuracy by choosing appropriate pairs of primers, so that smaller fragments are obtained after S$_1$ nuclease digestion.

12. We can selectively amplify fragments from the deleted mtDNA or from the normal-sized mtDNA without separating the two populations. For this purpose, we should chose appropriate pairs of primers from the list. We should adjust the length of extension time in PCR.

Short duration of extension period results in preferential amplification of short fragments over long ones.

13. Heteroduplex formation is dependent on cooling rate. Too slow cooling will yield formation of more homoduplexes and less heteroduplexes, resulting in a faint protected band after S_1 nuclease digestion. Programmed cooling in a DNA Thermal Cycler at a rate of 1°C/s is also recommended for heteroduplex formation.

14. The activity of S_1 nuclease differs from preparation to preparation. The amount of S_1 nuclease to be used and duration of incubation should be determined empirically. Too much enzyme and too long incubation will result in digestion of double-stranded portion of heteroduplex, and the deletion will be estimated to be larger than the actual size. Compare the sizes of homoduplexes before and after the S_1 nuclease treatment to examine whether the digestion is excessive or not.

Appendix

Nomenclature of Primers

Our nomenclature system for primers is as follows. Primers L and H are 20-mer oligonucleotides possessing sequences specific for the light (L) strand and the heavy (H) strand of mtDNA, respectively. The 5' end of primer L(n) corresponds to nucleotide position $(10n + 1)$ and its 3' end to nucleotide position $(10n + 20)$. Similarly, the 3' end of primer H(n) corresponds to nucleotide position $(10n + 1)$ and its 5' end to nucleotide position $(10n + 20)$. These sequences were selected so that no other homologous sequences (higher than 65–70% homology) of each primer were found in the entire mtDNA. Primers FL are 38-mer oligonucleotides possessing both the M13mpl8 forward universal sequence of 18 nucleotides (-21M13, 5'-TGTAAAACGACGGCC AGT-3') on the 5' side and a sequence of 20 nucleotides specific for the L strand of mtDNA at the 3' side. For example, FL4 has the following sequence: 5'-TGTAAAACGACGGCCAGTCTCCATGCATTT GGTATTTT-3'.

Determination of Primer Concentration

Concentration of oligonucleotide is determined by using the following formula.

$$C = A_{260} \times 100 / (1.54 n_A + 0.75 n_C + 1.17 n_G + 0.92 n_T)$$

Where: C = concentration (μM), n_x = number of residues of base x in the oligonucleotide, and A_{260} = absorbance at 260 nm.

References

1. Anderson, S., Bankier, A. T., Barrell, B. G., de Bruijn, M. H. L., Coulson, A. R., Drouin, J., Eperon, I. C., Nierlich, D. P., Roe, B. A., Sanger, F., Schreier, P. H., Smith, A. J. H., Staden, R., and Young, I. G. (1981) Sequence and organization of the human mitochondrial genome. *Nature* **290,** 457–465.
2. Chomyn, A., Mariottini, P., Cleeter, M. W. J., Ragan, C. I., Doolittle, R. F., Matsuno-Yagi, A., Hatefi, Y., and Attardi, G. (1985) Functional assignment of the products of the unidentified reading frames of human mitochondrial DNA, in *Achievements and Perspectives of Mitochondrial Research. vol. II: Biogenesis* (Quagliariello, E., Slater, E. C., Palmieri, F., Saccone, C., and Kroon, A. M., eds.), Elsevier, Amsterdam, NY, pp. 259–275.
3. Tanaka, M., Ino, H., Ohno, K., Ohbayashi, T., Ikebe, S., Sano, T., Ichiki, T., Kobayashi, M., Wada, Y., and Ozawa, T. (1991) Mitochondrial DNA mutations in mitochondrial myopathy, encephalopathy, lactic acidosis, and stroke-like episodes (MELAS). *Biochem. Biophys. Res. Commun.* **174,** 861–868.
4. Ino, H., Tanaka, M., Ohno, K., Hattori, K., Ikebe, S., Ozawa, T., Sano, T., Ichiki, T., Kobayashi, M., and Wada, Y. (1991) Mitochondrial leucine tRNA mutation in mitochondrial encephalomyopathy. *Lancet* **337,** 234–235.
5. Yoneda, M., Tanno, S., Horai, S., Ozawa, T., Miyatake, T., and Tsuji, S. (1990) A common mitochondrial DNA mutation in the t-RNA^{Lys} of patients with myoclonus epilepsy associated with ragged-red fibers. *Biochem. Int.* **27,** 789–796.
6. Shoffner, J. M., Lott, M. T., Lezza, A. M. S., Seibel, P., Ballinger, S. W., and Wallace, D. C. (1990) Myclonic epilepsy and ragged-red fibers disease (MERRF) is associated with a mitochondrial DNA tRNA^{Lys} mutation. *Cell* **61,** 931–937.
7. Wallace, D. C., Singh, G., Lott, M. T., Hodge, J. A., Schurr, T. G., Lezza, A. M. S., Eisas, L. J., and Nikoskelainen, E. K. (1988) Mitochondrial DNA mutation associated with Leber's hereditary optic neuropathy. *Science* **242,** 1427–1430.
8. Holt, I. J., Harding, A. E., Petty, R. K., and Morgan-Hughes, J. A. (1990). A new mitochondrial disease associated with mitochondrial DNA heteroplasmy. *Am. J. Hum. Genet.* **46,** 428–433.
9. Tanaka, M., Ino, H., Ohno, K., Hattori, K., Sato, W., and Ozawa, T. (1990) Mitochondrial tRNA^{Ile} mutation in fatal infantile cardiomyopathy. *Lancet* **336,** 1452.
10. Ozawa, T., Yoneda, M., Tanaka, M., Ohno, K., Sato, W., Suzuki, H., Nishikimi, M., Yamamoto, M., Nonaka, I., and Horai, S. (1988) Maternal inheritance of deleted mitochondrial DNA in a family with mitochondrial myopathy. *Biochem. Biophys. Res. Commun.* **154,** 1240–1247.
11. Linnane, A. W., Marzuki, S., Ozawa, T., and Tanaka, M. (1989) Mitochondrial DNA mutations as an important contributor to ageing and degenerative diseases. *Lancet* **i,** 642–645.
12. Ikebe, S., Tanaka, M., Ohno, K., Sato, W., Hattori, K., Kondo, T., Mizuno, Y., and Ozawa, T. (1990) Increase of deleted mitochondrial DNA in the striatum in Parkinson's disease and senescence. *Biochem. Biophys. Res. Commun.* **170,** 1044–1048.

13. Ozawa, T., Tanaka, M., Ikebe, S., Ohno, K., Kondo, T., and Mizuno, Y. (1990) Quantitative determination of deleted mitochondrial DNA relative to normal DNA in parkinsonian striatum by a kinetic PCR analysis. *Biochem. Biophys. Res. Commun.* **172,** 483–489.

14. Ozawa, T., Tanaka, M., Sugiyama, S., Hattori, K., Ito, T., Ohno, K., Takahashi, A., Sato, W., Takada, G., Mayumi, B., Yamamoto, K., Adachi, K., Koga, Y., and Toshima, H. (1990). Multiple mitochondrial DNA deletions exist in cardiomyocytes of patients with hypertrophic or dilated cardiomyopathy. *Biochem. Biophys. Res. Commun.* **154,** 830–836.

15. Hattori, K., Ogawa, T., Taizo, K., Mochizuki, M., Tanaka, M., Sugiyama, S., Ito, T., Satake, T., and Ozawa, T. (1991) Cardiomyopathy with mitochondrial DNA mutations. *Am. Heart J.* **122,** 866–869.

16. Hattori, K., Tanaka, M., Sugiyama, S., Obayashi, T., Ito, T., Satake, T., Hanaki, Y., Asai, J., Nagano, M., and Ozawa, T. (1991) Age-dependent increase in deleted mitochondrial DNA in the human heart: Possible contributing factor to "presbycardia." *Am. Heart J.* **121,** 1735–1745.

17. Higuchi, R., Beroldingen, C. H. V., Sensabaugh, G. F., and Erlich, H. A. (1988) DNA typing from single hairs. *Nature* **332,** 543–546.

18. Innis, M. A., Myambo, K. B., Gelfand, D. H., and Brow, M. A. (1988) DNA sequencing with Thermus aquatics DNA polymerase and direct sequencing of polymerase chain reaction-amplified DNA. *Proc. Natl. Acad. Sci. USA* **85,** 9436–9440.

19. Tanaka, M., Sato, W., Ohno, K., Yamamoto, T., and Ozawa, T. (1989) Direct sequencing of deleted mitochondrial DNA in myopathic patients. *Biochem. Biophys. Res. Commun.* **164,** 156–163.

20. Gyllensten, U. B. and Erlich, H. A. (1988) Generation of single-stranded DNA by the polymerase chain reaction and its application to direct sequencing of the HLA-DQA locus. *Proc. Natl. Acad. Sci. USA* **85,** 7652–7656.

21. Ohno, K., Tanaka, M., Sahashi, K., Ibi, T., Sato, W., Takahashi, A., and Ozawa, T. (1991) Mitochondrial DNA deletions in inherited recurrent myoglobinuria. *Ann. Neurol.* **29,** 364–369.

22. Sato, W., Tanaka, M., Ohno, K., Yamamoto, T., Takada, G., and Ozawa, T. (1989) Multiple populations of deleted mitochondrial DNA detected by a novel gene amplification method. *Biochem. Biophys. Res. Commun.* **162,** 664–672.

23. Tanaka-Yamamoto, T., Tanaka, M., Ohno, K., Sato, W., Horai, S., and Ozawa, T. (1989) Specific amplification of deleted mitochondrial DNA from a myopathic patient and analysis of deleted region with S_1 nuclease. *Biochim. Biophys. Acta* **1009,** 151–155.

CHAPTER 4

The Design and Use of Oligonucleotides

Alan N. Bateson and Mark G. Darlison

1. Introduction

Chemically synthesized oligonucleotides are powerful tools in the molecular biologist's repertoire. This chapter describes the design of oligonucleotides and their applications, with particular reference to their use in the isolation and characterization of recombinant DNA clones. Examples of their use are taken either from the literature or from work carried out in our laboratory. Two specific areas are covered: the design of oligonucleotides, based on peptide sequence derived from a purified protein, for use in the isolation of clones from recombinant DNA libraries, and the design of consensus oligonucleotide probes, corresponding to conserved sequences in multigene families, for use in the isolation and characterization of such related genes. We do not cover the use of oligonucleotides for the localization of mRNA transcripts by *in situ* hybridization or for the analysis of mRNA steady-state levels by Northern blot analysis since this has been described elsewhere *(1,2)*. A detailed discussion of all of the factors involved in the formation and melting of oligonucleotide/target DNA duplexes is beyond the scope of this treatise; for this, readers are referred to the work of Lathe *(3)*. However, formulae for the calculation of melting temperatures and the application of the base analog inosine are described in the Notes.

From: *Methods in Molecular Biology, Vol. 13: Protocols in Molecular Neurobiology*
Edited by: A. Longstaff and P. Revest Copyright © 1992 The Humana Press, Totowa, NJ

2. Oligonucleotides
Based on Known Peptide Sequence

Methods for the isolation of sequences from recombinant DNA libraries using oligonucleotide probes designed on the basis of peptide sequence information can be grouped into three alternative strategies. First, a mixture of short oligonucleotides can be synthesized that contain all possible codon choices for any particular amino acid in a peptide sequence (degenerate oligonucleotides). Second, a single (best-guess) oligonucleotide can be constructed that contains the preferred codon for any amino acid within a sequence. The third strategy involves the synthesis of two different (usually degenerate) oligonucleotide probes that are designed on the basis of two peptide sequences, as primers for the polymerase chain reaction (PCR).

The choice of method is largely dependent on the degree of certainty that can be placed on the accuracy of the available peptide sequence, which in turn is often dependent on the abundance of the protein in the tissue of origin. Where short peptide sequences have been determined with a high degree of accuracy, the first method has proven to be successful (*see,* for example, *4,5*). It is often the case, however, that the protein of interest can only be purified in very small amounts. This is particularly true when studying membrane-bound proteins specific to the nervous system, for example, neurotransmitter receptor subunit polypeptides, ion channels, and so forth. In such cases, where the derived peptide sequence would be expected to be less accurate than that for more abundant proteins, it is preferable to design best-guess oligonucleotides, especially if longer peptide sequences are generated (15 amino acids or so). These probes may contain regions of low homology to the DNA sequence of interest, but, by virtue of their increased length, should be sufficiently homologous overall to enable hybridization to target sequences to take place under the appropriate conditions.

The recently developed technique of PCR *(6)* allows the amplification of DNA fragments between two known sequences. Consequently, if two separate peptide sequences are available, degenerate oligonucleotides can be designed, and the DNA fragment encoding the intervening polypeptide sequence can be amplified. In addition to the requirement for accurate peptide sequence determination, it is essential that both amino acid sequences derive from the same polypeptide. Other workers *(7,8)* have employed the PCR technique to isolate novel, yet related, DNA sequences using short degenerate oligonucleotides designed on the basis of highly conserved regions of previously cloned members of a multigene family.

Although the three categories just outlined serve as a convenient means of classifying the various approaches to the design and use of oligonucleotides in the isolation of DNA sequences, there are no distinct divisions among them, and consequently, many applications may involve a combination of approaches. The three strategies will now be described in more detail using examples from published work. Before designing any oligonucleotide probe on the basis of a chemically determined peptide sequence, it is first important to search a suitable protein sequence data base for the possible presence of that sequence. Much time and effort can be saved by such analysis that would otherwise have been expended in screening libraries and analyzing false-positive clones, since peptide sequences have been known to derive from a contaminant or integral component of the preparation (e.g., a protease inhibitor), rather than the protein under study.

2.1. Degenerate Probes

A chemically determined peptide sequence of five or more amino acids can be used to design a mixture of oligonucleotides that contain all possible DNA sequences that code for that particular sequence. Probes designed in this manner have been used successfully to isolate both cDNA (e.g. *4,5*) and occasionally genomic clones *(9)*. The advantage of this approach is that, if the amino acid sequence used to design the oligonucleotides is correct, then there must be one oligonucleotide in the mixture that is an exact match to the target DNA sequence. Consequently, hybridization and wash conditions can be used such that only perfectly matched duplexes can form. This does not mean, however, that all hybridizing clones contain the required DNA sequence, since there may be oligonucleotides in the mixture that correspond to DNA sequences that are unrelated to the sequence of interest. These artifacts may be more apparent if the hybridization conditions used do not distinguish between perfect matches and one base pair mismatches. It is desirable, therefore, to counterscreen positive clones with a second pool of degenerate oligonucleotides. These can be designed on the basis of a second known peptide sequence. There is always the possibility, however, that the second amino acid sequence is not derived from the same polypeptide, but either from a contaminating polypeptide or from a different subunit of a hetero-oligomeric protein. Hence, if two pools of degenerate oligonucleotides are to be used, they should, where possible, be derived from one contiguous amino acid sequence. When designing these probes, regions rich in arginine, leucine, and/or serine residues should be avoided, since each of these amino acids has six possible codon choices. Designing oligonucleotide mixtures with as low a degeneracy as possible decreases the chance of isolating false positives; use of a highly degenerate mixture results in a decreased hybridization signal strength. The length of the probe is also important; too short a probe increases the frequency of random hybridization events, and too long a probe necessarily increases its degeneracy. Typically such oligonucleotides should be between 14 and 24 nucleotides in length; a degeneracy of up to 384-fold has been used successfully *(10)*.

A fine example of the use of degenerate oligonucleotide probes for screening libraries is provided by the isolation of the *Torpedo californica* nicotinic acetylcholine receptor (AChR) α-subunit cDNA by Numa and coworkers *(4)*. A cDNA library was screened with a mixture of 32 oligonucleotides representing all possible DNA sequences encoding a pentapeptide sequence contained within the amino terminus of the α subunit (Fig. 1A). Fifty-seven positive clones were identified, which were rescreened with a second mixture of 32 oligonucleotides corresponding to a hexapeptide sequence from a different portion of the amino terminus of the a subunit (Fig. 1B). Twenty clones that hybridized to both probes were further characterized; two of these were found to encode the entire AChR α subunit.

2.2. Best-Guess Probes

The approach described in the previous section is only feasible when the peptide sequence information can be relied on. It is often the case, however, that only a peptide sequence that contains some uncertainty can be obtained. This applies especially to proteins of low abundance. In this event, it is better to design oligonucleotide probes of 30 or more bases that have a low degeneracy (up to 10-fold) by making use of known codon preferences for each individual amino acid. Analysis of all cloned DNA sequences available in the GenBank data base has allowed Aota and colleagues *(11)* to compile codon-usage tables for individual genes and for individual species. Although, in general, codon preference varies little among, for example, different mammals, different types of genes can demonstrate particular preferences, e.g., histone genes *(12)*. Codon choices that generate CpG

A

	His	Thr	His	Phe	Val			amino-acid sequence
		U			U			
		A			A			
5'	CAU	ACC	CAU	UUU	GUC	3'		possible codons
	C	G	C	C	G			
		A						
		T						
3'	GTA	TGG	GTA	AAA	CA	5'		probe sequence
	G	C	G	G				

B

	Glu	Asn	Tyr	Asn	Lys	Val		amino-acid sequence
						U		
						A		
5'	GAA	AAU	UAU	AAU	AAA	GUC	3'	possible codons
	G	C	C	C	G	G		
3'	CTT	TTA	ATA	TTA	TTT	C	5'	probe sequence
	C	G	G	G	C			

Fig. 1. Degenerate oligonucleotide probes used to isolate the *T. californica* AChR α-subunit cDNA. Oligonucleotide sequences used to screen (**A**) and counterscreen (**B**) an electroplaque cDNA library are shown below the corresponding amino acid sequences and all of the possible codon choices for each residue (from ref. *4*).

dinucleotides among codons should, if possible, be avoided, since these have been shown to be underrepresented in mammalian DNA (*13*). Best-guess oligonucleotide probes are unlikely to form a perfect match with the target DNA sequence, but they should contain sufficient identity to allow hybridization under appropriate conditions. Such conditions cannot be predicted, however, and they therefore need to be determined empirically. When using best-guess oligonucleotide probes, we recommend that hybridization and washing should be performed at low stringency, followed by autoradiography. Filters should then be washed progressively more stringently, followed by autoradiography after each increase in stringency. This allows the background hybridization to be clearly distinguished from target DNA hybridization signals. As a general rule, the oligonucleotide sequence needs to be at least 70% identical to the target DNA sequence for specific hybridization to be readily detected. In addition, a short region of the probe having high homology to the target sequence appears to be necessary

for the formation of stable hybrids *(3)*. Once a best-guess oligonucleotide probe has been designed, we routinely search the EMBL DNA
sequence data base for sequences that have homology with the probe
sequence in order to assess whether the probe will fortuitously hybridize to known nontarget sequences.

One of the first examples of the successful use of a best-guess oligonucleotide probe was the cloning of human insulin receptor subunit cDNAs *(14)*. Amino-terminal peptide sequences were determined
for both the α and β subunits of this receptor, and these were used to
design the oligonucleotide probes that were used. Figure 2 shows the
peptide sequence determined for the α subunit and the corresponding oligonucleotide probe that was used. Sequence determination of
isolated receptor cDNA clones revealed that there was 84% identity
between the probe sequence and the target DNA sequence.

2.3. PCR Primers

The introduction of PCR has shortened the time needed to perform many molecular biological manipulations. Several laboratories
have recently used this technique to generate partial cDNA clones
encoding, for example, G proteins *(8)* and G-protein-coupled receptors *(7)*. By designing a pair (one sense and one antisense with respect
to mRNA) of degenerate oligonucleotide primers on the basis of two
peptide sequences, an intervening cDNA fragment can be generated
without the need for library construction. Furthermore, the sensitivity
of PCR allows the experimenter to use primers of higher degeneracy
than would normally be used for screening a library; primers of up to
262,144-fold degeneracy have been employed *(15)*.

The cloning of a cDNA encoding the rat luteal lutropin-choriogonadotropin receptor (LH-CG-R) is a good example of the use of
this approach. Seeburg and colleagues *(16)* used peptide sequence
from the amino terminus of the purified receptor, and from fragments
generated either by cyanogen bromide cleavage or by lysyl C-endo-
peptidase cleavage. A 32-fold degenerate oligonucleotide primer of
59 nucleotides (named ks) was designed based on part of the chemically
determined amino-terminal sequence. Similarly, primers of 44
nucleotides (64-fold degeneracy) and 47 nucleotides (fourfold
degeneracy) were designed on the basis of internal peptide sequences.
The latter two oligonucleotides (named fsrc and rsrc, respectively)
were constructed so as to hybridize to mRNA. PCR amplification
was performed using either ks and fsrc, or ks and rsrc, as primers. A

```
     Leu Tyr Pro Gly Glu Val XXX Pro Gly Met Asp Ile Arg Asn XXX Leu Thr Arg XXX His Glu

5' CTG TAC CCC GGC GAG GTG TGC CCC GGC ATG GAC ATC AGG AAC AAC CTG ACC AGG TAC CAC GAG 3'
3' GAC ATG GGG CCG CTC CAC ACG GGG CCG TAC CTG TAG TCC TTG TTG GAC TGG TCC ATG GTG CTC 5'
```

Fig. 2. Best-guess oligonucleotide probe used to isolate the human insulin receptor α-subunit cDNA. XXX indicates an uncertain residue in the peptide sequence; codons for these positions were chosen on the basis of tentative amino acid determinations. The best-guess sequence was designed and built as a double-stranded probe; this is shown below the amino acid sequence. Underlined nucleotides represent mismatches with the isolated cDNA sequence (from ref. *14)*.

622-bp DNA fragment was generated with the former set of primers, and this was shown to encode a polypeptide fragment that contained additional chemically determined peptide sequences. This partial cDNA was subsequently used as a probe to isolate full-length clones from a cDNA library in order to obtain the complete sequence of the LH-CG-R. It is not clear from this report why no product was generated using the second pair of oligonucleotides. Comparison of the rsrc primer sequence to that of the cDNA sequence reveals 74% identity, but the mismatches are distributed evenly along the length of the duplex. It is possible, therefore, that the low degeneracy of the rsrc primer compared to the fsrc primer resulted in the absence of a nucleus of identity in the former. This may have prevented hybridization of the rsrc primer (under the conditions used) to the target DNA sequence. Nevertheless, this example demonstrates the power of this approach for generating partial cDNA fragments, particularly those derived from rare mRNA transcripts.

3. Consensus Oligonucleotides

The advent of recombinant DNA technology has revealed the existence of multigene families that code for many classes of proteins. For example, the cloning of cDNAs encoding GABA$_A$ receptor subunit polypeptides *(2,17)* and a glycine receptor subunit polypeptide *(18)* demonstrated a superfamily of ion-channel receptors *(19)* that includes the previously characterized nicotinic acetylcholine receptor subunits (both muscle and neuronal types). Amino acid sequence alignments of subunits of a particular receptor, or of subunits of different receptors, have revealed extensive regions of homology within the superfamily. These similarities may reflect related functional properties and/or a common evolutionary origin. It is possible, therefore, to design oligonucleotide probes to such regions that will, under appropriate conditions, crosshybridize to DNA fragments that encode members of the same family. This procedure has been used to extend the GABA$_A$ receptor subunit gene family *(20)*. These workers used a 96-fold degenerate pool of 23-base oligonucleotides designed to correspond to an octameric amino acid sequence that is shared by all known GABA$_A$ and glycine receptor subunits. This sequence (Thr-Thr-Val-Leu-Thr-Met-Thr-Thr) occurs in the second proposed transmembrane domain, which is thought to form the lining of the ion channel *(19)*. In our laboratory, we have used consensus oligonucleotide probes, also corresponding to conserved regions (e.g., transmembrane or extracellular domains) of bovine $GABA_A$ receptor subunits, to analyze a series of clones that had been isolated from a 1-d-old chick whole brain cDNA library by crosshybridization using bovine cDNA probes (ref. *21*, and Bateson and Darlison, unpublished results). Fifteen cDNA clones encoding α and β subunits have been divided into four groups based on the following three criteria:

1. The combination of oligonucleotide probes that hybridized;
2. The strength of the hybridization signal obtained with each probe; and
3. The wash temperature at which each hybridization signal was removed.

Representative clones of each group were subsequently sequenced, confirming the validity of the subgroup classification. In theory, therefore, it should be possible to design consensus oligonucleotide probes that correspond to conserved regions of any multigene family, in order to sort large numbers of positive clones into subgroups, thereby aiding clone analysis.

4. Summary

This chapter has attempted to illustrate the different strategies applicable to the isolation and analysis of DNA sequences using oligonucleotide probes. It is not intended to be an exhaustive review of all possible approaches. It should, however, enable the reader to make a more informed decision as to the most suitable approach to his or her problem. It must also be stressed that, because of unknown factors (e.g., the abundance of the mRNA of interest, the accuracy of the chemically determined peptide sequence, and so on), no single strategy or experimental attempt can be guaranteed to result in success. Consequently, perseverance is just as important as the choice of approach.

5. Notes

1. The formula we use for the calculation of the melting temperature (T_m) of an oligonucleotide probe is:

$$T_m = 16.6 \log M + 0.41\ (\%\ G + C) + 81.5 - 820/L - 1.2\ (100 - h)$$

   where M is the monovalent cation concentration (molarity), $\%\ G + C$ refers to the base composition of the oligonucleotide, L is the probe length in nucleotides, and h is percentage identity between the probe and the target sequence. Hybridizations are performed between 5 and 25°C below T_m, and high stringency washing is carried out between 0 and 5°C below T_m. For a full discussion of these parameters, *see* ref. *3*.

2. Since inosine can form stable base pairs of approximately equal strength with adenosine, cytosine, guanine, or thymine *(12)*, one possible way to reduce the degeneracy of oligonucleotide probes is to incorporate inosine at mixed positions, e.g., where the third position of the amino acid codon is ambiguous (*see* ref. *22*). The successful use of this base in the design of PCR primers has also been reported in the isolation of G-protein-coupled receptor sequences *(7)*.

Acknowledgments

We would like to thank the members of our group for critical reading of this manuscript, and for helpful comments and advice concerning the work described herein. We also thank Lucy Howes for excellent secretarial assistance.

References

1. Wisden, W., Morris, B. J., and Hunt, S. P. (1990) *In situ* hybridization with synthetic DNA probes, in *Molecular Neurobiology. A Practical Approach,* (Chad, J. and Wheal, H., eds.), IRL Press, Oxford, UK, pp. 205–225.
2. Levitan, E. S., Schofield, P. R., Burt, D. R., Rhee, L. M., Wisden, W., Köhler, M., Fujita, N., Rodriguez, H. F., Stephenson, A., Darlison, M. G., Barnard, E. A., and Seeburg, P. H. (1988) Structural and functional basis for GABA$_A$ receptor heterogeneity. *Nature* **335,** 76–79.
3. Lathe, R. (1985) Synthetic oligonucleotide probes deduced from amino acid sequence data. *J. Mol. Biol.* **183,** 1–12.
4. Noda, M., Takahashi, H., Tanabe, T., Toyosato, M., Furutani, Y., Hirose, T., Asai, M., Inayama, S., Miyata, T., and Numa S. (1982) Primary structure of α-subunit precursor of *Torpedo californica* acetylcholine receptor deduced from cDNA sequence. *Nature* **299,** 793–797.
5. Noda, M., Takahashi, H., Tanabe, T., Toyosato, M., Kikyotani, S., Hirose, T., Asai, M., Takashima, H., Inayama, S., Miyata, T., and Numa, S. (1983) Primary structures of β- and δ-subunit precursors of *Torpedo californica* acetylcholine receptor deduced from cDNA sequences. *Nature* **301,** 251–255.
6. Saiki, R. K., Scharf, S., Faloona, F., Mullis, K. B., Horn G. T., Erlich, H. A., and Arnheim, N. (1985) Enzymatic amplification of β-globin genomic sequences and restriction site analysis for diagnosis of sickle cell anemia. *Science* **230,** 1350–1354.
7. Parmentier, M., Libert, F., Maenhaut, C., Lefort, A., Gerard, C., Perret, J., Van Sande, J., Dumont, J. E., and Vassart, G. (1989) Molecular cloning of the thyrotropin receptor. *Science* **246,** 1620–1622.

8. Strathmann, M., Wilkie, T. M., and Simon, M. I. (1989). Diversity of the G-protein family: Sequences from five additional a subunits in the mouse. *Proc. Natl. Acad. Sci. USA* **86,** 7407–7409.

9. Jacobs, K., Shoemaker, C., Rudersdorf, R., Neill, S. D., Kaufman, R. J., Mufson, A., Seehra, J., Jones, S. S., Hewick, R., Fritsch, E. F., Kawakita, M., Shimizu, T., and Miyake, T. (1985) Isolation and characterization of genomic and cDNA clones of human erythropoietin. *Nature* **313,** 806–810.

10. Whitehead, A., Goldberger, G., Woods, D. E., Markham, A. F., and Colten, H. R. (1983) Use of a cDNA clone for the fourth component of human complement (C4) for analysis of a genetic deficiency of C4 in guinea pig. *Proc. Natl. Acad. Sci. USA* **80,** 5387–5391.

11. Aota, S.-I., Gojobori, T., Ishibashi, F., Maruyama, T., and Ikemura, T. (1988) Codon usage tabulated from the GenBank Genetic Sequence Data. *Nucl. Acids Res.* **16,** r315–r402.

12. Sambrook, J., Fritsch, E. F., and Maniatis, T. (1989) *Molecular Cloning. A Laboratory Manual.* Cold Spring Harbor Laboratory, Cold Spring Harbor, NY.

13. Bird, A. P. (1980) DNA methylation and the frequency of CpG in animal DNA. *Nucl. Acids Res.* **8,** 1499–1504.

14. Ullrich, A., Bell, J. R., Chen, E. Y., Herrera, R., Petruzzelli, L. M., Dull, T. J., Gray, A., Coussens, L., Liao, Y.-C., Tsubokawa, M., Mason, A., Seeburg, P. H., Grunfeld, C., Rosen, O. M., and Ramachandran, J. (1985) Human insulin receptor and its relationship to the tyrosine kinase family of oncogenes. *Nature* **313,** 756–761.

15. Gould, S. J., Subramani, S., and Scheffler, I. E. (1989) Use of the DNA polymerase chain reaction for homology probing: isolation of partial cDNA or genomic clones encoding the iron-sulfur protein of succinate dehydrogenase form several species. *Proc. Natl. Acad. Sci. USA* **86,** 1934–1939.

16. McFarland, K. C., Sprengel, R., Phillips, H. S., Köhler, M., Rosemblit, N., Nikolics, K., Segaloff, D. L., and Seeburg, P. H. (1989) Lutropin-choriogonadotropin receptor: an unusual member of the G protein-coupled receptor family. *Science* **245,** 494–499.

17. Schofield, P. R., Darlison, M. G., Fujita, N., Burt, D. R., Stephenson, F. A., Rodriguez, H., Rhee, L. M., Ramachandran, J., Reale, V., Glencorse, T. A., Seeburg, P. H., and Barnard, E. A. (1987) Sequence and functional expression of the GABA$_A$ receptor shows a ligand-gated receptor superfamily. *Nature* **328,** 221–227.

18. Grenningloh, G., Rienitz, A., Schmitt, B., Methfessel, C., Zensen, M., Beyruether, K., Gundelfinger, E. D. and Betz, H. (1987) The strychnine-binding subunit of the glycine receptor shows homology with nicotine acetylcholine receptors. *Nature* **328,** 215–220.

19. Barnard, E. A., Darlison, M. G., and Seeburg, P. (1987) Molecular biology of the GABA$_A$ receptor: the receptor/channel superfamily. *Trends Neurosci.* **10,** 502–509.

20. Ymer, S., Schofield, P. R., Draguhn, A., Werner, P., Köhler, M., and Seeburg,

P. H. (1989) GABA$_A$ receptor β subunit heterogeneity: functional expression of cloned cDNAs. *EMBO J.* **8,** 1665–1670.

21. Bateson, A. N., Ultsch, A., Harvey, R. J., Sutherland, M. L., Lasham, A., Glencorse, T. A., Barnard, E. A., and Darlison, M. G. (1990) The chicken GABA$_A$ receptor: cDNA cloning and structural analysis of the α1-subunit gene promoter, in *Regulation of Gene Expression in the Nervous System* (Guiffrida Stella, A. M. Perez-Polo, J. R., and De Vellis, J., eds.), Alan R. Liss, NY, pp. 241–252.

22. Ohtsuka, E., Matsuki, S., Ikehara, M., Takahashi, Y., and Matsubara, K. (1985) An alternative approach to deoxyoligonucleotides as hybridisation probes by insertion of deoxyinosine at ambiguous codon positions. *J. Biol. Chem.* **260,** 2605–2608.

CHAPTER 5

The Use of Degenerate Oligonucleotides for Polymerase Chain-Reaction-Based Isolation of Related DNA Sequences

Tania E. Webb and Alan N. Bateson

1. Introduction

The polymerase chain reaction (PCR; *1*) employs a pair of oligonucleotide primers, one sense and one antisense, with respect to the template DNA. During repeated cycles of heat denaturation, annealing of the oligonucleotide primers to the template and extension of the primers by the enzyme *Taq* polymerase, the intervening fragment is amplified. The use of this thermostable polymerase has led to the automation and subsequent widespread use of this technique. This technique has a number of different applications *(2)* and, in particular, provides a powerful alternative approach to conventional library screening.

In Chapter 4, various examples of the design and use of oligonucleotides in the isolation and characterization of DNA sequences are described. Here, we describe in detail a protocol, utilizing PCR amplification, similar to that used by Libert and coworkers *(3)*, that we have used in our laboratory to isolate related members of the G-protein-coupled receptor superfamily. The technique is founded on the design of degenerate oligonucleotide primers using regions of amino acid and DNA sequence homology (*see* Note 1), and allows the isolation of related sequences. Many neuronal specific genes are members of families or superfamilies *(3,4)*, and therefore, this technique has wide applications in the field of molecular neurobiology.

From: *Methods in Molecular Biology, Vol. 13: Protocols in Molecular Neurobiology*
Edited by: A. Longstaff and P. Revest Copyright © 1992 The Humana Press, Totowa, NJ

2. Materials

1. TE buffer: 10 mM Tris-HCl and 1 mM EDTA, pH 7.5.
2. dNTP mix: 1.25 mM dATP, 1.25 mM dCTP, 1.25 mM dGTP, and 1.25 mM dTTP diluted in TE buffer, pH 7.5.
3. Low-density liquid paraffin (product number 29436, BDH Chemicals Ltd., Poole, UK)
4. 3M sodium acetate, pH 4.9.
5. 3M sodium acetate, pH 5.5.
6. Phenol, 0.1% hydroxquinoline equilibrated with several changes of 0.5M Tris-HCl, pH 8.0, and finally with 0.1M Tris-HCl, pH 8.0, 0.2% mercaptoethanol. (Phenol:chloroform means a 1/1 [v/v] mix of buffered, saturated phenol with chloroform and isoamylalcohol in a ratio of 24:1 [v/v]. Phenol can cause severe burns and should be handled accordingly.)
7. 20X KGB: 4M potassium glutamate, 1M tris-acetate (pH 7.5), 0.4M magnesium acetate, 2 mg/mL bovine serum albumin (Fraction V, Sigma), and 20 mM β-mercaptoethanol.
8. 2X TY: 1.6% (w/v) tryptone, 1% (w/v) yeast extract, and 0.5% (w/v) NaCl.
9. H-Top: 0.8% (w/v) bactoagar, 0.8% (w/v) bactotryptone, and 0.8% (w/v) NaCl.
10. H-Plates: 1% (w/v) bactoagar, 0.1% (w/v) bactotryptone, and 0.8% (w/v) NaCl.
11. 100 mM isopropylthio-β-D-galactoside (IPTG).
12. 20 mg/mL (w/v) 5-bromo-4-chloro-3-indolyl-β-D-galactoside (Xgal) in dimethylformamide.
13. 50 mM CaCl$_2$.
14. PEG solution: 20% (w/v) polyethylene glycol 6000, 2.5M NaCl.
15. Enzymes: *Taq* polymerase, T4 DNA ligase, *Sal*I, and *Sac*I.
16. Oligonucleotide primers for insert sizing, diluted in water to 0.2 µg/µL:LMB2: 5'-GTAAAACGACGGCCAGT-3', and LMB3: 5'-AACAGCTATGACCATGA-3'.
17. Sizing master mix for 10 amplifications: 32 µL of dNTP mix, 4 µL of LMB2, 4 µL of LMB3, 20 µL 10X *Taq* polymerase buffer (Promega Corporation, Madison, WI, USA), and 13.9 µL of water.
18. Degenerate oligonucleotide primer mixes for the PCR, diluted to 0.2 µg/µL.
19. 95% (v/v) ethanol.
20. 80% (v/v) ethanol.

21. Isopropanol.
22. Molecular-biology-grade agarose.
23. Molecular-biology-grade, low-melting-point agarose.

All solutions were made up with double-distilled deionized water. Solutions 1, 4, 5, 13, and 14 should be autoclaved and stored at room temperature. All media (materials 8, 9, and 10) should be autoclaved and stored indefinitely at room temperature. Phenol should be stored at 4°C in brown glass bottles. Solutions 2, 7, 11, 12, 17, and 18 should be stored at –20°C.

3. Methods

3.1. Amplification of First-Strand cDNA (see Note 2)

1. To a 500-μL Eppendorf tube, add 5 μL of 10X *Taq* Polymerase buffer, 8 μL of dNTP mix, 5.8 μL of each primer mix (0.2 μg/μL) (*see* Note 3), 1 μL of first-strand cDNA (approx 75 ng), and 29.7 μL of water. The appropriate control reactions should also be set up (*see* Note 4).
2. Overlay the reaction mixture with 50 μL of light paraffin oil and briefly spin (11,500*g*, 5 s, room temperature) to bring the contents to the bottom of the tube.
3. Incubate the tubes in a thermal cycle at 94°C for 5 min. Maintaining the temperature at 94°C, remove one tube at a time, and quickly add 0.5 μL of *Taq* DNA polymerase (5 U/μL), immediately returning it to the thermal cycler.
4. As soon as the *Taq* polymerase has been added to the last tube, start the PCR reaction: 94°C for 1 min, 1 min at the annealing temperature (*see* Note 5), and 5 min at 72°C (*see* Note 6) for 30 cycles. Finish the cycling reaction with a final extension of 7 min at 72°C.
5. Electrophorese 5 μL of each amplification reaction in a 1% (w/v) agarose gel (for details, *see* ref. *5*) in parallel with appropriate size markers. Visualize under UV light (*see* Notes 4 and 7). Store the remaining product at –20°C.

3.2. Cloning of Amplification Products

1. Digest 5 μg of M13mp18 *(6)* with 20 U of *Sac*I for 16 h at 37°C in 0.5X KGB *(7)* in a total vol of 50 μL.
2. Electrophorese 1 μL of the digestion on a 0.6% (w/v) agarose gel. If the DNA has cut to completion, only a single band of 7.25 kb will be visible. If further digestion is required, add 10 U *Sac*I, and incubate at 37°C for a further 16 h.

3. Once the DNA has cut to completion with *Sac*I, add 3.75 μL of 20X KGB, 6.25 μL of water, and 1 μL of *Sal*I (20 U/μL). Digest for 3 h (*see* Note 8).

4. Add 40 μL of water, and precipitate the DNA by the addition of 10 μL of 3*M* sodium acetate (pH 4.9) and 60 μL of isopropanol (*see* Note 9).

5. Place on wet ice for 15 min, and spin (11,500*g*, 15 min, 4°C). Wash the pellet with 80% (v/v) ethanol, dry, and resuspend in 100 μL water (*see* Note 10).

6. Remove all of the aqueous layer of the amplification to a 1.5-mL Eppendorf tube, and add 55 μL of water. Extract with 1 vol of phenol/chloroform, separating the phases by a 5-min spin (11,500*g*, 5 min, room temperature). Precipitate the DNA by the addition of 10 μL of 3*M* sodium acetate (pH 5.5) and 250 μL of 95% (v/v) ethanol.

7. Spin (11,500*g*, 15 min, 4°C). Wash the pellet with 80% (v/v) ethanol, dry, and resuspend in 20 μL of TE.

8. Electrophorese 10 μL of the DNA in a 0.6% (w/v) low-melting-point agarose gel with appropriate size markers. Stain the gel without shaking. Handle with care, since the gel is fragile. Visualize the DNA under UV light, and excise the region(s) of interest (*see* Note 11), making sure to recover the DNA in the minimum vol of agarose possible.

9. Collect the agarose at the bottom of the Eppendorf tube by a short spin (11,500*g*, 10 s, room temperature).

10. Melt the agarose at 65°C for 10 min, vortexing the tube periodically. Be sure that the agarose is completely molten. In a separate tube, incubate 13.5 μL of water and 0.5 μL of 20X KGB at 37°C while the agarose is melting.

11. Add 5 μL of the molten agarose to the water/KGB, and mix well by pipeting. Add 1 μL of *Sac*I (20 U/μL), mix again, and incubate at 37°C for 16 h. Any remaining DNA in agarose can be stored at –20°C.

12. Add 6.5 μL of water, 2.5 μL of 20X KGB, and 1 μL of *Sal*I (20 U/μL). Incubate for a further 2 h at 37°C.

13. While the digestion of the amplification product is still molten, take 5 μL (*see* Note 12) and add to 11 μL of water prewarmed to 37°C. Mix well by pipeting.

14. Add 1 μL of *Sac*I/*Sal*I cut M13mpl8 (50 ng/μL), 2 μL 10X ligase buffer, and 1 U of T4 DNA ligase (1 U/μL) to make the ligation reaction, and mix well.

15. Incubate at 4°C overnight. During this time, the agarose should settle to the bottom of the tube. Also set up the following control ligations under the same conditions: (1) *Sac*I/*Sal*I cut M13mpl8 vector without added insert in the absence of ligase, and (2) *Sac*I/*Sal*I cut M13mpl8 vector without added insert in the presence of ligase.

16. Inoculate 10 mL of 2X TY with a single bacterial colony (TG1; *8*) and incubate, with shaking (250 rpm), overnight at 37°C.
17. Inoculate 1 mL of the overnight culture into a 500-mL conical flask containing 100 mL of 2X TY, and incubate with shaking for 2–2.5 h at 37°C until the OD is between 0.6–0.8 at 595 nm.
18. Transfer the cells to 50-mL Falcon tubes that have been precooled on ice. Allow to cool for 10 min.
19. Harvest the cells by centrifugation (850*g*, 10 min, 4°C). Pour off the supernatant, and drain the tubes. Gently resuspend the pellets in 25 mL of ice-cold 50 m*M* CaCl$_2$. Incubate the cells on ice for 20 min.
20. Repellet the cells as before, and gently resuspend in 6.6 mL of ice-cold 50 m*M* CaCl$_2$. These competent cells can be stored at 4°C for up to 3 d, but the highest transformation efficiencies are obtained with 1-d-old cells.
21. Remove 7.5 µL of the ligation reaction mixture without remelting the solidified agarose in it. Add this to 300 µL of TG1 competent cells in precooled sterile Oxoid tubes. Mix the tubes gently by flicking, and incubate on ice for 45 min. Carry out transformations of both the ligation controls (step 15) and two further control transformations, namely competent cells on their own and competent cells with uncut M13mp18 (1 ng).
22. Melt the H-top agar, and pipet 3.5-mL aliquots into Oxoid tubes prewarmed to 55°C.
23. Heat-shock the transformations by incubating at 42°C for 2 min, and add 25 µL of 100 m*M* IPTG and 25 µL of 20 mg/mL X-gal.
24. Quickly mix each tube of competent cells with 3.5 mL of H top, and promptly plate out onto prewarmed H plates. When the plates have set, incubate them upside down at 37°C overnight.
25. After the overnight incubation period at 37°C, the number of individual blue plaques on the 1 ng of uncut M13mp18 control plate should be counted. This will give a measure of the transformation efficiency, which should be in the region of 5×10^6 plaques/µg of DNA. The ligation control plates (1 and 2) should have very few plaques present on them. The number of white plaques (false positives) present on the ligation control plate 2 should be subtracted from the total number of whites on each real (experiment) transformation plate. This gives an indication of the success of the cloning procedure. For example, if the number of white plaques on the real transformation plate is not significantly greater than on the ligation control plate 2, but the transformation efficiency is approx 1×10^6, the insert preparation and ligation should be repeated.

3.3. Clone Characterization
by PCR Insert Sizing (see Note 13)

1. Insert size can be estimated by using a pair of PCR primers (LMB2 and LMB3) that flank the polylinker of M13mp18 *(9)*. Amplification of recombinant clones is carried out in 20-µL aliquots of the sizing master mix.

2. DNA is added to the reaction mix by picking a region of the plaque with a sterile toothpick that is then swirled in the reaction mix, and removed and retained (*see* Note 14). Also set up the appropriate control reactions (*see* Note 3), and in this case, a positive control can also be included: a reaction set up with a blue phage plaque from one of the control transformation plates.

3. Overlay the aliquots with 20 µL of paraffin oil, and briefly spin (11,500*g*, 10 s, room temperature).

4. Incubate the tubes in a thermal cycler at 94°C for 5 min. Maintaining the temperature at 94°C, remove one tube at a time, quickly add 0.1 µL of *Taq* DNA polymerase (5 U/µL), and subject to 20 cycles of: 94°C for 1 min, 55°C for 1 min, and 72°C for 2 min, with a final extension of 7 min at 72°C.

5. Electrophorese 5 µL in a 1% (w/v) agarose gel with appropriate size markers. Visualize the DNA under UV light. The positive control should give a band at 115 bp. This protocol will give the sizes of the sequences isolated. Obviously, sequences worth pursuing will not include those that are the same size, or only slightly larger, than the polylinker region amplified in the positive control. It should be realized that the bands isolated from the initial amplification may possibly be cut internally by the enzymes used in the cloning procedure, and therefore, although the sizes of the isolated sequences may be smaller than anticipated, they may still be of interest.

3.4. Clone Analysis by Dideoxy Sequencing

Further characterization of individual clones requires the determination of their sequence using standard protocols. We routinely prepare a single-stranded template DNA *(5,6)* that is sequenced using the dideoxy method of Sanger and coworkers *(5,10)*.

When a large number of clones are generated, as is commonly the case with this PCR-based approach, we track the clones prior to carrying out a full sequencing reaction. This is achieved by sequencing each template with only one of the four deoxy/dideoxy nucleotide mixes (T-tracking), thereby allowing the initial characterization of up to 40 clones on one sequencing gel. As a positive control, single-stranded cloning vector (M13mpl8) should be T-tracked in parallel with the isolated clones to ensure that only clones with inserts are further characterized by a full sequencing reaction using all four nucleotide mixes.

4. Notes

1. Broadly speaking, there are two applications of this technique: first, to isolate new members of multigene families from a given species using the sequences of previously classified family members from that same species and, second, to use corresponding sequences from various species to isolate that same sequence from a new species. In many instances, a combination of these approaches is used.

   The amino acid sequences should be aligned. The length of the primer needs to be at least 24 bases; therefore, regions of homology of eight or more amino acid residues are required. Once these regions of homology have been defined, the corresponding DNA sequences are compared. Better yields are achieved if the 3' end of the primer is designed to correspond to either a methionine or a tryptophan, such that it is nondegenerate. The GC content of the primer mix should be between 40–60%. If the GC content is above 60%, this can lead to the primers annealing to nonspecific GC-rich sequences. If the GC content is below 40%, there may be an increase in nonspecific priming. Any totally conserved bases should be maintained in the final primer design. To decrease the degeneracy of the primer mix codon, preference can be taken into consideration *(11)* and inosine may be included at positions where all four bases occur *(12)*. At points of great sequence diversity, the primer mix may be synthesized in a number of groups. These can be recombined subsequently or used separately. Restriction enzyme sites should be included at the 5' end of the primer mix to allow directional forced cloning of the amplification products. Rare cutting enzymes should be chosen. *Not*I should not be used, however, since it requires a large overhang to cut to completion. If the region of homology extends in the 5' direction, a choice of enzyme can be made that makes use of any conserved bases. The 5' end of the primer mix should be further extended by five bases to ensure digestion by the enzyme of choice.

2. First-strand cDNA is used as the template source, in preference to genomic DNA, to remove the potential problem of the presence of introns in the intervening sequence between the primers. First-strand cDNA is synthesized using the method described by D'Alessio and Gerard *(13)*.

3. The amount of primers added to the reaction is equivalent to 1×10^{-10} mol. Therefore, for an oligonucleotide of 35 bases of a concentration of 0.2 μg/μL, the amount necessary to add to the reaction is calculated by $M \times M_r/[P]$,

$$1 \times 10^{-10} \times (35 \times 330)/0.2 = 5.8 \ \mu L \tag{1}$$

where M is the number of moles required, M_r is the mol wt of the primer, and [P] is the primer concentration.

4. Because of the nature of PCR, it is extremely important that control reactions be carried out in parallel with each "real" amplification, primarily from the consideration of contamination *(2)*. Controls that should be carried out are:
   a. The sense primer mix in the presence of template DNA.
   b. The antisense primer mix in the presence of template DNA.
   c. The sense primer mix in the absence of template DNA.
   d. The antisense primer mix in the absence of template DNA.
   e. Both primer mixes in the absence of template DNA

   Controls a and b will indicate if the individual primer mixes are able to give rise to spurious products. Controls c, d, and e will indicate if any of the solutions involved in the reaction are contaminated with DNA.

   The following precautions can be taken to decrease the chance of contamination. First, PCR reactions should be set up in an area away from further manipulations. Second, the solutions used should be kept in small aliquots solely for the purpose of PCR.

   Although optimization of the amplification reaction cannot be absolutely achieved for highly degenerate primer mix, an important variable is the magnesium concentration of the reaction. If the concentration is too far from the optimum, the product yield can be reduced or the reaction can even be totally inhibited, since *Taq* polymerase has been shown to possess an absolute requirement for Mg^{2+}. This ion is chelated by the template, primers, and nucleotides present in the reaction. For this reason, a Mg^{2+} titration should be performed for each new set of template and primers to establish a suitable Mg^{2+} concentration for the amplification.

5. The conditions of the annealing step of a PCR amplification dictate both the efficiency and specificity of the reaction. Obviously, the annealing conditions are chosen in an attempt to maximize the specificity while still maintaining the efficiency. For degenerate oligonucleotides, the optimum annealing temperature can be calculated using the computer program OLIGO *(14)* or as follows (adapted from ref. *15*):

$$T_a^{opt} = 0.3\ T_m^{primer} + 0.7\ T_m^{product} - 14.9$$

$$T_m^{primer} = 81.5 + 16.6\ (\log_{10}[Mg^{2+}]) + 0.41\ (\%G+C)^1 - 600/N$$

$$T_m^{product} = 81.5 + 16.6\ (\log_{10}[Mg^{2+}]) + 0.41\ (\%G+C)^2 - 675/L \quad (2)$$

where $[Mg^{2+}]$ is the magnesium concentration in the PCR, $(\%G+C)^1$ is the minimum GC content of the primers involved, $(G+C)^2$ is the GC content of the product (this can be taken as an average of 40%), N is the length of the primer, and L is the length of the product, which can be estimated from the known sequences that the design of the primers was based on. These parameters, once determined, should only be used as a guide to establish the optimal conditions for each specific experiment empirically.

6. The extension rate of *Taq* polymerase under optimum conditions is reported to be in the region of 60 nucleotides/s. Therefore, amplification of a 1-kbp product should occur in well under 1 min. We have found that product yield is greatly increased by using the longer extension time of 5 min.

7. The expected result will be determined, to a large extent, by the degree of heterogeneity of the DNA sequences targeted. Thus, although individual band(s) give a clear indication of the success of this step, a smear of DNA fragments may still contain sequences of interest.

8. The incubation time for the *Sac*I digestion is much longer than that for the *Sal*I digestion, because *Sac*I shows reduced activity in KGB buffer in our hands. Regardless of this, we prefer to use KGB as a buffering system, instead of those buffers recommended by the manufacturer, to avoid a DNA precipitation step prior to digestion with *Sal*I.

9. The region of the polylinker excised by the *Sal*I/*Sac*I will not be precipitated under these conditions.

10. At this point, the background of the vector should be tested by transforming vector with and without ligating it, as described in Section 3.2. steps 14, 15, and 21–24. If the vector has been incompletely cut, the background (number of false positives, i.e., white plaques) may be unacceptable. If this is the case, the vector preparation should be discarded and the preparation should be repeated.

11. The region(s) to be excised from the gel should be any band(s) that is(are) not present in control lanes. If a smear is present, running the gel for a shorter period of time allows for the recovery of size-selected regions in smaller volumes of agarose. In this case, the expected size of the product can be estimated roughly from the length of the DNA

sequence between the regions used to design the oligonucleotides. Slices of the gel should be excised as contiguous sections from this region.

12. If individual bands are present in the amplification, their concentration can be estimated by electrophoresis in parallel with known concentrations of DNA, and the appropriate molar ratio of vector to insert can be used (1:1). If a smear is visible, this ratio cannot be accurately calculated, and varying amounts of insert should be used (1–5 μL). We have found that the addition of more than 5 μL of the digestion to the ligation can lead to inhibition of the T4 DNA ligase.

13. This protocol works well when plaques are used directly as the source of template. Single-stranded template can also be used. In this case, dilute 1 μL of the preparation in 1 mL of water, and use 1 μL in the 20-μL sizing reaction.

14. The toothpick used for the insert sizing protocol should be retained for use in the template preparation. We routinely set up both experiments at the same time.

References

1. Saiki, R. K., Gelfand, D. H., Stoffel, S., Scharf, S. J., Higuchi, R., Horn, G. T., Mullis, K. B., and Erlich, H. A. (1988) Primer-directed enzymic amplification of DNA with a thermostable DNA polymerase. *Science* **239,** 487–491.
2. Erlich, H. A., Gelfand, D., and Sninsky, J. J. (1991) Recent advances in the polymerase chain reaction. *Science* **252,** 1643–1651.
3. Libert, F., Parmentier, M., Lefort, A., Dinsart, C., Van Sande, J., Maenhaut, C., Simons, M., Dumont, J. E., and Vassart, G. (1989) Selective isolation and cloning of four new members of the G-protein coupled receptor family. *Science* **244,** 569–572.
4. Barnard, E. A., Darlison, M. G., and Seeburg, P. (1987) Molecular biology of the GABA$_A$ receptor: the receptor/channel superfamily. *Trends Neurosci.* **10,** 502–509.
5. Sambrook, J., Fritsch, E. F., and Maniatis, T. (1989) *Molecular Cloning. A Laboratory Manual,* 2nd ed. Cold Spring Harbor Laboratory, Cold Spring Harbor, NY.
6. Messing, J. (1983) New M13 vectors for cloning. *Methods Enzymol.* **101,** 20–77.
7. McClelland, M., Hanish, J., Nelson, M., and Patel, Y. (1988) KGB: a single buffer for all restriction endonucleases. *Nucleic Acids Res.* **16,** 364.
8. Gibson, T. J. (1984) Studies of the Epstein-Barr virus genome, Ph.D. thesis, Cambridge University, UK.
9. Gussow, D. and Clackson, T. (1989) Direct clone characterisation from plaques and colonies by PCR. *Nucleic Acids Res.* **17,** 4000.
10. Sanger, F., Nicklen, S., and Coulson, A. R. (1977) DNA sequencing with chain termination inhibitors. *Proc. Natl. Acad. Sci. USA* **74,** 5463–5467.
11. Aota, I., Gojobori, T., Ishibashi, F., Maruyama, T., and Ikemura, T. (1988)

Codon usage tabulated from the Genbank Genetic Sequence Data. *Nucleic Acids Res.* **16,** r315–r402.

12. Knoth, K., Roberds, S., Poteet, C., and Tamkun, M. (1988) Highly degenerate, inosine containing primers specifically amplify rare cDNA using the polymerase chain reaction. *Nucleic Acids Res.* **16,** 10,932.

13. D'Alessio, J. M. and Gerard, G. F. (1988) Second strand cDNA synthesis with *E. coli* DNA polymerase I and RNase H: the fate of information at the mRNA 5' terminus and the effect of *E. coli* DNA ligase. *Nucleic Acids Res.* **16,** 1999–2014.

14. Rychlik, W. and Rhoads, R. E. (1989) A computer program for choosing optimal oligonucleotides for filter hybridisation, sequencing and *in vitro* amplification of DNA. *Nucleic Acids Res.* **17,** 8543–8551.

15. Rychlik, W., Spencer, W. J., and Rhoads, R. E. (1990) Optimisation of the annealing temperature for DNA amplification *in vitro. Nucleic Acids Res.* **18,** 6409–6412.

CHAPTER 6

Isolation of Photoreceptor Cell-Specific MEKA cDNA by Differential Hybridization

Che-Hui Kuo

1. Introduction

All cells have a large number of proteins; for instance, a single hepatocyte has about 10^4 different proteins, and a number of them are involved in tissue-specific functions. In neuronal cells, there are additional 10^4 protein species, which are predicted, from kinetic of DNA–RNA hybridization studies, to be involved in specialized neural functions and in intricate systems of communication *(1,2)*. At present, it is possible to separate and analyze numerous proteins using two-dimensional gel electrophoresis; however, only around 10^3 protein species in a given tissue can be detected by this method even when combined with the silver-staining technique. In contrast, if a cDNA library from mRNAs is screened, using molecular biological techniques, about 10^5–10^6 independent clones, each of which is expected to encode different proteins, can be easily isolated and screened. Recent advances in the techniques of genetic engineering encourage the neuroscientist to search for molecules underlying neuronal functions.

In retinal rod cell outer segments, a signal transduction mechanism occurs, which is mediated by retina-specific proteins, such as rhodopsin, transducin (T), phosphodiesterase, and a cyclic GMP-dependent Na channel *(3)*.

From: *Methods in Molecular Biology, Vol. 13: Protocols in Molecular Neurobiology*
Edited by: A. Longstaff and P. Revest Copyright © 1992 The Humana Press, Totowa, NJ

There are a large number of homologies between the photo-transduction system and the receptor G protein effector systems, such as the adenylate cyclase second messenger cascade *(4)*. Arrestin and opsin kinase, both of which are thought to be retina-specific proteins, affect not only the phototransduction system, but also the adrenergic β-receptor system, indicating that homologous proteins might be present in other tissues and function in similar signal transduction *(5,6)*. Isolated cDNA of the γ subunit of a G protein shows a 55% homology with cDNA of the γ subunit of transducin *(7)*. From this evidence, it is apparent that the discovery of a retina-specific protein(s) could produce a further understanding of other signal transduction systems, as well as the visual transduction mechanism.

In the search for other proteins thought to be involved in the visual transduction system, I have analyzed soluble proteins from bovine retina and found that the majority of retinal proteins were identical to those of brain. Similar results were also obtained when in vitro translational products derived from retinal and brain mRNAs were compared. This evidence led me to the use of the differential colony hybridization method *(8)* for the isolation of retina-specific cDNA clones.

The principle of differential hybridization is (1) to prepare three sets of replicated filters on which identical sets of clones from a retina-derived cDNA library are grown, and (2) to hybridize a replica filter with ^{32}P- labeled single-stranded (ss) cDNA probes synthesized from either brain or retinal mRNA. All clones that hybridize with the retinal probe, but not with brain probe, can be distinguished using auto-radiography. Candidates for retina-specific cDNA clones are isolated from the remaining filter (Fig. 1).

A second method, known as subtractive and differential hybridization, has also been reported and theoretically improves the probability of cDNA cloning, especially for the isolation of cDNA of low abundance (0.001%). Briefly, ^{32}P ss cDNA from retina mRNA and an excess of brain mRNA are hybridized. The unhybridized (enriched retina-specific) ss cDNA species is separated from the hybrid brain mRNA - ss cDNA by chromatography using a hydroxyapatite column. Recently, Bowes et al. *(9)* have cloned a cDNA-encoding molecule that might be involved in retinal regeneration. Their sophisticated approach was based on high-affinity binding between biotin and avidin immobilized onto an agarose column. Their method is also summarized in Fig. 1.

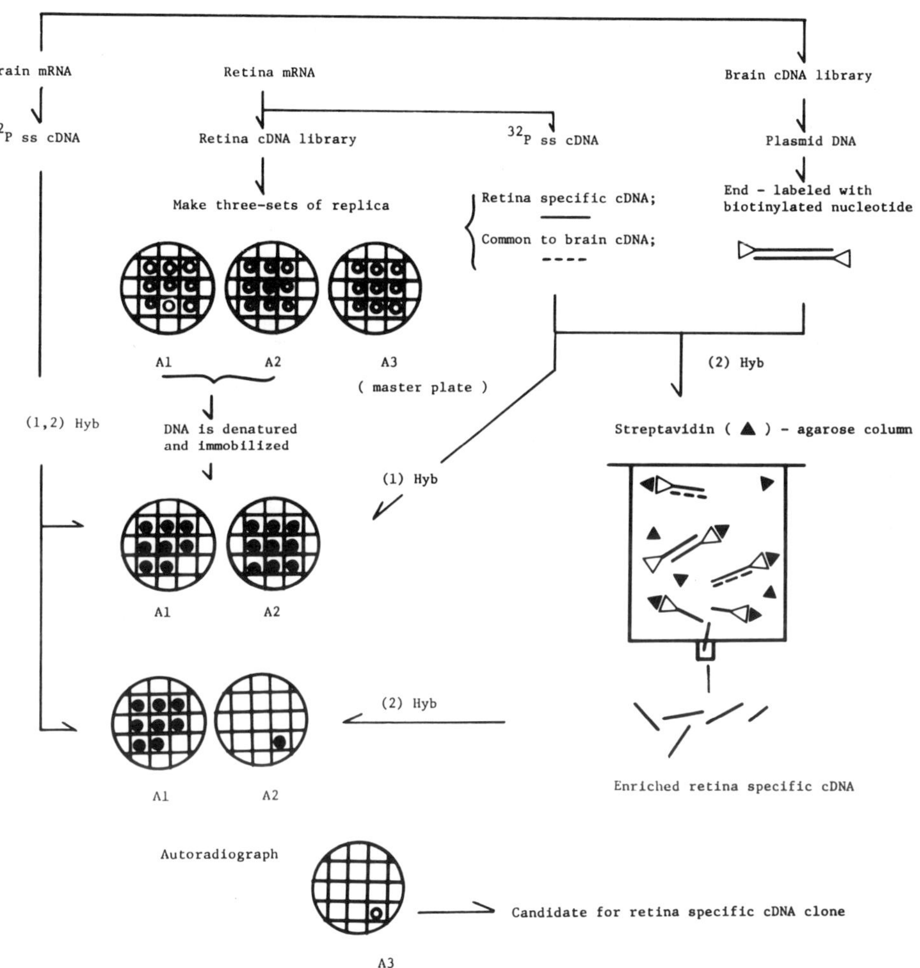

Fig. 1. An outline of the isolation of retina specific cDNA clone. Differential hybridization (1) and subtractive and differential hybridization (2) methods are summarized. *See* detail in Sections 1. and 3. Hyb indicates hybridization.

In this chapter, I describe the differential colony hybridization method by which I have cloned three cDNAs (opsin, transducin γ subunit, and MEKA), which are specific to photoreceptor cell as shown in Fig. 2 *(10–14)*. To isolate those clones, I screened only about 1000 colonies, suggesting that the retina is a highly differentiated tissue. I believe, therefore, that many more cDNAs specific to the

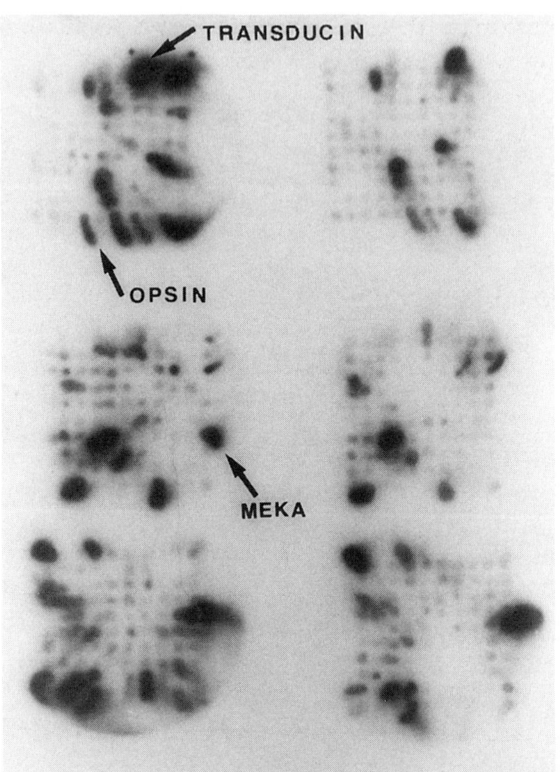

Fig. 2. Autoradiograph of differential colony hybridization. Colonies on the filters (two of three sets of replica filters) were hybridized with ss cDNA synthesized from either retinal (left lane) or brain (right lane) mRNA as described in Section 3.5. Arrows indicate photoreceptor cell-specific cDNA clones (transducin γ subunit, opsin, and MEKA).

photoreceptor cell can be isolated by simply scaling up the number of clones for screening.

2. Materials

2.1. Colony Hybridization

1. 5 mg/mL Diaminopimeric acid (DAP): DAP is dissolved in autoclave-distilled water, filtered through a 0.22-μm millex GV filter (Millipore, Yonezawa, Japan), and stored at 4°C.
2. 5 mg/mL Thymidine.
3. 5 mg/mL Tetracycline solution: To a 10 mg/mL tetracycline hydro-

chloride solution, add equal vol of absolute ethanol, mix, and wrap the container with aluminum foil. Store at –20°C.

4. 0.2M Dithiothreitol (DTT).
5. Chloramphenicol: 34 mg/mL in 100% ethanol, stored at –20°C.
6. L-broth: 10 g/L bacto-tryptone (Difco, Detroit, MI), 5 g/L of yeast-extract (Difco), and 5 g of NaCl (final concentration, 86 mM). Adjust pH to 7.4 with 4M NaOH, make up to 1 L with water, and autoclave for 20 min at 120°C.
7. LDTT (L-broth-DAP-thymidine-tetracycline) medium: add 100 µL of the DAP (5 mg/mL), 100 µL of thymidine (5 mg/mL), and 30 µL of tetracycline (5 mg/mL) to 10 mL L-broth.
8. 20X SSC solution: 3M NaCl, 0.3M citrate.
9. 10% Sarcosyl solution: 10 g sodium N-lauroyl sarcosinate/100 mL water.
10. NTE: 200 mM NaCl, 50 mM Tris-HCl, pH 7.4, 5 mM EDTA.
11. TE: 10 mM Tris-HCl, pH 7.4, 0.1 mM EDTA.
12. 5M NaCl: 292.2 g/L.
13. 0.5M EDTA: 186.1 g disodium ethylene diamine tetraacetate-dihydrate/L, pH 9.0.
14. 1M Tris-HCl, pH 7.4: 121.1 g Tris (hydroxymethyl) amino-methane/1 L. (*See* Note 1).
15. 50X Denhardt solution: Dissolve 2 g each of Ficoll 400 (Pharmacia, Uppsala, Sweden), Polyvinyl pyrrodidone 25 (Nakarai), and BSA F-V (Nakarai, Kyoto, Japan) in H$_2$O. Adjust the vol to 200 mL. Sterilize by filtration under sterile conditions, and divide into 40-mL aliquots in 50-mL Falcon tubes to store at 4°C.
16. 2X hybridization buffer: Add 2.0 mL of 5M NaCl, 2.0 mL of 50X Denhardt solution, 0.5 mL of 1M Tris-HCl, pH 7.4, 0.2 mL of 0.5M EDTA, 0.1 mL of 10% sarcosyl solution, and 0.2 mL H$_2$O to a 15-mL Falcon tube, and mix completely.

2.1.1. Preparation of Culture Plates

LDTT plates: Add 15 g bacto-agar (Difco) to 1 L L-broth. Autoclave for 20 min. Then (*see* Note 3) add 10 mL DAP, 10 mL thymidine, and 3 mL tetracycline, and mix completely. Dispense 20 mL of this LDTT solution into each sterile Petri dish (Falcon, 90 mm) under sterile conditions. Leave them for 1 h to harden. Invert the dishes with or without nitrocellulose filters on the dish and place at 37°C for 24 h to dry.

LDTT-C (LDTT plus chloramphenicol) plates: Prepare as above after addition of 5 mL chloramphenicol (34 mg/mL in 100% ethanol) to 1 L LDTT plate medium.

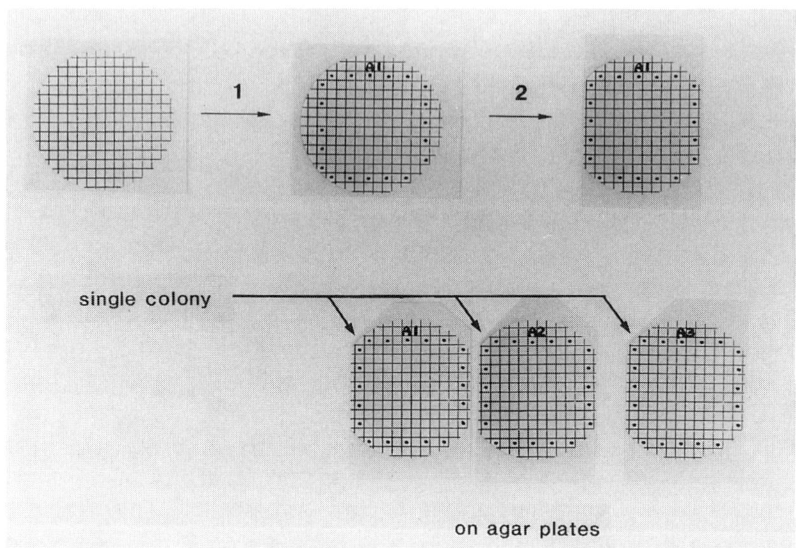

Fig. 3. Preparation of three sets of replicated filters. The nitrocellulose filter was marked and numbered (1), and two corners of the filter were cut (2). This makes it possible to grow two filter sets on a single agar plate (90 mm diameter). A single colony grown on an LDTT plate is transferred to identical positions on three sets of filters as in Note 10.

2.1.2. Preparation of Sheared DNA

1. HB101 genome DNA and pBR322 plasmid DNA: Both of these DNAs are prepared as described in ref. *15,* sheared with an Ultrasonic cell disrupter and checked on an agarose gel. The average size of the sheared DNA should be about 0.5–1.0 kb pairs.
2. Toothpicks (in a beaker) for transferring colonies. Autoclave for 20 min.
3. Whatman 3MM paper (wrapped with aluminum foil).
4. Nitrocellulose filters prepared as in Fig. 3 (*see* Note 2). Autoclaved for 5 min.

2.2. Reagents for ^{32}P ss cDNA Synthesis

1. NTE-saturated phenol (*see 2,3*).
2. CIAA solution: Mix 48 mL of chloroform and 2.0 mL of isoamylalcohol, and store in a glass bottle (*see* Note 4).
3. 10X cDNA synthesis buffer: Autoclave 10X buffer consisting of 0.5*M* Tris-HCl, pH 8.3 at 42°C, 0.1*M* MgCl$_2$, 0.5*M* KCl.
4. 10 m*M* dXTP solution: Make approx 50 m*M* solutions of dATP, dGTP, and dTTP in autoclaved 50 m*M* Tris-base, and check the actual concentrations by UV absorption (*see* Note 5). Dilute each of the

solutions with TE to 40 mM. Prepare 10 mM dXTP solution by mixing equal vol of 40mM dATP, dGTP, dTTP, and TE. Store at –80°C after dispensing into Eppendorf tubes (*see* Note 6).

5. 20 μM dCTP solution: Make 40 mM dCTP as in step 4 above. (*See* Notes 5 and 6). Dilute (1/2000) aliquots of the solution with TE to make 20 μM dCTP solution, and store at –80°C.
6. DNA-grade Sephadex G-50 (Pharmacia): Equilibrate approx 2 g of Sephadex G-50 with 200 mL of NTE and autoclave for 20 min.
7. Oligo (dT)$_{12-18}$: Prepare 500 μg/mL of oligo (dT)$_{12-18}$ (P-L Biochemicals), and store at –20°C (*see* Note 7).
8. Polyadenylic acid: Prepare 5 mg/mL of polyadenylic acid solutions, and store at –20°C (*see* Note 7).
9. Avian myeloblastosis virus reverse transcriptase (24 U/μL), stored at –80°C (*see* Note 8).
10. α-[^{32}P] dCTP (specific activity 3000 Ci/mmol).
11. X-ray film (XAR 5), developer, fixer, and replenisher.
12. Retina and brain poly(A) mRNAs (0.2 μg/μL of TE) stored at –80°C Poly(A) containing mRNA should be purified by oligo(dT) column chromatography from bovine retina and brain total RNAs as described *(10,15* and vols. 2 and 4 of this series*)*.

2.3. Preparation of NTE-Saturated Phenol (Note 4)

1. Dissolve 200 mL of phenol at 68°C, mix with 200 mL H$_2$O, add 200 mg of 8-Quinolinol (antioxidant), and let the mixture separate into two phases.
2. Aspirate the water (upper) phase, add an equal vol of 0.1M Tris-base to the phenol (lower) phase, and mix well.
3. Aspirate the water (upper) phase again.
4. Transfer the phenol phase (approx 100–150 mL) into a 50-mL Falcon tube, wrap with aluminum foil, and store at 4°C (neutralized phenol).
5. Transfer 5 mL of the neutralized phenol into a 15-mL Falcon tube.
6. Add 5 mL NTE, vortex, and leave until two phases are formed.
7. Remove upper phase, and repeat step 6.
8. Store the NTE-saturated phenol at 4°C in the dark.

3. Methods

Retina-specific clones can be isolated by differential colony hybridization derived from a bovine retinal cDNA library expressed in *E. coli*. strain λ 1776 stock in 30% glycerol. This is prepared with retina cDNA and the *Pst*I fragment of pBR322 plasmid vector according to conventional G:C tailing method *(10)*.

3.1. Preparation of Three Sets of Replicas

1. Take 10 µL of the cDNA library solution, and inoculate it into 10 mL of LDTT medium.
2. Incubate at 37°C with shaking for approx 8h. Cells will have grown to give an approx A_{600} of 1.0. This gives about 1×10^7 cells/mL.
3. Dilute the cells with LDTT medium to make 3×10^3 cells/mL and plate an aliquot of 100 µL with a bent Pasteur pipet onto an LDTT plate. This will produce approx 300 colonies.
4. Incubate the plate upside down at 37°C (*see* Note 9) until colonies become a suitable size (about 1 mm diameter).
5. Transfer a single colony grown on the LDTT plate to the same positions on three sets of grid-iron patterned filters as illustrated in Fig. 3 (*see* Note 10).
6. Incubate all the LDTT plates upside down at 37°C for approx 5 h.
7. Transfer two of the three sets of replicas from LDTT plates to LDTT-C (chloramphenicol) plates using dissecting forceps, and incubate at 37°C for 15 h to allow amplification of the plasmid DNA. Store the remaining filter on the LDTT plate at 4°C as a master plate.

3.2. Denaturation and Immobilization of Plasmid DNA on the Filters (16)

1. Remove the filter from one of LDTT-C plates (Section 3.1.; step 7), and place it colony-side up on a sheet of 3MM paper soaked in 0.5N NaOH for 10 min (*see* Note 11).
2. Transfer the filter to a paper towel to dry.
3. Rinse the filter on 3MM paper as in steps 1 and 2 with 1M Tris-HCl (pH 7.4), 0.5M Tris-HCl (pH 7.4)/1.5M NaCl, and then with 2X SSC (dilute 20X solution 10-fold) for 5 min.
4. Dry the filter on a paper towel for 10 min and bake at 80°C for 3h.
5. Keep the filters in a Hybri-bag (Cosmo Bio, Tokyo, Japan) at 4°C until used.

3.3. Hybridization and Washing (17)

1. Remove the filters from the Hybri-bag, and incubate in 250 mL of 2X SSC for 30 min at room temperature.
2. Boil the sheared HB101 genome DNA (5 mg/mL) and pBR322 DNA (0.5 mg/mL) for 10 min, and chill rapidly in ice-cold water for 10 min.
3. Add 200 µL of the HB101 genome DNA, 100 µL of the pBR322 DNA, and 20 µL of polyadenylic acid (5 mg/mL) to 5 mL of 2X hybridization buffer in a 15-mL Falcon tube, and then adjust the volume to 10 mL with H_2O to give the final hybridization buffer. (*See* Note 12).

4. Divide the filters (into two groups for the two probes) into two Hybri-bags, and add 5 mL of the final hybridization buffer.
5. Seal the bags with Polysealer (Fuji Impulse, Tokyo, Japan) after elimination of air, and incubate in a shaking water bath at 65°C for at least 5 h (>10 filters in a bag) or 2 h (<10 filters).
6. Cut off a corner of the Hybri-bag, and pour off the solution. Wear gloves, and add the boiled probe (prepared as in Section 3.5.) suspended with 5 mL of final hybridization buffer to the Hybri-bag.
7. Seal the two bags containing brain or retina ^{32}P ss DNA. Incubate at 65°C for 15 h.
8. The filters in each of two bags should be washed separately. Pour off radioactive solution into the appropriate waste container. Add 20 mL of 2X SSC to Hybri-bag, rinse, and pour off. Repeat once.
9. Transfer the filters from Hybri-bag to 2X SSC (250 mL) in a stainless-steel container, and incubate for 5 min at room temperature.
10. Discard the solution, replace by fresh 2X SSC (250 mL), and incubate as in step 9 for 30 min. Repeat once.
11. Discard the solution, replace by prewarmed 250 mL of 0.1X SSC/0.1% Sarcosyl, cover the container with Saran Wrap™, and incubate at 65°C for 30 min with constant agitation. Repeat at least three times (*see* Note 13).
12. Rinse the filter in 2X SSC (250 mL) at room temperature for 5 min. Repeat once.
13. Place on a paper towel to absorb the liquid, and allow to dry for 10 min.
14. Place the filters in rows on 3MM paper, tape down a corner of each filter, cover with Saran Wrap, and autoradiograph with Kodak XAR 5 film at −80°C for 20 h with Kodak intensifying screen (X-omatic regular type).
15. Develop the film for 5 min, stop development in 1% acetic acid for 1 min, and fix for 10 min under a safe light. An example of an autoradiograph can been seen in Fig. 2 *(10,11)*.

3.4. Preparation of Spun-Column

1. Remove the plunger from a 1-mL disposable syringe, put a small piece of sterile cotton wool in the barrel, and using the plunger, push the cotton wool over the noozle outlet.
2. Fill up the syringe with Sephadex G-50 resin equilibrated in NTE, and centrifuge with a swinging bucket rotor at 2500*g* for 2 min at 20°C.
3. Repeat step 2 until bed vol is reduced to approx 1.0 mL.
4. Wash the column with at least 1.0 mL of NTE, and spin as in step 2.
5. Add 100 µL of NTE, and spin as in step 2. Repeat once.

3.5. Synthesis of ^{32}P ss cDNA (10,18)

1. Add 10 µL solution containing 2 µg retina (brain) mRNA, 4 µL of 10X cDNA synthesis buffer, 1 µL of oligo (dT)$_{12-18}$, and 9.5 µL of H_2O to an Eppendorf tube, mix, incubate at 65°C for 5 min, chill rapidly in ice-cold water, and leave for 1 h at 4°C.
2. Spin for a few seconds in a microcentrifuge.
3. Add 2 µL of 0.2 *M* DTT, 2 µL of 10 m*M* dXTP mixture (dATP, dGTP, and dTTP), 1 µL of 20 µM dCTP, 10 µL (100 µCi) of ^{32}P dCTP, and 0.5 µL of reverse transcriptase solution (24 U/µL) to the tube. Total vol is 40 µL (*see* Note 14).
4. Gently mix with pipet. Do not introduce air bubbles. Incubate at 42°C for 2 h to synthesize ss cDNA.
5. Spin for a few seconds.
6. Add 20 µL of NTE-saturated phenol and 20 µL of CIAA. Make sure the tube is capped tightly, vortex for 1 min, and spin (12,000*g* for 5 min at 20°C).
7. Transfer water (upper) phase, containing cDNAs, to a new tube. Back-extract as follows: To the phenol (lower) phase, add 60 µL of NTE, vortex, spin as step 6, pipet out water phase, and add to the new tube (*see* Note 15).
8. Take the spun column as prepared in Section 3.4.
9. Put an uncapped Eppendorf tube as a reservoir into a 15-mL Falcon tube, and then place the spun column in the Falcon tube.
10. Apply 100 µL of the ss cDNA solution obtained in step 7 to the spun column, and centrifuge with swinging bucket rotor at 2500*g* for 2 min (the effluent will be collected in the reservoir).
11. Transfer the effluent to a new Eppendorf tube.
12. Mix with 250 µL of absolute ethanol, and store at –20°C over night.
13. On the next day, centrifuge the tube at 12,000*g* for 10 min at 4°C.
14. Carefully pour off the ethanol, add 500 µL of cold 100% ethanol to the tube, and centrifuge as step 13. Pour off the ethanol, and place the tube upside down on 3MM filter to drain complete.
15. Dry the cDNA pellet under vacuum, and resuspend with 35 µL H_2O (*see* Note 16).
16. Add 15 µL of 1*N* NaOH, incubate for 30 min at 60°C to hydrolyze the mRNA in the tube, and spin for a few seconds.
17. Add 15 µL of 1*N* HCl and then 10 µL of 1*M* Tris-HCl, pH 7.4 to neutralize the solution, and mix. Total vol is 75 µL.
18. Add 190 µL of ethanol, mix, and store at –20°C over night.
19. Centrifuge at 12,000*g* for 20 min at 4°C, and drain ethanol.

20. Add 500 µL of 75% ethanol (do not mix). Centrifuge as in step 19, and drain ethanol.
21. Add 500 µL of 100% ethanol, spin as in step 19, pour off solution, place tube upside down on 3MM paper and allow to dry.
22. Suspend the cDNA pellet with 100 µL of TE (it usually includes $0.3–1 \times 10^7$ cpm), boil for 10 min, and cool down rapidly in ice water for 10 min.
23. Use as an ss cDNA probe in Section 3.3., step 6.

4. Notes

1. Dissolve Tris in 700 mL of H_2O. Adjust the pH to approx 7.5 by addition of concentrated HCl, leave for at least 5 h, adjust it to pH 7.4, and make up to 1 L with H_2O.
2. Mark and number each cross-striped filter (HAWG, 04700, HA, 0.45 µm; Millipore) with a waterproof pencil (e.g., Al, A2, and A3), and cut two corners of a filter as in Fig. 3. Float it for at least 30 min, and then immerse in a beaker half filled with H_2O.
3. Leave to cool until you can touch the container for at least 20 s.
4. Chromatography-grade phenol and CIAA should be handled with gloves and in a chemical hood. Do not pipet the reagents by mouth. Wash your hands with soap and water thoroughly after use. Check the color of the NET-saturated phenol. Do not use pinkish or reddish phenol.
5. The optical coefficients for deoxy-adenosine-triphosphate at 259 nm, guanosine at 253 nm, thymidine at 260 nm, and cytidine at 271 nm are 1.54, 1.37, 0.74, and 0.91×10^4, respectively. Use a cell with a 1 cm path length.
6. This solution is stable for at least 1 yr.
7. Do not filter.
8. Repeated freezing and thawing cause a reduction in the enzyme activity.
9. Usually small colonies appeared after 11 h at 37°C.
10. Be careful not to tear the filters. Touch a single colony with the toothpick only once, and transfer it by only one touch to the surface of the filter (this is enough to form a colony). It takes about 3 h to transfer 1000 colonies on the three sets of filters. Tilt and examine the filters on the LDTT plate. You can easily judge which positions you have already replicated.
11. Do not leave any air bubbles between the filter and the paper.
12. The final constituents of the hybridization buffer consist of 50 mM Tris-HCl, pH 7.4, 1M NaCl, 10 mM EDTA, and 10X Denhardt con-

taining polyadenylic acid (final concentration 10 µg/mL), HB101 genome DNA (100 µg/mL), and pBR322 DNA (5 µg/mL).

13. Monitor the radioactivity on the filter with a Geiger counter (keep the filter wet). If the equation *Tm* (melting temperature) = 16.6 log (Na⁺) + 0.41 (% G + C) + 81.5 *(17)* is used, the probe will dissociate from a <97% homologous nucleotide sequence under these washing conditions, if the G + C content is estimated to be 40%.

14. Final concentrations for this incubation are 50 m*M* Tris-HCl, pH 8.3, 50 m*M* KCl, 10 m*M* MgCl₂, 10 m*M* DTT, 0.5 m*M* dATP, dGTP and dTTP, and 0.5 µ*M* dCTP containing 0.5 µg of oligo (dT) and 12 U of reverse transcriptase in a tube.

15. Check the activity with a Geiger counter. More than 80% of the added ³²P dCTP should be recovered in this tube.

16. Avoid overdrying the pellet; otherwise, it will be difficult to resuspend it. Pipet the 35 µL of H₂O into a tip, monitor the radioactivity in the tip and tube to check the extent of resuspension, and then return the 35 µL of H₂O to the tube.

References

1. Hahan, W. E. and Laird, C. D. (1971) Transcription of nonrepeated DNA in mouse brain. *Science* **173**, 158–161.
2. Chaudhari, N. and Hahan, W. E. (1983) Genetic expression in the developing brain. *Science* **220**, 924–928.
3. Stryer, L. (1986) Cyclic GMP cascade of vision. *Ann. Rev. Neurosci.* **9**, 87–119.
4. Stryer, L. (1986) G protein: A family of signal transducers. *Ann. Rev. Cell. Biol.* **2**, 391–419.
5. Benovic, J. L., Kuhn, H., Weyand, I., Codina, J., Caron, M. G., and Lefkowitz, R. J. (1987) Functional desensitization of the isolated β-adrenergic receptor by the β-adrenergic receptor kinase: Potential role of an analog of the retinal protein arrestin (48-kDa protein). *Proc. Natl. Acad. Sci.* **84**, 8879–8882.
6. Benovic, J. L., Mayer, F., Jr., Somers, R. L., Caron, M. G., and Lefkowitz, R. J. (1986) Light-dependent phosphorylation of rhodopsin by β-adrenergic receptor kinase. *Nature* **321**, 869–872.
7. Gautam, N., Baetscher, M., Aebersold, R., and Simon, M. I. (1989) A G protein gamma subunit shares homology with *ras* proteins. *Science* **244**, 971–974.
8. Taniguchi, T., Fujii-Kuriyama, Y., and Muramatsu, M. (1980) Molecular cloning of human interferon cDNA. *Proc. Natl. Acad. Sci.* **77**, 4003–4006.
9. Bowes, C., Danciger, M., Kozak, C. A., and Farber, D. B. (1989) Isolation of a candidate cDNA for the gene causing retinal degeneration in the rd mouse. *Proc. Natl. Acad. Sci.* **86**, 9722–9726.
10. Kuo, C.-H., Yamagata, K., Moyzis, R. K., Bitensky, M. W., and Miki, N. (1986) Multiple opsin mRNA species in bovine retina. *Mol. Brain Res.* **1**, 251–260.

11. Kuo, C.-H., Akiyama, M., and Miki, N. (1989) Isolation of a novel retina-specific clone (MEKA cDNA) encoding a photoreceptor soluble protein. *Mol. Brain Res.* **6,** 1–10.
12. Kuo, C.-H. and Miki, N. (1989) Translocation of a photoreceptor-specific MEKA protein by light. *Neurosci. Lett.* **103,** 8–10.
13. Kuo, C.-H., Watanabe, Y., Yamagata, K., and Miki, N. (1989) Developmental changes of MEKA protein and opsin in normal and rd mice. *Develop. Brain Res.* **50,** 139–141.
14. Kuo, C.-H., Taniura, H., Watanabe, Y., Fukada, Y., Yoshizawa, T., and Miki, N. (1989) Identification of a retina-specific MEKA protein as a 33 k protein. *Biochem. Biophys. Res. Comm.* **162,** 1063–1068.
15. Maniatis, T., Fritsch, E. F., and Sambrook, J. (1982) *Molecular Cloning, A Laboratory Manual,* Cold Spring Harbor Laboratory, Cold Spring Harbor, NY.
16. Grunstein, M. and Hogness, D. S. (1975) Colony hybridization: A method for the isolation of cloned DNAs that contain a specific gene. *Proc. Natl. Acad. Sci.* **72,** 3961–3965.
17. Meinkoth, J. and Wahl, G. (1984) Hybridization of nucleic acids immobilized on solid supports. *Anal. Biochem.* **138,** 267–284.
18. Nabeshima, Y., Fujii-Kuriyama, Y., Muramatsu, M., and Ogata, K. (1982) Molecular cloning and nucleotide sequences of the complementary DNAs to chicken skeletal muscle myosin two alkali light chain mRNAs. *Nucleic Acid Res.* **10,** 6099–6110.

Generation of Isoform-Specific Antisera from Cloned cDNAs

Application to Multiple Forms
of the Na/K-ATPase Expressed in Rat Brain

Andrew W. Shyjan and Robert Levenson

1. Introduction

Many proteins that are expressed in the brain, in particular membrane-associated proteins, are the products of multigene families. The HT (serotonin) receptor *(1)*, GABA receptor *(2)*, glutamate receptor *(3)*, Na$^+$ channel *(4)*, muscarinic acetylcholine receptor *(5)*, and glycine receptor *(6)* are examples of neural proteins that are encoded by multiple genes. A central issue in cell and molecular biology involves characterization of the functional differences between closely related members of a family of proteins. One approach to this problem involves biochemical analysis of individual protein isoforms, either *in situ,* or after transfection of cognate cDNAs into suitable recipient cell lines. A second, intersecting line of approach is localization of individual members of a multigene family within specific anatomical compartments or cell types. Structure–function analysis, in conjunction with subcellular localization studies, represents a powerful approach for understanding the physiological significance of isoform diversity.

Na/K-ATPase (EC 3.6.1.3) is an example of a plasma membrane-associated protein that is expressed in brain and is encoded by a multigene family. Na/K-ATPase is the enzyme responsible for the

From: *Methods in Molecular Biology, Vol. 13: Protocols in Molecular Neurobiology*
Edited by: A. Longstaff and P. Revest Copyright © 1992 The Humana Press, Totowa, NJ

active transport of sodium and potassium ions in virtually all animal cells. Na/K-ATPase plays a central role in mediating the electrical activity of the brain. In neurons, the activity of the Na/K-ATPase maintains the ion gradients that provide the driving force for the action potential *(7)*. The uptake of neurotransmitters and the efflux of calcium from neurons are also coupled to the activity of the Na/K-ATPase *(8)*. In glia, the Na/K-ATPase plays a critical role in potassium uptake during periods of intense neuronal activity *(9)*.

Na/K-ATPase has been shown to consist of two subunits. The α or catalytic subunit is a polypeptide with $M_r \sim 100$ kDa that contains the site for ATP hydrolysis *(10)*. The β subunit is a glycosylated polypeptide ($M_r \sim 55$ kDa) whose function has not yet been established. Molecular cloning has revealed that the α and β subunits of the Na/K-ATPase are each encoded by multiple genes. Three α and two β subunit genes have been localized to different chromosomes in the mouse *(11,12)*, and cDNA clones encoding three distinct rat α (α_1, α_2, α_3) and two distinct rat β (β_1,β_2) subunit isoforms have been characterized *(13–16)*. Substantial differences in the tissue and developmental specificity of expression have been found for the genes encoding each of the α and β subunits *(14–18)*. Chromosomal dispersion and tissue-specific expression of the α and β subunit genes suggest that the enzyme encoded by each gene may have properties selected in response to different physiological demands.

Several lines of evidence suggest that isoforms of the Na/K-ATPase differ with respect to their functional properties. Gene transfer experiments indicate that the rat α_1 subunit is much less sensitive to cardiac glycoside inhibition than the rat α_2 and α_3 subunits *(19–23)*. Examination of the enzymatic properties of the α_3 isoenzyme suggests that this ATPase exhibits a lower apparent K_m for Na^+ than the α_1 enzyme and does not show the positive cooperative Na^+ activation that is a characteristic feature of the α_1 isoform *(24)*. In adipocytes, insulin has been shown to stimulate the activity of the α_2 subunit *(25)*, suggesting differential regulation of α subunit isoforms. Expression of β_1 subunits has been detected in rat kidney, heart, and brain *(18)*, whereas expression of β_2 subunits appears to be restricted to brain, pineal gland, and thymus *(26)*. These results suggest that β_1 and β_2 subunits may play specialized roles in different tissues and cell types. However, the physiological significance for α and β subunit isoform diversity has not been clearly explained.

To gain further insight into possible physiological differences between Na/K-ATPase isoforms, we have initiated a series of experiments designed to localize expression of the isoforms within the brain. Initial experiments indicate that mRNA transcripts encoding all three α and two β subunit isoforms are expressed in rat brain *(14–16)*. *In situ* hybridization histochemistry employing cDNA probes specific for each α subunit isoform indicates that the pattern of expression of α_1, α_2, and α_3 subunit mRNAs is complex, and varies markedly within distinct anatomical regions and cell types *(26a)*. However, characterization of the cellular and subcellular distribution of Na/K-ATPase α and β subunit polypeptides within the brain has not yet been accomplished.

The ability to localize expression of Na/K-ATPase α and β subunit polypeptides is a challenging problem, primarily because antibodies capable of distinguishing between the different α and β subunit isotypes have not been available. From a general perspective, isolation and purification of antisera capable of distinguishing between closely related members of a family of proteins continues to be technically difficult goals to achieve, especially in situations in which such proteins are present in low abundance or when purified preparations of the proteins cannot be obtained. Generation of antisera specific for each of the Na/K-ATPase α subunit isoforms has been limited for three principal reasons. First, tissues that express high levels of either the α_2 or α_3 subunit have not yet been identified. Second, the α_2 and α_3 subunits exhibit similar mobilities on SDS-containing polyacrylamide gels and, therefore, cannot be separated and purified by conventional protein chemistry techniques. Third, the three α subunit isoforms exhibit a high degree of amino acid sequence homology. The deduced amino acid sequence for each rat α subunit isoform indicates a very high degree of evolutionary conservation. Complete identity is observed at 825 of 1013 positions in the amino acid sequence for all three isoforms *(13,14)*. Pairwise comparison of the amino acid sequence of the isoforms reveals that most of the differences are represented by single amino acid changes that occur at random positions along the length of the polypeptide. We believed that the amino acid sequence homology between the three α subunit isoforms was so high as to preclude the use of synthetic peptides as antigens for generating isoform-specific antisera.

In this chapter, we describe a strategy for generating a panel of antisera specific for each of the Na/K-ATPase α and β subunit isoforms. Our approach utilizes cloned cDNAs as starting material. Specific fragments of cDNAs encoding portions of the rat Na/K-ATPase α_1, α_2, α_3, β_1 and β_2 subunits were separately fused to the *E. coli* trpE gene, and used to produce trpE-α and -β subunit fusion proteins in *E. coli*. Purified fusion proteins were then used as antigens to produce polyclonal antisera in rabbits. Antisera produced from a specific α or β subunit fusion protein contain antibodies that are monospecific for that isoform, and can be purified by immunoabsorption to the other α or β subunit fusion proteins. This approach has a number of advantages for the development of isoform-specific antisera. First, antigen is derived from a fusion protein expressed in bacteria. Isolation and purification of a specific isoform as starting material are therefore not required. Second, antibodies are raised against a segment of the polypeptide specified by the cDNA sequence. Antisera can thus be generated that are reactive with defined domains of a particular polypeptide. Third, it is possible to generate monospecific antisera against highly conserved members of a multigene family. Monospecific antisera can be produced in situations where amino acid sequence homology between polypeptides is so high as to preclude the use of synthetic peptides as antigens. Our approach therefore has wide applicability in the design of immunologic reagents that can discriminate among closely related members of a family of proteins.

2. Materials

2.1. Plasmid Constructions

1. Restriction endonucleases (*see* Section 3.2.1.).
2. Calf intestinal phosphatase.
3. 1% Agarose gel.
4. TE: 1 m*M* EDTA, 10 m*M* Tris-HCl, pH 7.6.

2.2. Fusion Protein Production

1. M9CA medium: This medium (*27*) is used for the growth of bacteria expressing fusion proteins. A stock solution of 5X M9 salts is made by dissolving the following salts in deionized water to a final vol of 1 L.

$Na_2HPO_4 \bullet 7H_2O$	64 g
KH_2PO_4	15 g
NaCl	2.5 g
NH_4Cl	5.0 g

Divide the salt solution into 200-mL aliquots, and sterilize by autoclaving. Stock solutions of M9 salts can be stored indefinitely at room temperature. To 750 mL of sterile deionized water, add:

5X M9 salts	200 mL
20% Glucose	20 mL
Casamino acids	0.5% (w/v)

2. Ampicillin stock solution: 25 mg/mL of the sodium salt of ampicillin in water. Sterilize by filtration, and store in aliquots at –20°C. Use at a final concentration of 100 µg/mL. (Stock solution can be stored indefinitely at –20°C.)

3 Tryptophan: Stock solution is 10 mg/mL. Store at 4°C. Use at a final concentration of 40 µg/mL. (Stock solution can be stored indefinitely at 4°C.)

4. 3β-indoleacrylic acid: Stock solution is 5 mg/mL in 95% EtOH. Store at –20°C. Use at a final concentration of 50 µg/mL. (Stock solution can be stored indefinitely.)

5. Cracking buffer: $6M$ urea, 2% SDS, 150 mM NaCl, 30 mM Tris-HCl, pH 7.5. Store at room temperature. Adjust to 1% β-mercaptoethanol just prior to use.

2.3. Antibody Purification

1. Ponceau-S stain: 0.5% Ponceau-S (w/v) in 0.5% trichloroacetic acid (TCA).

2. Glycine hydrochloride buffer: Working solution is $0.2M$ glycine, pH to 2.5 with concentrated hydrochloric acid. (Can be stored indefinitely at room temperature.)

3. Tris-buffered saline (TBS): Working solution is 150 mM NaCl and 30 mM Tris-HCl, pH 7.5. Can be stored indefinitely at 4°C.

2.4. Microsomes and Western Blotting

1. Homogenization buffer: Working solution is $0.25M$ sucrose, 150 mM NaCl, and 30 mM Tris-HCl, pH 7.5. Store at 4°C. (Working solution can be stored indefinitely.)

2. Transfer buffer: 20% methanol, $0.2M$ glycine, and 25 mM Tris base (do not adjust pH).

3. Blotto: Working solution is 5% dry milk (milk protein solids, available in the supermarket), 0.5% Tween 20, 150 mM NaCl, and 30 mM Tris-HCl, pH 7.5. Store at 4°C. (Working solution can be stored indefinitely.)

4. Wash buffer: Working solution is 0.5% Tween 20, 150 mM NaCl, and 30 mM Tris-HCl, pH 7.5. Store at 4°C. (Working solution can be stored indefinitely.)

3. Methods

3.1. Strategy for Antibody Production

The strategy we adopted for the production of isoform-specific antisera is shown in Fig. 1. This approach utilizes cloned cDNAs for each α subunit isoform as starting material. We chose cDNA fragments

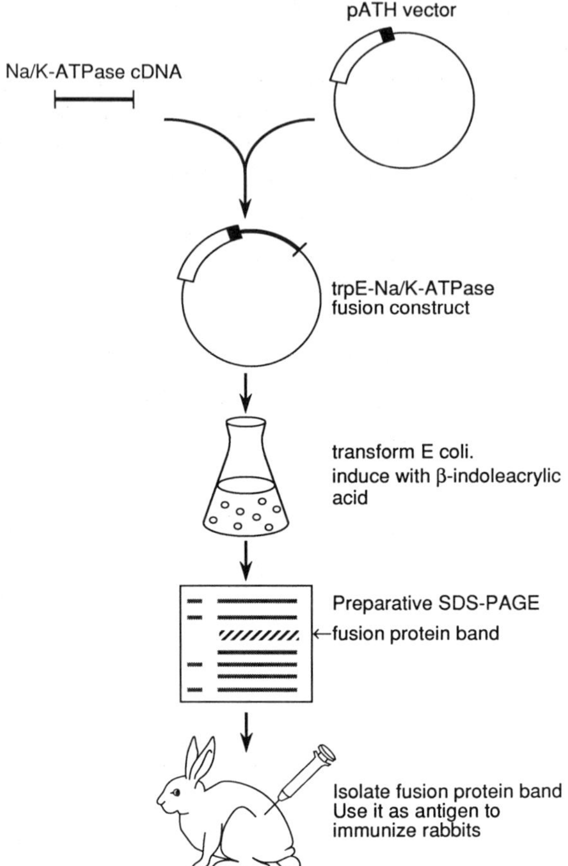

Fig. 1. Strategy for production of fusion protein-derived antisera. To produce isoform-specific antisera, cDNAs encoding portions of the Na/K-ATPase α and β subunits were ligated into the pATH expression vector in-frame with the trpE gene. The open box in the drawing of the pATH vector construct represents the trpE gene, the closed box represents the polylinker, and the shaded line depicts the cDNA insert. Transcription occurs in the clockwise direction. *E. coli* harboring a pATH vector construct express the fusion protein following tryptophan starvation and addition of 3β-indoleacrylic to the culture medium. To purify fusion protein, bacteria were lysed, total bacterial protein solubilized, and the protein size fractionated by preparative SDS-PAGE. Proteins were visualized by immersing the gel in ice-cold 2.5M KCl. As shown in the drawing, the fusion protein band is the predominant band in the gel. Mol-wt markers are depicted on the left side of the gel. The fusion protein band was then excised from the gel, emulsified in Freund's adjuvant, and used to immunize rabbits.

spanning a common, overlapping portion of the H4–H5 intracellular domain of the α subunit (each α subunit contains seven putative membrane-spanning domains designated H1–H7). Each of the cDNA fragments was ligated to the *E. coli* trpE gene using the expression vector pATH *(see* details in Section 3.2.) and used to generate fusion proteins in *E. coli.* The amino acid residues encoded by the cDNA fragments share >90% identity. Antisera directed against the fusion proteins should therefore contain antibodies that crossreact with each of the three α subunit isoforms. We reasoned that antisera produced from a specific α subunit fusion protein should contain antibodies that react uniquely with that isoform and that crossreacting antibodies could be removed from each antisera by absorption against heterologous fusion proteins.

3.2. Plasmid Constructions

Segments of the cDNAs encoding each of the three rat Na/K-ATPase α subunits and two β subunits are fused to the 3' terminus of the *E. coli* trpE gene using pATH expression vectors. The pATH vectors were originally developed by Dieckmann and Tzagoloff *(28)*. They contain a pBR-derived backbone, a selectable marker (the ampicillin resistance gene), and a polylinker containing multiple cloning sites. The utility of the vectors derives from two principal considerations. First, the polylinker is constructed so that, by choosing appropriate restriction sites, cDNAs can be inserted in-frame with the trpE gene. Second, the trpE promoter can be induced to high levels by tryptophan starvation and addition of 3β-indoleacrylic acid to the culture medium. Under these conditions, the trpE fusion protein is the predominant protein expressed in *E. coli.* Plasmid constructions employing α subunit cDNAs are shown in Fig. 2, whereas constructions utilizing β subunit cDNAs are shown in Fig. 3. The steps involved in plasmid constructions are outlined in the following.

1. Digest pATH vector DNA with appropriate restriction endonucleases *(see* Section 3.2.1.) according to the manufacturer's instructions.
2. Add 1 μL of calf intestinal phosphatase to the restriction digest reaction, and incubate for 30 min at 37°C.
3. Electrophorese DNA through a 1% agarose gel (10 V, 20 h).
4. Isolate the vector DNA from the gel and resuspend in TE at a final concentration of 1 μg/mL. Store DNA at 4°C.
5. Digest the plasmids containing Na/K-ATPase α or β subunit cDNA with appropriate restriction endonucleases so that the DNA fragments have ends compatible with those of the vector *(see* Section 3.2.1.).

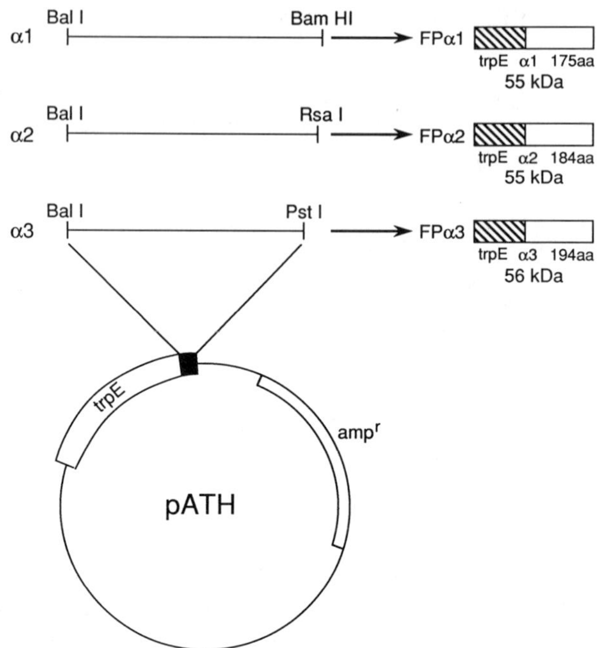

Fig. 2. Structures of the trpE–Na/K-ATPase α subunit fusion proteins. Segments of the cDNAs encoding each of the three rat Na/K-ATPase α subunits (top left) were fused to the trpE gene using the pATH expression vector (bottom left). This plasmid contains the trpE gene (trpE), a polylinker with multiple cloning sites (shaded box), and a selectable marker, the ampicillin-resistance gene (amp^r). The restriction endonuclease sites used in constructions are depicted above each cDNA fragment. Structures of the trpE–Na/K-ATPase fusion proteins are shown at the right. Each fusion protein consists of the 37 kDA trpE protein fused to: 175 amino acids of the α_1 subunit to yield the 55-kDa fusion protein FPα_1; 184 amino acids of the α_2 subunit to yield the 55-kDA fusion protein FPα_2; 194 amino acids of the α_3 subunit to yield the 56-kDA fusion protein FPα_3.

6. Electrophorese DNA through a 1% agarose gel (10 V, 20 h).
7. Isolate the DNA restriction fragment from the gel, and resuspend in TE at a final concentration of 1 μg/mL. DNA can be stored at 4°C.
8. Ligate 200 ng of insert DNA to 200 ng of vector DNA. To the ligation reaction, add 1 μL of 10X ligation buffer (provided by manufacturer), 1 μL of 10 m*M* ATP, H_2O to 10 μL, and 0.5 Weiss U of bacteriophage T4 DNA ligase.
9. Incubate 8–12h at 16°C.

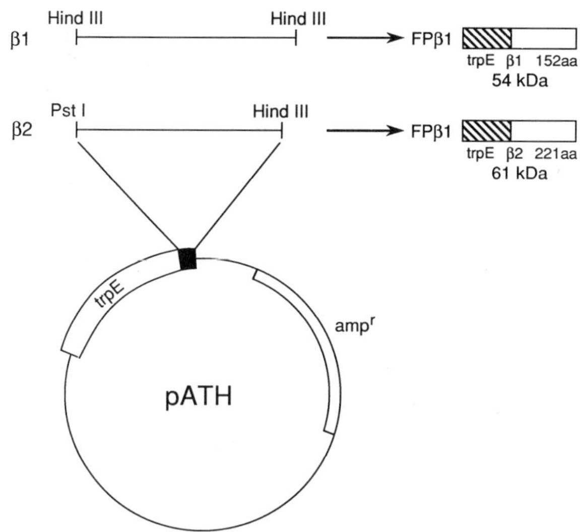

Fig. 3. Structures of the trpE–Na/K-ATPase β subunit fusion proteins. Segments of the cDNAs encoding the rat Na/K-ATPase β_1 and β_2 subunits (top left) were inserted in-frame with the trpE gene using the pATH expression vector (bottom left). Structures of the trpE–Na/K-ATPase fusion proteins are shown at the right. Each fusion protein consists of the 37 kDA trpE protein fused to: 152 amino acids of the β_1 subunit to yield the fusion protein FPβ_1; 221 amino acids of the β_2 subunit to yield the fusion protein FPβ_2.

10. Transform competent *E. coli* with 0.0, 0.01, 0.1, and 1 µL of the ligation reaction.
11. Pick several transformed colonies, and check for presence and proper orientation of the cDNA insert (*see* Note 1).

3.2.1. Appropriate Restriction Endonucleases

Plasmid constructions for Na/K-ATPase α and β subunits are described in the following.

1. α_1 Subunit: A cDNA clone encoding the entire coding region of the rat α_1 subunit *(19)* is digested with *Bal*I and *Bam*HI. The 526-bp fragment is ligated to pATH11, which has been digested with *Sma*I and *Bam*H1. The resulting plasmid construction (pFPα_1) fuses the trpE protein to 175 residues of the α_1 subunit protein (Fig. 2), which correspond to residues 338–513 of the sequence deduced from cDNA.

2. α_2 Subunit: A full-length rat α_2 subunit cDNA *(20)* is digested with *Bal*I and *Rsa*I. The 553-bp fragment is ligated to pATH11, which has been digested with *Sma*I. The resulting plasmid construction (pFPα_2) fuses the trpE protein to 184 residues of the α_2 subunit protein (Fig. 2), which correspond to residues 335–519 of the sequence deduced from cDNA.

3. α_3 Subunit: Full-length rat α_3 subunit cDNA *(18)* is digested with *Bal*I and *Pst* I. The 583-bp fragment is ligated to pATH11, which has been digested with *Sma* I and *Pst* I. The resulting plasmid construction (pFPα_3) fuses the trpE protein to 194 amino acids of the α_3 subunit protein (Fig. 2), which correspond to residues 320–514 of the sequence deduced from cDNA.

4. β_1 Subunit: A full-length rat β_1 subunit cDNA *(15)* is digested with *Hind* III. The 739-bp fragment is ligated to pATH3, which has been digested with *Hind* III. The resulting plasmid construction (pFPβ_1) fuses the trpE protein to 152 residues of the β_1 subunit (Fig. 3), which correspond to residues 152–304 (carboxyl-terminus) of the sequence deduced from cDNA.

5. β_2 Subunit: A full-length rat β_2 subunit cDNA *(16)* is digested with *Pst* I and *Hind* III. The 663-bp fragment is ligated to pATH11, which has been digested with *Pst* I and *Hind* III. The resulting plasmid construction (pFPβ_2) fuses the trpE protein to 221 residues of the β_2 subunit (Fig. 3), which correspond to residues 63–284 (carboxyl-terminus) of the sequence deduced from cDNA.

3.3. Production of Fusion Proteins

3.3.1. Mini-Induction of Bacteria
Expressing Plasmid Constructs

After transformation of *E. coli* with a fusion protein-producing plasmid, it is important to verify that the trpE gene can be induced to high levels and that the cells produce a novel fusion protein *(see* Note 2). This can be accomplished by gel electrophoretic analysis of total cellular proteins from bacteria producing the fusion protein of interest. A small culture of bacteria will produce large amounts of the fusion protein after tryptophan starvation and growth in 3β-indoleacrylic acid. To analyze fusion protein production, the bacteria are lysed, total bacterial protein solubilized, and the proteins size-fractionated on an SDS-containing polyacrylamide gel (SDS-PAGE). Uninduced bacteria and bacteria expressing the unfused trpE gene are run in control lanes. Visualization of the proteins with Coomassie blue stain will reveal whether the bacteria produce a fusion protein of the expected size. The steps involved in this procedure are outlined in the following.

1. Pick a single bacterial colony transformed with the fusion protein construct. Incubate overnight in 10 mL of M9CA medium containing ampicillin (100 µg/mL) and tryptophan (40 µg/mL).
2. Inoculate 0.5 mL of each overnight culture into duplicate 250-mL Ehrlenmyer flasks containing 10 mL of fresh M9CA medium (with ampicillin). Add tryptophan (40 µg/mL) to one culture, and grow the second culture in the absence of tryptophan. Incubate the cells for 2 h at 37°C with vigorous shaking, and then add 3β-indoleacrylic acid (50 µg/mL) to the culture lacking tryptophan to induce high levels of fusion protein. Grow the cultures for an additional 4 h at 37°C.
3. Pellet the bacteria by centrifugation (500g, 15 min, 4°C). Discard the supernatant and resuspend the pellet in 300 µL of cracking buffer. Incubate at 37°C for 3 h.
4. Add 3 µL of 1% bromophenol blue to each sample and centrifuge (6000g, 15 min, room temperature). Carefully remove supernatant and discard pellet (*see* Note 3a).
5. Load 50 µL of each sample onto an SDS gel containing 10% polyacrylamide.
6. Visualize the proteins by staining with Coomassie brilliant blue.

3.3.2. Large-Scale Induction of Bacteria

In order to produce sufficient quantities of fusion protein for use as antigen, large-scale induction of bacteria harboring the trpE–cDNA plasmid construct is required. The procedure for large-scale inductions is described in the following.

1. Inoculate the bacteria (shown by mini-induction to produce fusion protein) into two 50-mL tubes containing 10 mL of M9CA medium (with ampicillin and tryptophan), and incubate overnight at 37°C.
2. Inoculate 5 mL of the overnight cultures into each of four 4-L flasks containing 100 mL of M9CA medium (with ampicillin and no tryptophan). Incubate the cultures at 37°C for 2 h with vigorous shaking (*see* Note 2).
3. Add 1 mL of 3β-indoleacrylic acid (stock solution) to each of the cultures. Grow the cultures for an additional 6 h at 37°C.
4. Store the cultures at 4°C overnight.
5. Pellet the bacteria by centrifugation (500g, 15 min, 4°C). Discard the

supernatant, and resuspend the pellet in cracking buffer (2 mL of cracking buffer/100 mL of bacterial culture).

6. Incubate the samples for 3 h at 37°C. Add 10 μL of 1% bromophenol blue/1 mL of sample and centrifuge (500*g*, 15 min, 4°C). Carefully remove the supernatant, and discard the pellet (*see* Note 3a).

7. Electrophorese the entire sample (8 mL from 400 mL of original culture) onto a preparative (6 mm thick) SDS gel containing 10% polyacrylamide. Run the gel at 30 mA for 3 h.

8. Visualize the fusion protein by immersing the gel in ice-cold 2.5*M* KCl for 10 min (*see* Note 3b). Excise the fusion protein band from the gel.

9. Rinse the gel slice three times for 5 min in distilled water. Homogenize the gel slice in a tissue grinder, and then pass the slurry through a 27-g needle.

3.3.3. Immunoreactivity of Fusion Proteins

It is important to verify that a fusion protein represents the product of its encoding cDNA. This can be accomplished by immunoblotting the fusion protein with an antiserum reactive with the polypeptide. In the case of a multigene family, a polyclonal antiserum that is reactive with all or most members of the gene family is most useful in this regard. To establish that Na/K-ATPase α and β subunit fusion proteins represent the products of α and β subunit cDNAs, they are immunoblotted with antisera raised against Na/K-ATPase purified from rat kidney (K3) or rat brain stem axolemma (Ax2) *(29)*. These antisera were kindly provided by Kathleen Sweadner (Massachusetts General Hospital, Boston, MA). As shown in Fig. 4A, K3 antiserum reacts specifically with α_1 subunit fusion protein. Ax2, on the other hand, is reactive with the α_2 and α_3 fusion proteins (Fig. 4B). K3 antiserum also reacts with β_1 subunit fusion protein (Fig. 4C). These results indicate that the trpE–cDNA constructs encode portions of the Na/K-ATPase α and β subunits.

3.4. Rabbit Immunizations

Rabbits are immunized according to the following schedule (*see* Note 4):

1. Emulsify 0.5 mL of gel slurry with 0. 5 mL of Freund's complete adjuvant.

2. Inject the gel slurry into multiple back sites of female New Zealand White rabbits. Each of the fusion proteins is used to immunize a set of two rabbits.

3. Rabbits receive additional intradermal boosts every 2 wk thereafter of 0.5 mL of gel slurry emulsified with 0.5 mL of incomplete Freund's adjuvant.

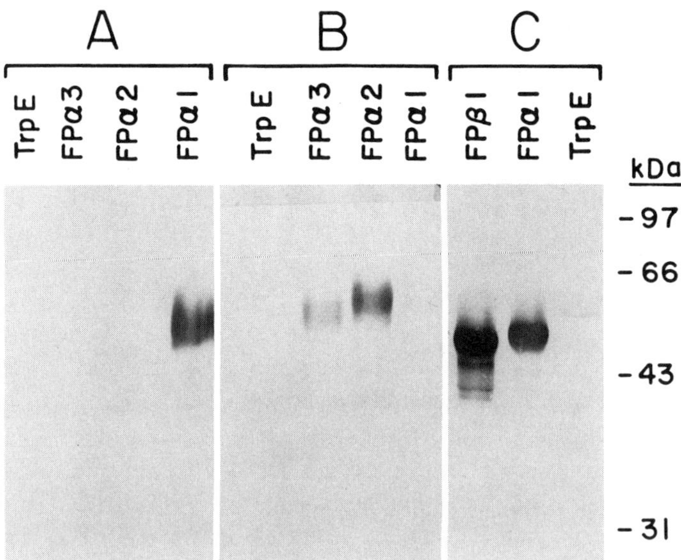

Fig. 4. Immunoreactivity of fusion proteins. Fusion proteins produced from trpE–α (panels A and B) and β (Panel C) subunit constructs were electrophoresed through SDS-containing 10% polyacrylamide gels, transferred to nitrocellulose filters, and probed with Na/K-ATPase antisera. Total protein from bacteria expressing the unfused trpE gene was run in control lanes. The filters shown in panels A and C were probed with K3 antiserum (raised against rat kidney Na/K-ATPase), whereas the filter shown in panel B was reacted with Ax2 antiserum (raised against Na/K-ATPase from rat brain stem axolemma). The positions of the mol-mass markers are shown at the right.

4. After the third boost, aliquots of serum are assayed for immunoreactivity on Western blots containing microsomal proteins prepared from rat kidney and brain.

3.5. Antibody Purification

The strategy for purifying antibodies specific for each Na/K-ATPase α and β subunit isoform is shown schematically in Fig. 5. Antiserum raised against a trpE–α or -β subunit fusion protein is immunoabsorbed against heterologous fusion proteins in order to remove nonspecific, crossreacting antibodies. For example, α_1 subunit antiserum is absorbed against nitrocellulose strips containing α_2 and α_3 subunit fusion proteins. The procedure for antibody purification is outlined in the following.

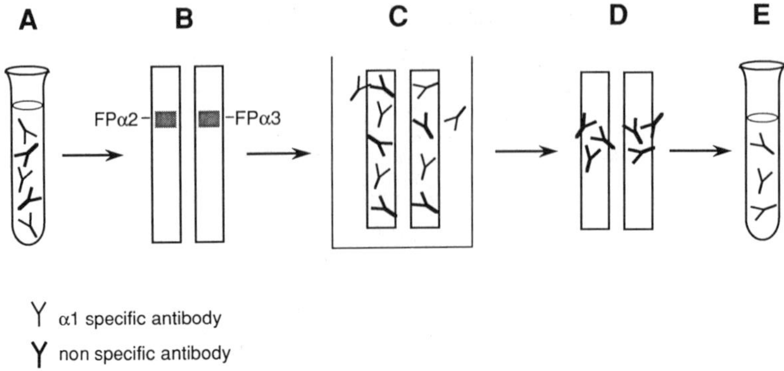

Fig. 5. Strategy for antibody purification. Antiserum produced against α_1 subunit fusion protein (A) contains antibodies specific for the α_1 (Y) subunit as well as nonspecific antibodies (Y). An α_1 specific antiserum can be purified by immunoabsorbing the antiserum onto nitrocellulose strips containing α_2 and α_3 subunit fusion proteins (FPα_2 and FPα_3 respectively; B). Nonspecific antisera are absorbed onto the filter (C and D), whereas α_1 specific antibodies remain unbound and in solution (E). The identical protocol procedure was used to purify antisera specific for the α_2, α_3, β_1, and β_2 subunits.

1. Follow steps 1–6 of the induction protocol described in Section 3.3.2.
2. Size fractionate the proteins on a 0.5 mm thick SDS-containing polyacrylamide gel. Run the gel at 30 mA for 3 h.
3. Transfer the proteins to a nitrocellulose filter (as in ref. *30)* using a semidry transfer chamber (24 V, 30 min, 4°C). Immerse the nitrocellulose filter in Ponceau-S stain for 5 min at room temperature. Wash the filter for 10 min in double-distilled water in order to visualize protein bands.
4. Remove the strip of nitrocellulose containing the fusion protein band (Fig. 5B), and quench the filter in blotto (*see* Section 2.4., step 3) for 20 min at room temperature.
5. Dilute 1 mL of fusion protein-derived antiserum in 10 mL of blotto.
6. Allow strips to absorb antiserum for 1 h (Fig. 5C). Remove antiserum and keep (Fig. 5D).
7. Wash the strips in glycine hydrochloride buffer for 2 min.
8. Neutralize the strips by washing three times for 2 min each in TBS.
9. Reapply strips to antiserum. Repeat steps 6–8 seven times.

3.5.1. Specificity of Fusion Protein-Derived Antisera

After affinity purification, it is essential to test the fusion protein-derived antisera for isoform specificity. Initially, this can be accomplished by immunoblotting antisera produced from one of the fusion

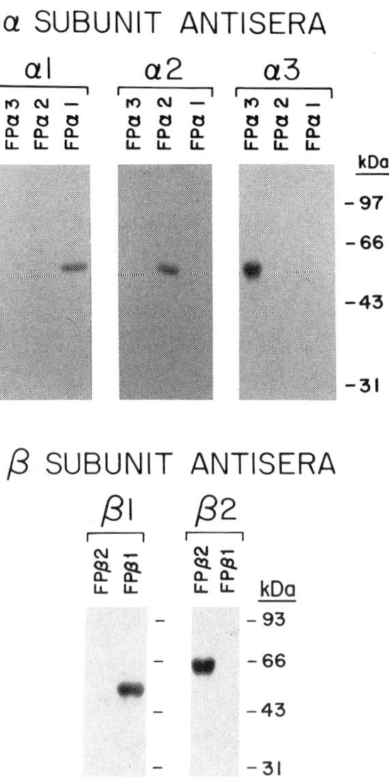

Fig. 6. Specificity of antisera generated against α and β subunit fusion proteins. Total cellular protein from induced bacteria expressing α (top panels) or β (bottom panels) subunit fusion protein genes was size-fractionated on SDS-containing 10% polyacrylamide gels, transferred to nitrocellulose filters, and probed with immunoabsorbed antisera generated against $FP\alpha_1$ (α_1 panel), $FP\alpha_2$ (α_2 panel), $FP\alpha_3$ (α_3 panel), $FP\beta_1$ (β_1 panel), or $FP\beta_2$ (β_2 panel). Bound antibody was detected by incubating nitrocellulose filters with ^{125}I-Protein A. Mol-mass markers are at the right.

proteins onto Western blots containing total protein from bacteria expressing each of the fusion proteins. An example of such an analysis is shown in Fig. 6. Bacteria harboring each of the trpE-α and -β subunit fusion protein genes are induced and processed according to the mini-induction protocol described in Section 3.3.1. Total cellular protein is size fractionated on SDS containing 10% polyacrylamide gels, transferred to nitrocellulose, and probed with immunoabsorbed antisera generated against α or β subunit fusion proteins. As shown in Fig. 6A (α_1 panel), affinity-purified α_1 antiserum reacts only with the α_1

fusion protein FPα_1. Likewise, affinity-purified α_2 antiserum (α_2 panel) reacts only with the α_2 fusion protein FPα_2, whereas α_3 antiserum (α_3 panel) reacts only with the α_3 fusion protein FPα_3. The specificity of the β subunit antisera is presented in Fig. 6B. Affinity-purified β_1 antiserum (β_1 panel) is found to react only with the β_1 fusion protein FPβ_1, whereas β_2 antiserum (β_2 panel) reacts only with the β_2 fusion protein FPβ_2. These results establish that affinity-purified, fusion protein-derived α and β subunit antisera are specific for their respective isotypes.

3.6. Expression of Na/K-ATPase α and β Subunit Isoforms in Brain

To evaluate expression of Na/K-ATPase polypeptides, antisera specific for each of the α and β subunits are used to probe Western blots of rat tissue microsomal membrane fractions. Our methods for the preparation of microsomes (adapted from the procedure described by Jorgenson *[31]*), and Western blotting *(30)* are outlined in the following.

1. Remove the tissue, and place in a tube containing ice-cold homogenization buffer (10 mL of homogenization buffer/g of tissue).
2. Homogenize in a rotary-blade tissue grinder at full power for 1 min. Keep tissue on ice during this procedure.
3. Centrifuge (6000g, 15 min, 4°C).
4. Save the supernatant, and place on ice. Resuspend the pellet in homogenization buffer, homogenize in tissue grinder, and centrifuge the homogenate (6000g, 15 min, 4°C).
5. Combine the supernatants, and centrifuge (45,000g, 45 min, 4°C).
6. Resuspend the pellet and homogenize in ice-cold homogenization buffer (0.5 mL buffer/g of tissue).
7. Determine the protein concentration. Adjust vol to 2 mg protein/mL with homogenization buffer.
8. Solubilize microsomes in SDS-PAGE sample buffer. Fractionate the samples (40 µg protein/lane) by SDS-PAGE, and then transfer to a nitrocellulose filter as described earlier.
9. Quench the nitrocellulose filter in blotto for 20 min.
10. Incubate the filter with fusion protein-derived antisera, diluted in blotto (1:1000), for 16 h.
11. Wash the filter four times for 15 min each in wash buffer.
12. Incubate the filter with Protein A (5 µCi/blot, diluted in blotto) for 1 h.
13. Wash the filter four times for 15 min each in wash buffer.
14. Air-dry the filter and expose to X-ray film (Kodak XAR-5) at –70°C for at least 24 h. Reexpose the filter for longer times if necessary.

The results of our Western blotting experiments are presented in Fig. 7. Blots exposed to α subunit antisera are shown in panel A, whereas blots exposed to the β subunit antisera are shown in panel B. Affinity-purified α_1 antiserum reacts with a polypeptide of $M_r \sim 100$ kDa in kidney, brain, adult pineal gland, and 5-d-old pineal gland. The α_2 and α_3 antisera react with a polypeptide of $M_r \sim 105$ kDa in brain, adult pineal gland, and 5-d-old pineal gland. No α_2 subunits are detectable in 5-d-old or adult pineal glands, or in kidney microsomes. Antiserum specific for the β_1 subunit reacts with two bands having $M_r \sim 50$ and ~ 54 kDa in kidney, and a broad band with $M_r \sim 48$ kDa in brain. No β_1 subunit polypeptides are detectable in either adult or 5-d-old rat pineal glands. Antiserum specific for the rat β_2 subunit reacts with polypeptides of $M_r \sim 46$ kDa in brain and $M_r \sim 50$ kDa in adult pineal glands. No β_2 subunits are detectable in 5-d-old pineals or in kidney microsomes. These results are consistent with the view that the Na/K-ATPase subunit antisera we have developed are specific for their respective isotypes.

4. Notes

1. Verification of fusion protein constructs. In many instances, insertion of a cDNA fragment into the expression vector can occur in the correct (5'→3') or reverse (3'→5') orientation relative to the trpE gene (i.e., when the restriction sites at both ends of the cDNA insert are the same as the cloning site within the polylinker). Insertion of cDNA in the reverse orientation is likely to produce stop codons or other errors in the cDNA sequence, thus causing premature termination of fusion protein synthesis or production of spurious peptide fragments. It is therefore important to determine the orientation of the cDNA following ligation to the vector. This can be accomplished by mapping of diagnostic restriction enzyme sites within the cDNA relative to sites within the pATH vector or by production of a novel fusion protein of the correct size employing the mini-induction protocol.

2. Optimizing conditions for fusion protein production. There are several key points to keep in mind for obtaining maximal induction of fusion protein synthesis. These are enumerated in the following.

    a. When setting up overnight cultures of bacteria for mini- or large-scale inductions, always start by picking bacteria from a freshly streaked plate or from a permanent freeze-down. If bacteria are picked from an old plate (>2 or 3 d) or liquid culture (>2 d), they

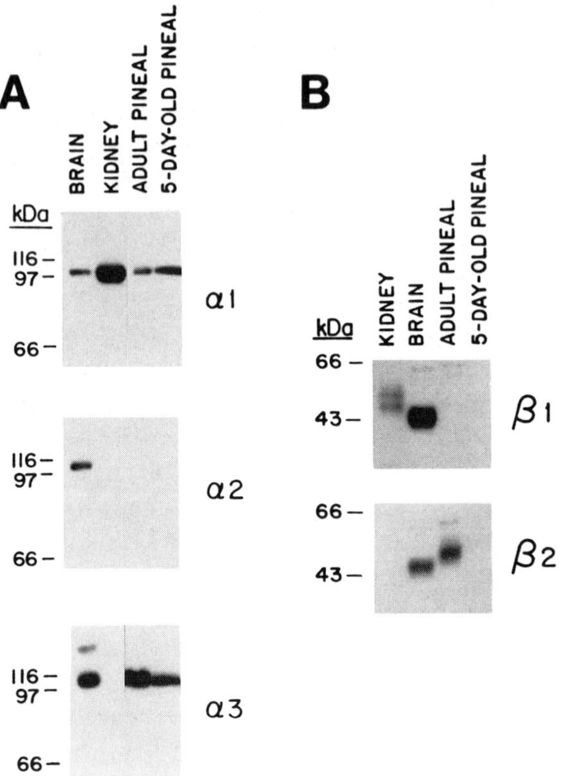

Fig. 7. Expression of Na/K-ATPase α and β subunits in rat tissues. Microsomes were prepared from 5-d-old and adult rat tissues. Solubilized microsomal proteins were size fractionated by electrophoresis through SDS containing 7.5% (α subunits, panel A) or 10% (β subunits, panel B) polyacrylamide gels, transferred to nitrocellulose filters, and probed with the α or β subunit isoform-specific antisera indicated at the right. Bound antibody was detected by incubating the filters with [125]I Protein-A. The positions of mol-mass markers are on the left.

often will not produce large amounts of fusion protein. In some cases, such cultures may be completely refractory to induction.

b. Proper aeration is critical for maximal induction of fusion protein synthesis. Always use small amounts of media in a large, flat-bottomed vessel to obtain best results. Shake vigorously. We employ a rotary shaker set at top speed (250 rpm).

c. Optimal conditions for induction of fusion protein synthesis will differ from construct to construct. It is therefore essential to assess the level of fusion protein production at various times following addition of 3b-indoleacrylic acid to the culture medium.

   d. Inclusion of long hydrophobic (membrane-anchoring) sequences within a trpE–cDNA construct may prevent optimal production of the fusion protein. For example, we have found that a trpE-β subunit construct containing a hydrophobic membrane-spanning domain cannot be induced to express fusion protein (A. W. Shyjan and R. Levenson, unpublished results). Hydrophobic sequences have also been found to inhibit production of several trpE–yeast fusion proteins (P. Novick and S. Ferro-Novick, personal communication). It is possible that long hydrophobic sequences may cause the fusion protein to become associated with the bacterial membrane and targeted for degradation by the cellular machinery. In contrast, a trpE-human poliovirus receptor fusion protein containing a putative membrane anchoring domain has been expressed as soluble protein in *E. coli* (V. Racaniello, personal communication). It is thus possible that certain types of foreign membrane-spanning sequences are not efficiently inserted into bacterial membranes.

3. Hints for processing fusion proteins.

   a. Processing bacterial lysates (Section 3.3.1.): Bacteria are lysed and solubilized with cracking buffer and then spun at high speed to pellet nucleic acid. The pellet contains DNA and is translucent, viscous, and not tightly compacted. The pellet can be more easily visualized by addition of bromophenol blue, and then removed by sliding it up the side of the tube with a toothpick or spatula.

   b. Visualization of fusion protein in large-scale inductions: For large-scale inductions, visualize the proteins by immersion of the gel in ice-cold $2.5M$ KCl. Cold KCl precipitates SDS, and the protein bands will appear milky white. The fusion protein band should be the predominant band in the gel. To avoid confusion, note positions of mol-wt markers in order to identify the position of the fusion protein band precisely. Alternatively, the preparative gel can be stained with Coomassie Blue. However, because of the large size of the gel, staining and destaining may take several hours. Visualization of proteins with KCl can be accomplished within 10 min, and the gel need not be destained.

4. Use of fusion proteins as antigens: For rabbit immunizations, fusion protein can be injected as a gel slurry. The gel slice is simply homogenized, passed through a syringe, and the slurry emulsified in Freund's adjuvant. The advantage of this method is that acrylamide is a potent

irritant, and the rabbit will develop a strong immune response to the antigen. However, it is important to note that, in this method, the fusion protein is in a denatured form, and antibodies produced against denatured antigen may not immunoreact with native protein. Further, antigen in the form of gel slurry cannot be used to raise monoclonal antisera. Injection of gel slurry in mice results in macrophage proliferation. Macrophages in the spleens of immunized mice will phagocytize hybridomas before they can be expanded into cell lines. To avoid these problems, fusion protein can be electroeluted from the gel by the procedure of Hunkapillar et al. *(32)*. Electroeluted protein is at least partially renatured. Antisera raised against electroeluted fusion protein is therefore more likely to immunoreact with native protein. Another advantage of electroelution is that the antigen can be injected intravenously into rabbits and mice to produce antisera of high titer.

Acknowledgments

This work was supported by Public Health Service grant CA-38992 from the National Cancer Institute and Public Health Service grant HL-39263 from the National Institutes of Health to R. Levenson. A. W. Shyjan was supported in part by Training Grant GM-07223 from the National Institutes of Health. R. Levenson is an established investigator of the American Heart Association.

References

1. Julius, D., Huang, K. M., Livelli, T. J., Axel, R., and Jessel, T. M. (1990) The 5HT2 receptor defines a family of structurally distinct but functionally conserved serotonin receptors. *Proc. Natl. Acad. Sci. USA* **87,** 928–932.
2. Schofield, P. R., Darlison, M. G., Fujita, N., Burt, D. R., Stephenson, F. A., Rodriguez, H., Rhee, L. M., Ramachandran, J., Reale, V., Glencourse, T. A., Seeburg, P. H., and Barnard, E. A. (1987) Sequence and functional expression of the $GABA_A$ receptor shows a ligand-gated receptor super-family. *Nature* **328,** 221–227.
3. Hollmann, M., O'Shea-Greenfield, A., Rogers, S.W., and Heinemann, S. (1989) Cloning by functional expression of a member of the glutamate receptor family. *Nature* **342,** 643–648.
4. Noda, M., Ikeda, T., Kayano, T., Suzuki, H., Takeshima, H., Kurasaki, M., Takahashi, H., and Numa, S. (1986) Existence of distinct sodium channel messenger RNAs in rat brain. *Nature* **320,** 188–192.
5. Bonner, T. I., Buckley, N. J., Young, C., and Brann, M. R. (1987) Identification of a family of muscarinic acetylcholine receptor genes. *Science* **237,** 527–532.

6. Grenningloh, G., Reinitz, A., Schmitt, B., Methfessel, C., Zensen, M., Beyreuther, K., Gundelfinger, E. D., and Betz, H. (1987) The strychnine-binding subunit of the glycine receptor shows homology with nicotinic acetylcholine receptors. *Nature* **328,** 215–220.

7. Thomas, R. C. (1972) Electrogenic sodium pump in nerve and muscle cells. *Physiol. Rev.* **52,** 563–594.

8. Iverson, L. L. and Kelly, J. S. (1975) Uptake and metabolism of gamma-aminobutyric acid by neurones and glial cells. *Biochem. Pharmacol.* **24,** 933–938.

9. Hertz, L. (1977) Biochemistry of glial cells, in *Cell, Tissue, and Organ Culture in Neurobiology* (Federoff, S. and Hertz, L., eds.), Academic, New York, pp. 37–71.

10. Cantley, L. C. (1981) Structure and mechanism of the (Na,K)-ATPase. *Curr. Top. Bioenerg.* **11,** 201–237.

11. Kent, R. B., Fallows, D. A., Geissler, E., Glaser, T., Emanuel, J. R., Lalley, P. A., Levenson, R., and Housman, D. E. (1987) Genes encoding α and β subunits of the Na,K-ATPase are located on three different chromosomes in the mouse. *Proc. Natl. Acad. Sci. USA* **84,** 5369–5372.

12. Malo, D., Schurr, E., Levenson, R., and Gros, P. (1990) Assignment of the Na,KATPase $\beta 2$ subunit gene *(Atpb-2)* to mouse chromosome 11. *Genomics* **6,** 697–699.

13. Shull, G. E., Greeb, J., and Lingrel, J. B. (1986) Molecular cloning of three distinct forms of the Na^+,K^+-ATPase α subunit from rat brain. *Biochemistry* **25,** 8125–8132.

14. Herrera, V. L., Emanuel, J. R., Ruiz-Opazo, N., Levenson, R., and Nadal-Ginard, B. (1987) Three differentially expressed Na,K-ATPase α subunit isoforms: structural and functional implications. *J. Cell Biol.* **105,** 1055–1065.

15. Mercer, R. W., Schneider, J. W., Savitz, A., Emanuel, J. R., Emanuel, J. R., Benz, E. J., Jr., and Levenson, R. (1986) Rat brain Na,K-ATPase β chain gene: primary structure, tissue-specific expression, and amplification in ouabain-resistant HeLa C^+ cells. *Mol. Cell. Biol.* **6,** 3884–3890.

16. Martin-Vasallo, P., Dackowski, W., Emanuel, J. R., and Levenson, R. (1989) Identification of a putative isoform of the Na,K-ATPase β subunit: primary structure and tissue specific expression. *J. Biol. Chem.* **264,** 4613–4618.

17. Emanuel, J. R., Garetz, S., Stone, L., and Levenson, R. (1987) Differential expression of Na,K-ATPase α- and β-subunit mRNAs in rat tissues and cell lines. *Proc. Natl. Acad. Sci. USA* **84,** 9030–9034.

18. Shyjan, A. W. and Levenson, R. (1989) Antisera specific for the $\alpha 1$, $\alpha 2$, $\alpha 3$, and β subunits of the Na,K-ATPase: differential expression of α and β subunits in rat tissue membranes. *Biochemistry* **28,** 4531–4535.

19. Emanuel, J. R, Schulz, J., Zhou, X.-M., Kent, R. B., Housman, D., Cantley, L., and Levenson, R. (1988) Expression of an ouabain-resistant Na,K-ATPase in CV-1 cells after transfection with a cDNA encoding the rat Na,K-ATPase $\alpha 1$ subunit. *J. Biol. Chem.* **263,** 7726–7733.

20. Emanuel, J. R., Graw, S., Housman, D., and Levenson, R. (1989) Identification of a region within the Na,K-ATPase α subunit that contributes to differential ouabain sensitivity. *Mol. Cell. Biol.* **9,** 3744–3749.

21. Fallows, D., Kent, R. B., Nelson, D. L., Emanuel, J. R., Levenson, R., and Housman, D. E. (1987) Chromosome-mediated transfer of the murine Na,K-ATPase alpha subunit confers ouabain resistance. *Mol. Cell. Biol.* **7,** 2985–2987.

22. Canfield, V., Emanuel, J. R, Spickofsky, N., Levenson, R., and Margolskee, R. F. (1990) Ouabain-resistant mutants of the rat Na,K-ATPase α2 isoform identified by using an episomal expression vector. *Mol. Cell. Biol.* **10,**1367–1372.

23. Hara, Y., Nikamoto, T., Matsumoto, A., and Nakao, M. (1988) Expression of sodium pump activities in BALB/c 3T3 cells transfected with cDNA encoding α3 subunits of rat brain Na,K-ATPase. *FEBS Lett.* **238,** 27–30.

24. Shyjan, A.W., Cena, V., Klein, D. C., and Levenson, R. (1990) Differential expression and enzymatic properties of the Na^+,K^+-ATPase α3 isoenzyme in rat pineal glands. *Proc. Natl. Acad. Sci. USA* **87,** 1178–1182.

25. Lytton, J., Lin, J. C., and Guidotti, G. (1985) Identification of two molecular forms of (Na^+K^+)-ATPase in rat adipocytes: relation to insulin stimulation of the enzyme. *J. Biol. Chem.* **260,** 1177–1184.

26. Shyjan, A.W., Gottardi, C., and Levenson, R (1990) The Na,K-ATPase β2 subunit is expressed in rat brain and copurifies with Na,K-ATPase activity. *J. Biol. Chem.* **265,** 5166–5169.

26a.Watts, A. G., Sanchez-Watts, G., Emanuel, J. R., and Levenson, R. (1991) Cell-specific expression of mRNAs encoding Na^+, K^+- ATPase α- and β-subunit isoforms within the rat central nervous system. *Proc. Natl. Acad. Sci. USA* **88,** 7425–7429.

27. Maniatis, T., Fritsch, E. F., and Sambrook, J. (1982) *Molecular Cloning: A Laboratory Manual.* Cold Spring Harbor Laboratory, Cold Spring Harbor, NY.

28. Dieckmann, C. L. and Tzagoloff, A. (1985) Assembly of the mitochondrial membrane system. *J. Biol. Chem.* **260,** 1513–1520.

29. Sweadner, K. J. and Gilkeson, R. C. (1985) Two isozymes of the Na,K-ATPase have distinct antigenic determinants. *J. Biol. Chem.* **260,** 9016–9022.

30. Towbin, H., Staehlin, T., and Gordon, J. (1979) Electrophoretic transfer of proteins from polyacrylamide gels to nitrocellulose sheets: procedure and some applications. *Proc. Natl. Acad. Sci. USA* **76,** 4350–4354.

31. Jorgenson, P. L. (1974) Purification and characterization of $(Na^+ + K^+)$-ATPase III. Purification from the outer medulla of mammalian kidney after selective removal of membrane components by sodium dodecylsulphate. *Biochim. Biophys. Acta* **356,** 36–52.

32. Hunkapillar, M. W., Lujan, E., Ostrander, F., and Hood, L. E. (1980) Isolation of microgram quantities of proteins from polyacrylamide gels for amino acid sequence analysis. *Methods Enzymol.* **91,** 227–240.

CHAPTER 8

Determination of Transcriptional Initiation Sites and Their Usage in the Nervous System

Gary Weisinger, Joseph D. DeCristofaro, and Edmund F. La Gamma

1. Introduction

The appearance and maintenance of differentiated neural cell types in eukaryotic organisms can be largely attributed to characteristic spatial and temporal regulatory cascades evoking cell-specific gene expression. Since gene expression causes information passage from DNA to RNA to proteins and since most mammalian brain protein coding DNA is expressed as messenger RNA (mRNA) *(1,2)*, it appears that the control of mRNA levels is a critical intermediate step in producing brain-specific patterns of gene expression. Of the many transcriptional mechanisms known, RNA initiation is recognized as an important modulator of expression *(3–5)*.

In eukaryotes, RNA is transcribed from DNA by RNA polymerase II after the formation of a transcription initiation complex *(1,2)*. In the nervous system, the formation of this complex as well as its ability to initiate a particular species of RNA transcripts, can be mediated through several second messenger systems *(2)*. In turn, second messenger pathways modify binding properties of different *trans*-acting factors, which may affect different classes of neuronal genes, for example, those of the neuroendocrine system. Neuroendocrine genes are precisely regulated at the level of neuropeptide biosynthesis, as well as release, by many afferent signals, which act individually or in combination. As a consequence of recent advances in molecular biology, one can now examine various subtle aspects of the control of RNA synthesis of neuroendocrine genes, such as preproenkephalin.

As a first step in the determination and characterization of the RNA initiation site(s) for a particular transcription unit, one of two general approaches can be used. One approach, the nuclease digestion assay (Fig. 1A), involves the hybridization of a radiolabeled (minus strand only) polynucleotide probe, with either total cellular RNA or poly A^+ mRNA. The polynucleotide probe fragment must overlap the putative 5' end of the gene of interest. Following this hybridization step, the labeled probe corresponding to the exact 5' end of the gene should be double stranded. Any portion of the probe that is upstream of the RNA start site(s) will remain single stranded as does any portion of the RNA of interest that does not complement the probe. Those remaining single strands are then removed with either S1 nuclease or by RNAse digestion. The remaining radiolabeled fragment(s) is fractionated on a $7M$ urea polyacrylamide sequencing gel next to high resolution mol-wt markers. Finally, autoradiography is used to visualize the result.

The other major approach for defining a gene's transcriptional start site(s) is generally termed primer extension (Fig. 1B). In this case, a small (30–40 bp) radio-end-labeled, single-stranded complementary oligonucleotide is chosen from sequences within the putative first exon of the gene's transcription unit. This oligonucleotide is hybridized to the same type of RNA used for the above nuclease assays. Following hybridization (which is less temperature dependent than the nuclease assays), the enzyme reverse transcriptase extends the oligomer primer to the start of the RNA transcript, where it stops for 90% or more of the transcripts. The newly formed double-stranded radiolabeled fragment is then denatured, sized by polyacrylamide gel electrophesis and visualized by autoradiography. The size of the major band(s) should correspond to that found by the nuclease assays and, hence, be a direct reflection of the position of any individual or multiple start site arrangement for any gene.

Once defined, these two complementary assays can be used to determine treatment, cell type, or tissue-specific usage of any start site for any cloned gene. In this way, one can begin to dissect one of the various controls of transcriptional regulation for neuroendocrine or

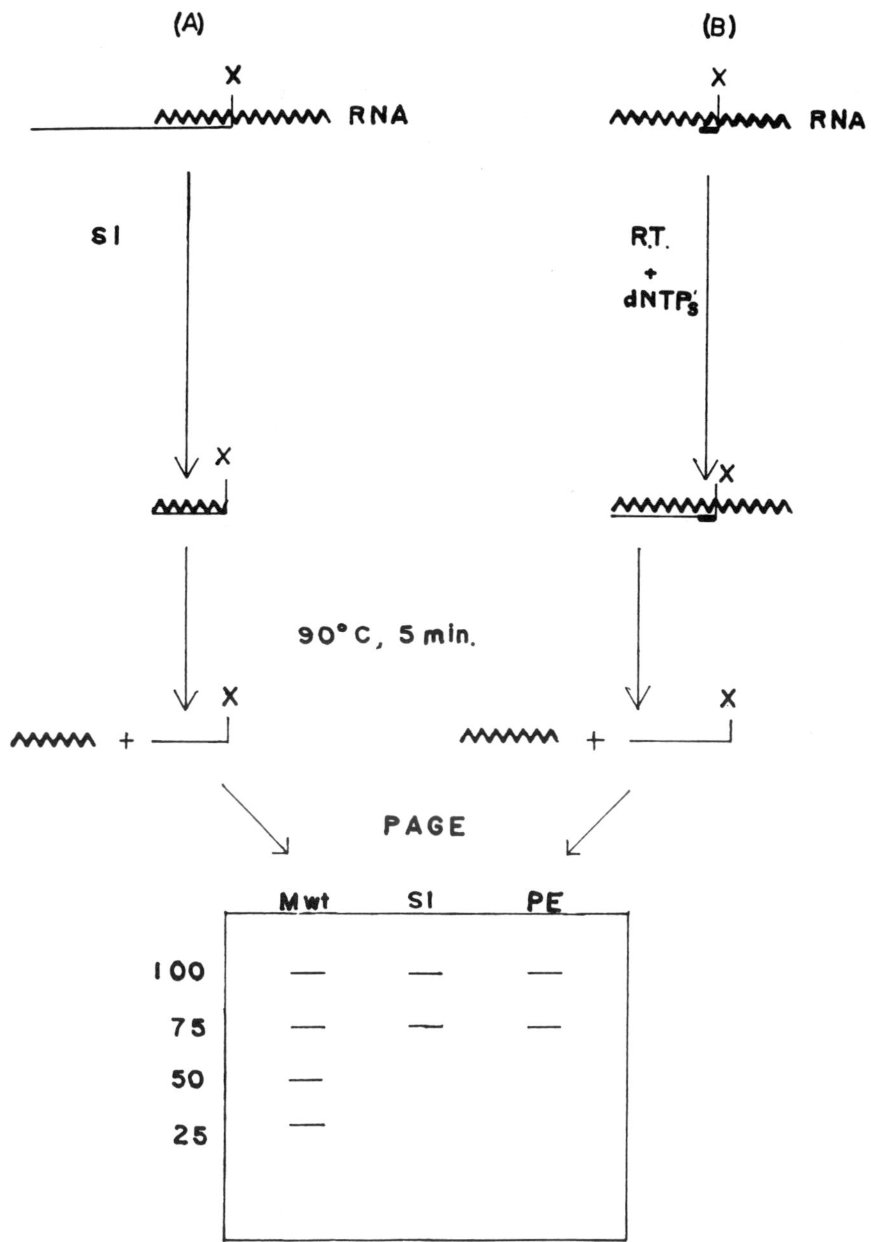

Fig. 1. Flow diagram of an S1 nuclease digestion assay **(A)** and primer extension (PE) **(B)** analysis to determine RNA initiation site(s) of a particular RNA species using urea polyacrylamide gel electrophesis (x marks the radiolabel).

other neuronal genes. In this chapter, we describe the preparation of total tissue RNA (from the rat), as well as the S1 nuclease and primer extension assays.

2. Materials

2.1. Total Tissue RNA Preparation

All reagents should be made with autoclaved (×2) diethyl pyrocarbonate (DEPC) treated double-distilled water (100 μL DEPC/L water; DEPC is HAZARDOUS, dispense wearing gloves in a fume hood).

1. Lysis solution: $4M$ guanidinium thiocyanate (HAZARDOUS: Use in fume hood), 25mM sodium citrate, pH 7.0, 0.5% sarcosyl, and 0.1M β-mercaptoethanol (BME). This reagent should be prepared by mixing all the components except the BME at 65°C. It can be stored at room temperature in a brown bottle for up to 3 mo. Add BME just prior to use.
2. Chloroform: isoamylalcohol (24:1, v/v). Store in a brown bottle.
3. $2M$ sodium acetate, pH 4.5.
4. Phenol (redistilled) saturated with 10 mM Tris-HCl, pH 7.5 in DEPC water. This solution should either be used directly or frozen in aliquots at −20°C. Storage of the frozen phenol, for up to 3 mo, is possible.
5. Absolute isopropanol.
6. A solution of 75% ethanol and 25% DEPC-treated water.
7. Absolute ethanol and 70% aqueous ethanol will be used as −20°C solutions through out these procedures.

2.2. S1 Nuclease Assay

1. Appropriate restriction enzymes with their reaction cocktails.
2. ANE (10X): 100 mM sodium acetate, pH 6.0, 100 mM NaCl, and 1 mM EDTA.
3. Chloropane: Made by mixing 300 mL of freshly melted (65°C), recrystalized phenol with 300 mL $CHCl_3$, 150 mg 8-hydroxyquinoline, and 250 mL ANE (1X). After allowing the phases to separate, the top aqueous phase is discarded, and the bottom organic phase is transferred into a brown bottle and stored at 4°C.
4. TE: 10 mM Tris-HCl, pH 8.0 and 1 mM EDTA.
5. DN buffer (10X): 0.2M Tris-HCl, pH 9.5, 10 mM spermidine, and 1 mM EDTA.
6. Linker kinase buffer (10X): 700 mM Tris-HCl, pH 7.4, 100 mM $MgCl_2$, and 50 mM dithiothreitol (DTT).
7. [γ-^{32}P] ATP: 10 mCi/mL, 3000 Ci/mmol.
8. TBE (10X): 0.89M Tris-borate, 0.89M boric acid, and 20 mM EDTA.

9. 40% Acrylamide-1.3% bisacrylamide (w/v): Use molecular biology grades. (HAZARD: neurotoxic.)
10. 10% Dimethyldichlorosilane in carbon tetrachloride: (HAZARD: extremely flammable, corrosive, causes burns, and mutagenic). Handle in a fume hood with heavy rubber gloves.
11. PIPES buffer (5X): 0.2 M sodium PIPES (piperazine-*N*-*N'*-*bis* [2-ethane sulfonic acid]), pH 6.4, 5 mM EDTA, and 2.0 M NaCl. To make a 1X solution, dilute stock 5X buffer fivefold with a high-purity formamide (we use Fluka).
12. S1 nuclease reaction buffer (10X): 0.3 M sodium acetate, pH 4.6, 0.5 M NaCl, 10 mM $ZnSO_4$, and 50% glycerol.
13. Sephadex G50 slurry: Add 1 g Sephadex G50 to 15 mL double-distilled water and stand for 2 d at room temperature. Replace the water with TE. This material should be stored at 4°C. Other investigators also add 0.02% sodium azide to the TE to stop bacterial growth, but we have not added this and have not had bacterial contamination.
14. 1% Dye mix: A 1% mixture of bromophenol blue and xylene cyanol in double-distilled water.
15. Ficoll gel-loading buffer: 25% Ficoll (Type 400) in double-distilled water and 25% of 1% dye mix.
16. Formamide denaturing gel-loading buffer (formamide dye mix): 80% Formamide (high purity), 10% of 1% dye mix, and 10% of TBE (10X). We have found that this reagent can be kept frozen with occasional freeze-thawing for up to 1 mo.
17. DMS buffer: 50 mM sodium cacodylate, pH 8.0 and 1 mM EDTA. It is usually unnecessary to adjust the pH (HAZARDOUS: irritant and toxic).
18. DMS stop buffer: 1.5 M sodium acetate, pH 7.0, 1 M BME, and 0.1 mg/mL tRNA.
19. Bacterial alkaline phosphatase (BAP): Store at –20°C.
20. Proteinase K: 5 mg/mL double-distilled water. Store at –20°C.
21. T4 polynucleotide kinase: Store at –20°C.
22. S1 nuclease: (1000 U/µL). Store at –20°C. This enzyme is particularly sensitive to temperatures above 0°C.
23. Yeast tRNA: 10–25 mg/mL DEPC-treated water. Store at –20°C.
24. 0.3 M sodium acetate.
25. 7.5 M ammonium acetate: This reagent should be filtered through a 0.45-µm filter and stored at room temperature.
26. Dry ice.
27. Ammonium persulfate.
28. Urea: Molecular biology grade.
29. TEMED: *N,N,N',N'*-Tetramethyl-ethylenediamine.
30. Dimethyl sulfoxide (DMS): (VERY HAZARDOUS).

31. Piperidine: The material sold by Sigma (St. Louis, MO) is 11*M*. Just before use, dilute 1:10 with double-distilled water.
32. X-ray film: Both 8 × 10" and 35 × 43 cm formats, Kodak XAR 5.
33. Siliconized glass wool.
34. The following list of separate reagents are prepared by autoclaving at 15 psi for 25–30 min: 1*M* Tris-HCl, pH 8.0, 3*M* sodium acetate, pH 4.5, mineral oil, and glycerol.
35. Preparative polyacrylamide gel electrophoresis (PAGE), sequencing PAGE, electroelution, and autoradiography equipment should be available.

2.3. Additional Materials for the Primer Extension Assay

1. Oligonucleotide primer: 0.1 m*M*/mL of a 30–40 bp sequence made by DNA synthesizer and further purified by polyacrylamide gel electrophoresis or HPLC. This primer should correspond to the minus strand within the first exon of the gene of interest.
2. Oligonucleotide kinase buffer (10X): 700 m*M* Tris-HCl, pH 7.5, 100 m*M* MgCl$_2$, 50 m*M* DTT, 1 m*M* spermidine, and 1 m*M* EDTA.
3. 4 m*M* dNTPs: 4 m*M* each of dATP, dCTP, dGTP, and dTTP mixed together in double-distilled water. Store as frozen aliquots to extend their half-lives.
4. M-MLV reverse transcriptase reaction cocktail (5X): 250 m*M* Tris-HCl, pH 8.3, 375 m*M* KCl, 15 m*M* MgCl$_2$, and 50 m*M* DTT.
5. 75% Ethanol: 25% 0.1*M* sodium acetate, pH 5.2 stored at −20°C.
6. 4*M* and 2.5*M* ammonium acetate.
7. 0.5*M* EDTA, pH 8.0: After autoclaving, this solution may be stored at ambient temperature.
8. M-MLV reverse transcriptase: Store at −20°C. (M-MLV is the molony murine leukemia virus).
9. Pancreatic ribonuclease A (RNAse A): (1 mg/mL double-distilled water); store frozen.
10. Placental ribonuclease inhibitor: e.g., RNAs in from Promega Biotec (Madison, WI). Store at 4°C.

3. Methods

3.1. Total Tissue RNA Preparation

This RNA preparation method was largely adapted from the acid guanidinium thiocyanate-phenol-chloroform method as previously described *(6,17)*. All the following procedures require baked glassware

(250°C for 4 h) as well as a Teflon™ homogenizer, micropipet tips, and 1.5-mL microcentrifuge tubes, which have been autoclaved (*see* Note 1).

1. Adult Sprague-Dawley rats are sacrificed by CO_2 narcosis/asphyxiation; the adrenal medulla, brain striatum, and other tissues are microdissected, and frozen immediately in liquid nitrogen or dry ice and stored at −70°C in 1.5 mL Eppendorf tubes.
2. Prepare the lysis solution as outlined in Section 2.1. by adding stock BME to the stored lysis solution minus BME to a final concentration of 0.1*M* (e.g., 0.35 mL BME to 50 mL lysis solution) (*see* Note 2).
3. Homogenize approx 100 mg of tissue (6–8 rat medullae, 2 striata, and so on) in 1 mL lysis solution using a glass Teflon™ (4 mL) homogenizer. Homogenize with four to six complete strokes (up and down) at 2100 rpm until the tissue appears to be completely homogenized. To avoid excessive bubbling from the sarcosyl, always keep the top of the pestle below the surface of the solution. (We use the highest speed on the Thomas six-speed tissue grinder, the 4-mL grinding vessel, and a smooth-tipped Teflon™ pestle.)
4. Transfer the homogenate to a siliconized, baked glass 15-mL corex (or pyrex) round-bottomed centrifuge tube, add the following reagents sequentially, and mix (per 1 mL lysis solution):
   0.1 mL 2*M* sodium acetate (pH 4)
   1 mL DEPC water/Tris-saturated phenol
   0.2 mL chloroform:isoamylalcohol
   Mix for 10 s by inversion or low-speed vortexing, and cool on ice for 15 min.
5. Centrifuge (10,000*g*, 20 min, 4°C), transfer the upper aqueous phase (containing the RNA) to another similarly prepared 15-mL corex tube, and add an equal vol of 100% isopropanol. Vortex this mixture, and incubate at −20°C for >90 min.
6. This suspension of RNA crystals is then recentrifuged (10,000*g*, 20 min, 4°C), and the resulting RNA pellet is resuspended in 1/3 of the original vol using lysis solution (e.g., 0.33 mL for 1 mL).
7. Transfer this solution to a preautoclaved microcentrifuge tube (1.5 mL), and reprecipitate by adding an equal vol of 100% isopropanol and vortex. Incubate again at −20°C for 60 min or overnight.
8. Centrifuge the Eppendorf tubes (10,000–14,000*g*, 15 min, 4°C), and wash the resulting RNA pellet by adding 1 mL of cold 75% ethanol/ 25% DEPC-treated water, followed by another quick vortex and centrifugation (10,000–14,000*g*, 10 min, 4°C).

9. The remaining RNA pellet is dried in a vacuum for 2 min. Resuspend the pellet in a small vol of DEPC-treated double-distilled water (e.g., 0.1 mL), and keep on ice. Take an aliquot of diluted RNA solution (e.g., 2 µL into 0.5mL) and measure the absorbance at 260 nm (an A_{260} nm of 25 is equivalent to 1mg RNA/mL solution), and calculate the yield/mg original tissue.

10. The RNA can then either be used directly, or as is more usual, stored for later use in 1/10 vol of $3M$ sodium acetate ($0.3M$ final aqueous concentration) and 2.5 vol absolute ethanol at −80°C. (This method of RNA storage helps preserve the RNA in an undegraded form.) (*See* Notes 3 and 4.)

3.2. S1 Nuclease Assay

There are a number of component procedures involved in performing S1 nuclease experiments. In the following, we divide the procedure into four sections. First, we describe the preparation of T4 kinase end-labeled DNA probes; second, we describe the preparation of bp resolution mol-wt markers (Maxam and Gilbert's G > A methylation and cleavage reaction). Third, we describe the RNA-probe hybidization step together with the S1 assay itself. Finally, we describe the preparation of a 0.2 mm thick urea polyacrylamide gel.

3.2.1. Preparation of ^{32}P End-Labeled DNA Probes (7–9)

As can be seen in Fig. 1, a DNA fragment is required that is radiolabeled at the 5' end, and which clearly contains the proposed transcription initiation site of the gene of interest. Here we describe 5' labeling with T4 polynucleotide kinase. To cut out the required fragment it is preferable to use a restriction enzyme that has only single (or at least few) internal sites within the plasmid and that the enzyme should optimally leave a 5' overhang. This last point is not essential, but it is helpful. If this is not possible, a flush or small 3' overhang will also label, however less efficiently.

1. Digest 10 µg of plasmid (containing the fragment of interest) with an appropriate restriction enzyme that cuts at the point to be labeled (point X in Fig. 1). Ideally, the reaction vol should be 50–100 µL.

2. When the digestion is complete (as determined by analyzing a fraction by agarose minigel electrophoresis), increase the reaction vol to 100 µL with TE and 3 µL of BAP. Adjust the pH to 8.0 if necessary (if the restriction reaction in step 1 was not at pH 8, 2 µL of $1M$ Tris-HCl, pH 8, may be added). Although there is variation in the standard

BAP unit definition, the BAP concentration is comparable among most companies; we suggest using 3 µL. Incubate at 60–65°C for 60 min. This step dephosphorylates the exposed free DNA ends, preparing them to receive radiolabeled phosphates in their place.

3. Add 1 µL of proteinase K, and incubate the mixture at 37°C for 30 min. This step digests the hardy BAP enzyme.
4. Add an equal vol of chloropane, vortex well, and microcentrifuge (10,000–14,000*g*, 3–5 min, 4°C).
5. Transfer the DNA containing upper phase to a fresh microcentrifuge tube, and add 1/10 vol of 3*M* sodium acetate and 2.5 vol of absolute ethanol (–20°C), vortex the mixture, and place it at either –20°C for 2 h (or more) or into a dry-ice/ethanol bath for 10–15 min.
6. Microcentrifuge (10,000–14,000*g*, 15 min, 4°C) the resulting crystaline DNA suspension, add 1 mL of cold 70% ethanol to wash the resulting DNA pellet, vortex briefly, and recentrifuge using the same conditions for 5 min. This last wash removes residual salt from the DNA. The resulting pellet is vacuum-dried and resuspended in 15 µL of TE.
7. Resuspended the DNA (10 µg) pellet in 15 µL TE, add 2.5 µL DN buffer (10X), heat to 75°C for 2 min, and then quench on ice. This step melts the ends of the linearized plasmid allowing easy access for the enzyme in the next step.
8. Next, add 5 µL of linker kinase buffer (10X), 25 µL [γ^{32}P] ATP, and 2 µL of T4 polynucleotide kinase, and incubate at 37°C with appropriate shielding for 45 min.
9. During this incubation period, prepare a G-50 Sephadex minicolumn. Remove the shaft of a 1-mL syringe, and plug the nozzle end with some siliconized glass wool. This minicolumn is then filled with the Sephadex slurry, placed in a 15-mL Falcon tube, and centrifuged (1000 rpm, 30 s). After this spin, more Sephadex slurry should be added, everything recentrifuged, and this sequence repeated until 90% of the tube has been filled with Sephadex. The Sephadex is washed four times with 60 µL of double-distilled water and 30-s centrifugations. When this has been completed, place a 0.5-mL uncapped Eppendorf tube at the mouth of the minicolumn in the 15-mL tube. (The tube should not fit tightly over the end of the column.)
10. After the 45-min kinase incubation is complete, place this reaction mix into the top of the minicolumn, and centrifuge (1000 rpm, 90 s). The resulting radioactive fraction that passes through the column into the Eppendorf tube should contain the radiolabeled DNA, and the radioactive portion retained in the column will contain unreacted [γ^{32}P] ATP, which should be discarded safely.

11. To this radiolabeled DNA solution, an appropriate second restriction enzyme and buffer are added (this cut should delimit the other end of your fragment of interest [i.e., 5' of the transcription unit]). Allow for fragment sizes clearly separable by PAGE. Plan on a 5–10-fold overdigestion of the plasmid to be certain that the restriction digest is complete. To overdigest a plasmid 5–10-fold, add 5–10 U of enzyme/1ug of DNA.

12. Add 1/10 vol of Ficoll gel-loading buffer (10X), and load the sample onto a nondenaturing 5% PAGE system. Briefly, to make the appropriate 5% PAG in TBE, the following are mixed:

    5 mL 40% acrylamide-1.3% *bis*-acrylamide (w/v)
    8 mL TBE (10X),
    27 mL double-distilled water
    40 mg ammonium persulfate

    Once in solution, the mixture is vacuum filtered through a double layer of Whatman #1 and degassed. Add 50 µL TEMED, quickly pour this solution into a 1 mm thick preparative gel glass assembly, and then insert the gel former ("comb"). Allow 10 min for polymerization to be complete. The sample, now containing Ficoll gel-loading buffer, is loaded into the formed wells with the running buffer in place. Under these conditions of electrophoresis (5% PAGE), the bromophenol blue marker will comigrate with a DNA fragment of 65 bp and the xylene cyanol as a fragment of 260 bp.

13. When the gel fractionation is complete, take the gel down, separate the plates, cover the gel on one glass plate with plastic wrap, and mark the gel with radioactive ink spots or "glogos" (available from Statagene, CA, a good nonradioactive alternative). Expose the gel to X-ray film (Kodak XAR-5) for 3–10 min at room temperature, and develop the film.

14. The desired radiolabeled fragment may be obtained from the gel by first completely cutting out the band image from the exposed X-ray film. Then using this film as a template, i.e., lay it over the the plastic-wrap-covered gel to cut out the labeled fragment of interest using a razor blade or scalpel.

15. Electroelute the labeled DNA from the gel. We use an IBI electroeluter, which we modified such that the electrodes are at the far opposite sides of the tank. The DNA fragment is electroeluted into a 7.5*M* ammonium acetate salt bridge, according to the manufacturer's directions. (We use 140 V for between 30–60 min, and use a minimonitor counter to follow the radioactive counts. It is of major importance to avoid disturbing the bench during the electroelution.)

16. Before removing the DNA–salt solution, gently remove the TBE electroelution buffer to below the sample level by low suction. Then collect the 7.5M ammonium acetate salt bridge, which contains the radiolabeled DNA, using a long-nosed Pasteur pipet. Add an equal vol of chloropane, vortex, and microcentrifuge for 3–5 min. (This can be followed by two chloroform steps to remove all traces of phenol.) Transfer the aqueous top layer to another Eppendorf tube, and add 1 mL of cold absolute ethanol with 25 µg tRNA (as a carrier). Vortex this mixture, and allow it to sit at –20°C for more than 2 h or in a dry-ice/ethanol bath for 10–15 min.
17. Microcentrifuge (10,000–14,000g, 15 min, 4°C), wash the pellet with 1 mL cold 70% ethanol, and remicrocentrifuge for 5 min.
18. This final probe pellet is vacuum dried and resuspended in TE at 50 cps/µL (based on readings of a hand-held Geiger counter).

3.2.2. Maxam-Gilbert "G > A" Methylation and Cleavage Reaction (7–9)

1. Add 100 cps (2 µL of solution in step 18 above) or more (<7.5 µL) to 0.3 mL DMS buffer, chill in ice water for 5 min, and then add 1.5 µL stock DMS (CAUTION HAZARDOUS: Work in fume hood).
2. Float the tube on a 20°C water bath for 3 min; then stop the reaction by adding 75 µL of DMS stop buffer and 1.12 mL cold absolute ethanol. This should just fit into a 1.5-mL microcentrifuge tube.
3. Incubate this mixture in a dry-ice/ethanol bath for 10–15 min, and microcentrifuge (10,000–14,000g, 15 min, 4°C).
4. Resuspend the resulting DNA pellet in 0.2 mL of 0.3M sodium acetate (4°C) and 0.6 mL absolute ethanol/tube.
5. Repeat step 3, and then wash the pellet with 1 mL of cold 70% ethanol. Vacuum dry the pellet, and resuspend it in 0.1 mL of 1.1M aqueous piperidine. Completely submerge this solution into a 95°C water bath, and incubate for 45–60 min; then quench the tubes on ice for 30 s.
6. Add 25 µg tRNA, 10 µL 3M sodium acetate, and 2.5 vol cold absolute ethanol to each tube.
7. Vortex and incubate in a dry-ice/ethanol bath for 10–15 min, and centrifuge (10,000–14,000g, 10 min, 4°C).
8. Resuspend the pellet in 0.2 mL of 0.3M sodium acetate (4°C), and add 0.5 mL absolute ethanol; then repeat step 7 (second precipitation). Wash the resulting pellet with 1 mL of cold 70% ethanol (–20°C), and microcentrifuge (10,000–14,000g, 5 min, 4°C). Finally, vacuum dry the pellet for 5–10 min, and resuspend in formamide gel-loading buffer at 1–5cps/µL.

3.2.3. RNA–DNA Hybridization
and S1 Nuclease Digestion (9–13)

1. As a first step, the hybridization temperature must be determined. A good bottom of the range temperature may be estimated by the DNA melting temperature (T_m) for a fragment *(11–13)*.

$$T_m = 81.5 - 16(\log_{10}[Na^+]) + 0.41(\%G + C) - 0.6\,(\%\text{formamide}) - 500/N \quad (1)$$

where N is the chain length in bp, $[Na^+]$ is the molar sodium concentration, and ($\%G + C$) is the approximate $G + C$ content of the region of the probe to the complementary RNA expressed as a percentage (e.g., 25% not 0.25). This equation can be simplified assuming the current procedure and buffers are used.

$$T_m = 27.1 + 0.41(\%G + C) - 500/N \quad (2)$$

2. To 0.25 mL of 0.3 M sodium acetate add the purified probe (50–200 cps) and RNA (2 µg poly A$^+$ or 10–50 µg total RNA). For an optimized annealing temperature (T_a), you should hybridize at least four samples. First, yeast tRNA as a negative control should be hybridized at T_m + 4°C. The other three samples are the experimental RNA hybridized at a minimum of three temperatures: T_m + 1°C, T_m + 4°C, and T_m + 7°C. (*See* Note 5.)
3. Add 0.75 mL of cold absolute ethanol to each sample, vortex, and incubate in a dry-ice/ethanol bath for 10–15 min. Microcentrifuge (14,000*g,* 15 min, 4°C) using a horizontal rotor (if possible). Discard the supernatant, and vacuum dry the pellet containing the coprecipitated RNA and probe, for 2–3 min.
4. Redissolve the dry pellet in 50 µL of 1X Pipes buffer, and then overlay with 35 µL sterile mineral oil.
5. Samples within their respective microcentrifuge tube should then be placed within larger cutdown plastic 15-mL test tubes ("boats") filled with water at 80°C and suspended in a 80°C water bath. Allow the samples to incubate in this manner for 12 min, then quickly transfer each "boat" with the Eppendorf sample to its respective annealing temperature. This allows a slow transition in temperature.
6. Annealing is allowed to proceed for at least 5 h or more (typically) overnight. The larger the relative abundance of the RNA of interest, the shorter the necessary hybridization time. To stop the hybridization, add 0.45 mL of 1X S1 buffer to each sample, vortex mix, and then microcentrifuge for 10 min. This step is necessary to make the mineral oil easily separable (*see* Note 6).

7. Remove 25 µL of each sample, and mix in a fresh tube with 25 µg tRNA, 0.375 mL of 0.3*M* sodium acetate, and 1 mL cold absolute ethanol. Hold at –20°C for the moment. These samples are the S1 untreated controls and are used to define the full-length probe for comparison to the S1-treated experiments. (If desired, this step need only be done with the tRNA control sample.)

8. Transfer the remaining 0.45 mL of the sample to a fresh tube and add 500–1000 U S1 (diluted to 100 U S1/µL S1 buffer (1X) immediately before use, and kept on ice) to 10 µg total RNA. Ideally, an S1 titration experiment, focusing on the amount of S1 nuclease necessary to digest the hybridized probe/RNA, should be performed after the optimal hybridization temperature is determined (e.g., 500, 750, and 1000 U). Allow the reaction to continue for 1 h at room temperature.

9. Add 25 µg of tRNA to each sample with 1 mL of cold absolute ethanol, vortex mix, and place (together with S1-untreated controls) in a dry-ice/ethanol bath for 10–15 min. Microcentrifuge (14,000*g*, 15 min, 4°C) these samples, wash the pellets with 1 mL of 70% ethanol twice (this includes a 5-min microcentrifugation, each time), and finally vacuum dry the pellets for 2–3 min.

10. Redissolve each pellet in 3 µL of formamide denaturing gel-loading buffer (fresh), denature at 90–95°C for 5–10 min, and then quench on ice.

11. The samples are then electrophoresed on a 7*M* urea polyacrylamide gel of the appropriate gel concentration and run in 1X TBE at 60 W constant power (*see* Section 3.2.4.). For example, to resolve the four rat preproenkephalin start sites, which are all within 42 bp of each other (and the largest fragment size is 92 bp), a 10% 0.2 mm thick sequencing gel is run at 60 W constant power until the xylene cyanol has run 25 cm *(9)*. The samples are run with the high-resolution mol-wt markers for accurate size determination (Section 3.2.2.). It is also necessary to prerun the gel at the running power for at least 30 min to preheat the gel so the polynucleotides remain denatured during the course of the electrophoresis.

12. After the gel has run its course, the plates are gently separated. The gel should have adhered to the nonsiliconized plate. Transfer the gel to an oversized piece of developed dry X-ray film (this is done preferably while the plates are still hot) by massaging the film over all the gel contact area with tissues. Then the gel on the film backing is covered with plastic wrap. Any air bubbles are gently massaged out of the sandwich, and the edges of the plastic are completely sealed with

cellotape to the backing film. (This prevents freeze-drying of the gel during exposure and hence, prevents the urea from coming out of solution, which would quench count from the sample.)

13. The film is now subject to autoradiography at −80°C with one intensifying screen (we find the Cronex Lightning Plus from Dupont (Wilmington, DE) quite satisfactory) for the necessary period of time. The length of this exposure is dependent on the number of counts loaded onto the gel (anytime between 1 d and 1 mo) (*see* Notes 7–10).

3.2.4. Preparation of 0.2 mm Thick Urea Polyacryamide Sequencing Gel

Both the relative amount and the accuracy of start site placement are determined by the quality of the polyacrylamide gel used to display the nuclease digested or primer extended fragments. The high degree of resolution necessary can be reproducible only if care is taken in the preparation and handling of the gel and gel plates.

The concentration of acrylamide used to prepare the gel depends on the size of the DNA fragments to be analyzed. To determine the size of fragments < 50 bp long accurately, gels should be cast with 10 or 12% acrylamide *(9,12)*. Fragments between 50–400 bp should be resolved on 6 or 8% gels *(9,12)*. The methods and apparatus used in our laboratory will now be described.

1. A gel length of 40–50 cm is standard. We use apparatus with a metal heat diffuser from the company American Bionuclear Inc. (Hayward, CA). We routinely use 42 × 33 cm glass plates. The thickness of the polyacrylamide gels is defined by thin plastic strips that are used as spacers between the front and back glass panels. Usually, we use 0.2 mm and occasionally 0.4 mm thick spacers. The 0.2 mm thick gels are more fragile, but give higher resolving power. Finally, we use conventional loading slot formers (combs), which generally allow either 36 or 45 samples (8 and 5 μL sample vol respectively) to be run together on each gel.

2. Before the plates are assembled, the shorter glass panel should be doubly siliconized with 10% dimethyldichlorosilane in CCl_4 (HAZARDOUS: Work in a fume hood) and washed with acetone followed by water. We then store our panels vertically in a dust-free cabinet. This siliconization procedure is necessary so that, after completion of the electrophoresis, the glass panels will be easily separable, with the gel always sticking entirely to the unsiliconized plate.

Resiliconization should occur after five to eight uses of the glass panels. At this stage the glass panels can be either stored or assembled as follows:

a. Each glass panel is washed with a scratch-proof soap (2X) (we use Dow brand [Indianapolis, IN] Scrubbing Bubbles™). This involves applying the soap over the whole surfaces to be used with a soft brush or sponge, rinsing the soap off, and repeating the soap application and removal. The wet plates are then mounted on plastic (nonscratching) supports with the washed side facing up.

b. With a fresh single-sided razor blade, a drop of double-distilled water is wiped over both glass faces to be used. This helps remove grit. Ethanol (70%) is sprayed over the surfaces and mostly wiped off with a Kimwipe™. This is repeated with 95% ethanol; this time the surface is wiped until the plates are dry.

c. The spacers are placed on either side of one of the panels, and the second panel overlayed so that the bottom of both plates are flush with each other. Four binder clips are then applied, two to either side to clamp the assembly together. The assembled glass panels are then stood at an angle against a wall with a plastic tray underneath, such that the bottom of the panels is standing in the corner at the edge of the plastic tray to form a trough at the base of the glass.

3. To prepare the gel solution, mix the following:

Reagents	6% gel	10% gel
40% Acrylamide-		
1.3% *bis*-acrylamide (w/v)	15 mL	25 mL
TBE (10X)	20 mL	20 mL
Urea	50 g	50 g
Ammonium persulfate	100 mg	100 mg

Make up to 100 mL with double-distilled water.

Place the suspension on a stirring block in a 40°C water bath until the suspension is dissolved (but not for too long or the solution will polymerize; note that the greater the amount of acrylamide present, the quicker the polymerization rate). Vacuum filter the clear solution through a double layer of Whatman #1 paper and degas thoroughly.

4. The next step is to form a polyacrylamide plug at the base of the assembled panels. Mix 12 mL of the gel solution by gentle inversion with 20 µL TEMED and pour over the base of the glass panels (from the outside) into the trough. Move the plates to allow for some of this solution to be taken up by capillary action. Allow the plug to polymerize over 2–3 min.

5. Once the plug has formed, position the plates at a 45° angle (we mount them on pipet tip racks). Add TEMED (18 µL) to the remaining gel solution, gently swirl (avoid bubbles), and carefully pour it in between the glass panels from one side, while tilting the plates toward that side. Keep a constant flow, or bubbles will form. Some authors suggest a 60-mL syringe to inject the gel solution; however, we find this unnecessary. Once the cavity is filled with the gel solution, insert the slot former, and lay the whole assembly on plastic supports as a near-horizontal arrangement. Make sure the end with the slot former is about 5° slightly elevated above the end with the plug.

6. Allow 15–30 min for the gel to polymerize. The progress of the polymerization can usually be followed by noting the Schlieren refraction pattern developing around the slot edges.

7. When all the slots have been formed, squirt some 2X TBE over the slot former and gently remove it. The gel is now ready for immediate use. We, as a rule, do not store these gels before use, since they easily dry out. Note: Clear the lanes of urea before loading any samples by flushing with 1X TBE.

8. Migration rate of marker dyes through denaturing PAGE:

% Polyacrylamide	Bromophenol blue	Xylene cyanol
5	35	130
6	26	106
8	19	76
10	12	55
20	8	28

The numbers in the dye lanes represent the approximate sizes of DNA (in bases) with which the marker dyes will comigrate *(12)*.

3.3. Primer Extension Assay

This technique can be used to measure both the amount and distance from an oligomeric primer to the 5' end of a RNA species. Thus, if the appropriate oligomeric-primer is used (*see* Fig. 1, labeled at point X), primer extension analysis should give identical bands as S1 analysis. This assay is based on the discovery that reverse transcriptase, an enzyme encoded by retrovirus, can transcribe primed RNA into DNA.

This assay involves hybridizing an end-labeled oligonucleotide that is complementary to a 30–40 base stretch of a specific RNA, to either total RNA or poly A$^+$ RNA. The primer is then extended using cold deoxynucleotides to form a cDNA copy of the RNA template. The resultant cDNA is then analyzed on a denaturing polyacrylamide gel. The length of the extended primer maps the position of the 5' end of the RNA, and the amount of the extended primer reflects the amount of steady-state initiated RNA present. For the sake of clarity, the following protocol is divided into two sections. The first section deals with the preparation of the end-labeled oligomer, and the second section the hybridization and primer extension reactions.

3.3.1. End Labeling of the Oligomeric Primer (14)

1. To end-label the oligomer primer using T4 kinase, the following reagents should be added sequentially:
     20 µL [γ^{32}P]ATP (10 mCi/mL, 3000 mCi/mmol)
     1 µL oligomeric primer (100 µg/mL) (chemically sythesized and purified by size)
     2.5 µL oligonucleotide kinase buffer (10X)
     1 µL T4 polynucleotide kinase (10 U/µL)
   Incubate at 37°C for 30 min.
2. Heat the reaction mixture at 65°C for 5 min to denature the enzyme. Next, add 25 µL of 4M ammonium acetate with 0.25 mL cold ethanol (–20°C), and mix by vortexing. Place the samples in a dry-ice/ethanol bath for 30 min, and then microcentrifuge (14,000g, 15 min, 4°C).
3. Redissolve the pellet in 25 µL double-distilled water, 25 µL 4M ammonium acetate, and 0.25 mL cold ethanol, and mix by vortexing. Place the samples in a dry-ice/ethanol bath again for 30 min, then microcentrifuge (14,000g, 15 min, 4°C).
4. Repeat step 3 again. Rinse the final pellet with 1 mL of 70% cold ethanol, and vacuum dry the pellet for 1–2 min. These three precipitations should have removed most of the free [^{32}P]ATP from the phosphorylated oligomer.
5. Resuspend this oligomer pellet in 0.1 mL of 0.3M sodium acetate, and dry count 1 µL by Cerenkov counting.

Ausubel et al. *(14)* claim that 2–4 × 10^7 cpm is a usual Cerenkov yield using this protocol. For this to be true, the isotope must be used immediately after receiving it from the company. It is permissible to use older batches of isotope, and sometimes obtain specific activity below this estimate. Nevertheless, these lower specific-activity probes can also be used for these experiments, but require longer exposure times.

3.3.2. Hybridization and Primer Extension Reaction (9)

1. As was the case for the S1 assay, an approximate determination of the hybridization temperature is necessary as a first step. For this protocol, a good estimate for the annealing temperature (T_a) is given by the melting temperature (T_m) if the oligomer is at least 30 bases long.

A simplified formula to calculate the T_m (*see* Section 3.2.3. for more details) assuming the current procedure and buffers are used, is as follows:

$$T_m = 27.1 + 0.41(\%G + C) - 500/N \qquad (3)$$

where N is the oligomer chain length. For $N = 30$, then: $T_m = 10.4 + 0.41\ (\%G + C)$.

2. To 0.25 mL of 0.3M sodium acetate, add the purified oligomer primer (1–2 µL, 10–20 ng) and RNA (2 µg poly A$^+$ or 10—50 µg total RNA). Since T_a is less critical, one sample of tRNA negative control and one of each test sample is adequate.

3. Add 0.75 mL of cold absolute ethanol to each sample, mix by vortex, incubate in a dry-ice/ethanol bath for 10 min, and then microcentrifuge (14,000g, 15 min, 4°C). Vacuum dry the pellet (consisting of both the RNA coprecipitated with the oligomer primer) for 2–3 min.

4. Redissolve the dry pellet in 50 µL of 1X Pipes buffer, and then overlay with 35 µL sterile mineral oil.

5. Place each microcentrifuge tube into a "boat" filled with water at 80°C floating in an 80°C water bath (as in Section 3.2.3., step 5). Incubate the samples in this manner for 12 min, and then transfer them quickly in their "boats" to a second water bath set at the calculated T_m.

6. Allow annealing to proceed overnight. Stop the hybridizations by adding 5 µL 3M sodium acetate/tube, and then 0.350 mL 0.3M sodium acetate. Mix the samples by vortex, and microcentrifuge (14,000g, 10 min, 4°) to separate the oil. Following this, transfer 0.4 mL from each sample to a fresh tube, mix by vortex with 1 mL of cold absolute ethanol, place in a dry-ice/ethanol bath for 10 min, and microcentrifuge (14,000g, 10 min, 4°C). Wash the pellet with 75% ethanol/25% 0.1M sodium acetate, recentrifuge for 5 min, and vacuum dry for 2–3 min.

7. Prepare the following reverse transcription cocktail on ice immediately before use (per reaction):
   3.5 µL 4 mM dNTPs
   5.0 µL 5X RT buffer
   1.25 µL RNAsin T_m (Promega Biotec, Madison, WI)
   15.5 µL double-distilled water
Resuspend the dried RNA/primer pellet in 25 µL of the above mixture.

8. Add 400 U M-MLV-reverse transcriptase (2 µL) to each sample, and allow the reaction to continue at 42°C for 90 min (*see* Notes 11 and 12).

9. At the end of this interval, add 1 µL of 0.5*M* EDTA and 1 µL RNaseA (1 mg/mL) to each sample, and incubate for a further 15 min at 37°C. (If working with ≤10 µg RNA, then the RNase A treatment may be omitted.)

10. Following the RNase treatment, add 0.1 mL of 2.5*M* ammonium acetate and 0.125 mL of chloropane to each sample, mix by vortex, microcentrifuge (14,000*g*, 3 min, 4°C), and transfer the aqueous supernatants to fresh tubes.

11. Precipitate these supernatants with 0.3 mL cold absolute ethanol, mix by vortex, place in a dry-ice/ethanol bath for 10 min, microcentrifuge (14,000*g*, 15 min, 4°C), wash with 0.5 mL cold 70% ethanol, recentrifuge for 5 min using the same conditions, and vacuum dry the pellet for 5–10 min.

12. Resuspend the resulting pellet in 3–6 µL of formamide denaturing gel-loading dye mix (1X), denature at 95°C for 3–5 min, and then quench on ice. This sample is now ready to be run on an appropriate percentage acrylamide sequencing gel (*see* Section 3.2.3., steps 11, 12, and 13 and Section 3.2.4.) (*see* Note 13).

4. Notes

4.1. RNA Extraction

1. The most important factor in RNA extraction is avoiding RNase contamination. Therefore, attention must be paid to keeping your hands gloved, baking all glassware, and using autoclaved microcentrfuge tubes (untouched by ungloved hands). Another source of RNase contamination is from freeze-thawing of the tissue to be homogenized. This potential source can be minimized by homogenizing frozen tissues immediately upon adding the lysis buffer. The original protocol included an additional incubation with SDS (0.5%) at 65°C to inactivate contaminating RNase. We have not included this step, since degradation of RNA has not been a problem following this procedure.

2. There are many methods to extract RNA from tissues or cultures. The acid guanidinium/phenol/chloroform method described here is a rapid, reliable, and efficient procedure for extracting RNA from small amounts of tissues (e.g., rat adrenal medulla). Filter sterilization of the lysis buffer prior to adding the β-mercaptoethanol using a 0.22-µm filter is performed in some protocols. This step may be necessary when a low-purity guanidinium thiocyanate is used. However, some filters dissolve in the lysis solution. Using ultrapure guanidinium thiocyanate can avoid this problem.

The guanidinium thiocyanate-ultracentrifugation method of Chirgwin et al. *(15)* results in a lower yield, but a more pure RNA. The cesium chloride step and overnight ultracentrifugation makes for a more expensive and prolonged procedure. Another method, the hot phenol method *(10)*, produces toxic vapors, which may lead to corrosive skin burns. Using the technique outlined in this chapter, one may process multiple samples in a single day, since it takes <5 h.

3. There are disadvantages to this method, however. In the original description, tRNA was shown to contribute significantly to the total RNA in only some of the samples (i.e., tRNA extraction seems variable). In addition, an isolated report *(16)* claiming the presence of a contaminating material that has a UV light absorbance at 260 nm was postulated to be a polysaccharide. These authors found that this material could be removed by an additional lithium chloride precipitation. Both of these problems can be overcome for poly (A^+) RNA by incorporating affinity chromatography using poly (T)-matrix columns into the procedure.

4. According to the original protocol of Chomozynski and Sacchi *(6)*, this method takes as little as 4 h. However, the time it takes to homogenize the tissues as well as quantify the RNA is not included in this estimate. In general, we begin RNA extractions in the afternoon using frozen tissues, incubate the sample overnight in isopropanol at −20°C (step 7, Section 3.1.), and do the final microcentrifugation and absorbance spectroscopy the following day. Certainly, one could extract RNA from at least 16 different samples in a single day.

4.2. S1 Nuclease Assay

For accurate quantitation we will now discuss several key issues. It is necessary that: the probe is in excess; the hybridization reaction is complete; the RNA:DNA hybrid complexes are equally stable; the S1 nuclease reaction proceeds properly; and that certain artifacts, like the S1 stutter, are avoided (*see* ref. *9)*.

5. The first issue can be dealt with by setting up a titration of different amounts of RNA sample, hybridized to a fixed amount of end-labeled probe. There should be a linear increase in band intensity with an increase in the amount of RNA.

6. The second concern is only an issue when a prolonged time (more than several hours) is required for complete hybridization or there is a limiting amount of probe. Using our current protocol, this should not be a problem, since the overnight hybridization should allow the reaction to go to completion. For situations where this is a concern,

samples from a given hybridization reaction can be removed at various times after combining the RNA to the probe, treated with S1 nuclease, and run on a gel. Hybridization is complete when the band intensity does not increase after further incubation.

7. The problem of complex stability is demonstrated in Fig. 2. Note, first, that the apparent optimal hybridization temperature for the distal two start sites is 54°C. The E2 start site is stronger at 57°C. At this higher temperature, the E3 and E4 start sites have all but disappeared. This phenomenon is based on relative (G + C) concentration, which directly influences the T_m (*see* Section 3.2.3., step 1). This problem is dealt with by performing a temperature calibration as described in the text (Section 3.2.3.) and confirming the start-site usage by primer extension (*see* Section 3.3. and Fig. 2).

8. This particular concern (i.e., the effective S1 nuclease concentration in the reaction) is simply dealt with by titrating the amount of S1 nuclease used, while keeping the amount of probe, RNA, the incubation time, and temperature of digestion all constant. In this way, one optimizes for optimal signal with complete degradation of unrelated bands. Additionally, we have found that, if the phenol used for the RNA extractions is slightly impure, the S1 nuclease may appear to be of much lower concentration. This can be avoided by using high molecular biology grade phenol and, after aqueous saturation, storing it at −20°C in small aliquots, which may be thawed only two to three times.

9. Other artifacts: Occassionally, the signal detected may be unexpected or nonexistent. This could be owing to either the degradation of the RNA itself or the fact that the RNA/probe annealed product may not have solublized in the original hybridization buffer. The former problem may be validated by either Northern analysis, or more simply by running the RNA on an agarose gel and staining with either ethidium bromide or acridine orange. The latter problem is solved by being overzealous in the resuspension step and using a minimonitor to follow it.

10. Allow one full working day to prepare the end-labeled probe. It takes about 2 h to set up the hybridization, followed by an overnight hybridization (16 h) The bench time involved in performing the S1 digest and preparing and running the gel should be no more than 6 h. Finally, the autoradiogaphy can be any time from overnight to a month.

4.3. Primer Extension Assay

11. The primary difficulty encountered in this type of analysis is associated with regions of RNA that cause the reverse transcriptase to pause or terminate. These show up as low-mol-wt bands on the gel. This can

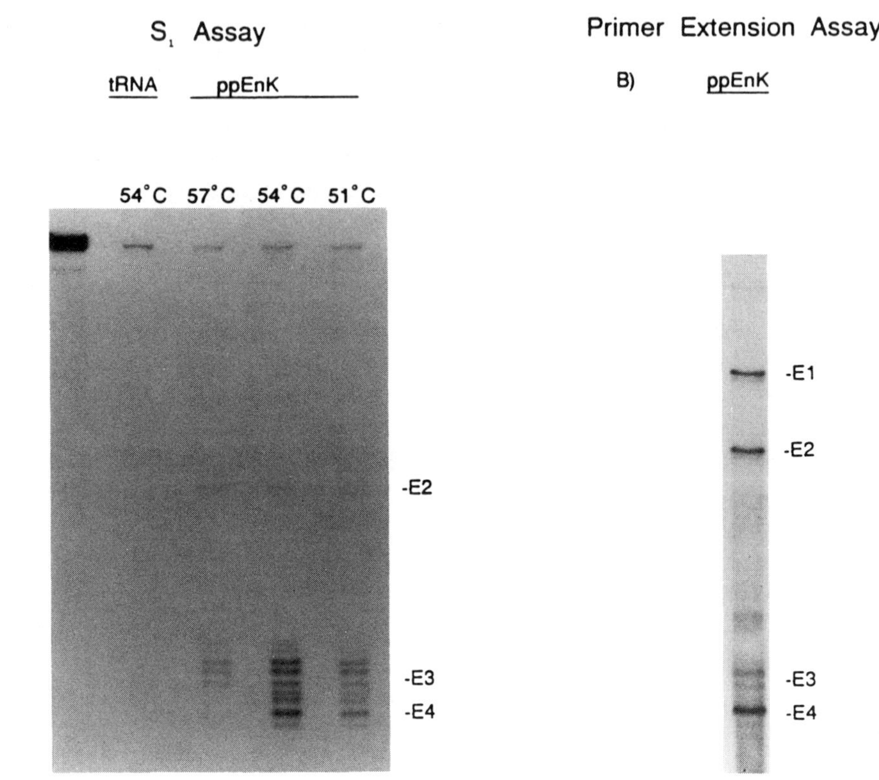

Fig. 2. Demonstration of rat basal striatal preproenkephalin RNA start sites by S1 nuclease and primer extension analysis. Lane 1 (left-hand panel) shows full-length (133 bp) S1 end-labeled probe. Lanes 2–5 show samples that have been treated with 750 U S1/10 µg total cellular striatal RNA. Lane 2, tRNA hybridized with probe at 54°C (T_m + 4); lane 3, rat striatal RNA hybridized at 57°C (T_m + 7); lane 4, rat striatal RNA at 54°C (T_m + 4); lane 5, rat striatal RNA at 51°C (T_m + 1); and the right-hand panel, a primer extension analysis of rat striatal RNA after hybridization at 48°C to a 30-bp oligomer primer. The S1 figure (**A**) demonstrates that, although 54°C is the optimal annealing temperature for E3 and E4 start sites, 57°C is better for E2 in an S1 experiment. Additionally, not only are all the sites more clearly seen using primer extension (PE) analysis, but also the band stuttering seen around E3 and E4 by S1 analysis is resolved. Occasionally, additional bands are visible by PE as well as those that can only be resolved by more complicated procedures (*see text* for combined PE/S1 analysis).

be caused by extensive G–C rich stretches and hairpin structures. Use of a primer within 100 bp of the start site(s) of the RNA will help alleviate this problem, since the reverse transcriptase will not have to travel very far.

12. Another difficulty arises out of the fact that most commercial supplies of reverse transcriptase (AMV or M-MLV) have a contaminating RNase H activity. This can result in "pseudo" upstream initiation sites seen as high-mol-wt bands on a gel. This is owing to hairpin looping over the freshly extended cDNA onto itself, displacing the RNase H-digested RNA. To control for this possibility, a series of primer extension experiments can be performed, followed by a titration of S1 nuclease under limiting conditions (e.g., 50 U over 1–5 min). This should allow the determination of the most 5' end of the RNA (*see* ref. *9* for details).

13. Allow 2–3 h to label the oligomer primer. This labeled oligonucleotide can be used for up to a period of a month. The hybridization takes about 2 h to set up, and this can easily be followed by the overnight hybridization (16 h). The bench time involved performing the extension, setting up the gel, and running it should be about 8 h. Finally, the autoradiography can last any time from overnight to a month.

Acknowledgments

This work was supported by grants from the National Science Foundation (#BNS8719872) and the March of Dimes Foundation.

References

1. Chaudhari, N. and Hahn, W. E. (1983) Genetic expression in the developing brain. *Science* **2**, 924–928.
2. Sutcliffe, J. E. (1988) mRNA in the mammalian central nervous system. *Ann. Rev. Neurosci.* **11**, 157–198.
3. Edlund, T., Walker, M. D., Barr, P. J., and Rutter, W. J. (1985) Cell specific expression of the rat insulin gene: evidence for role of two distinct 5' flanking elements. *Science* **230**, 912–916.
4. Maniatis, T., Goodbourn, S., and Fischer, J. A. (1987) Regulation of inducible and tissue-specific gene expression. *Science* **236**, 1237–1245.
5. Maire, P., Wuarin, J., and Schibler, U. (1989) The role of *cis*-acting promoter elements in tissue-specific albumin gene expression. *Science* **244**, 343–346.
6. Chomozynski, P. and Sacchi, N. (1987) Single-step method of RNA isolation by acid guanidinium thiocyanate-phenol-chloroform extraction. *Anal. Biochem.* **162**, 156–159.

7. Maxam, A. M. and Gilbert, W. (1977) A new method for sequencing DNA. *Proc. Natl. Acad. Sci. USA* **74,** 560–564.

8. Weisinger, G., Remmers, E. F., Hearing, P., and Marcu, K. B. (1988) Multiple negative elements upstream of the murine *c-myc* gene share nuclear factor binding sites with SV40 and polyoma enhancers. *Oncogene* **3,** 635–646.

9. Weisinger, G., DeCristofaro, J. D., and La Gamma, E. F. (1990) Multiple preproenkephalin transcription start sites are induced by stress and cholinergic pathways. *J. Biol. Chem.* **265,** 17,389–17,392.

10. Julius, M. A., Street, A. J., Fahrlander, P. D., Yong, J. G., Eisenman, R. N., and Marcu, K. B. (1988) Translocated *c-myc* gene produces chimeric transcripts containing antisense sequences of the immunoglobulin heavey gene locus in mouse plasmacytomas. *Oncogene* **2,** 469–476.

11. Sharp, P. A., Berk, A. J., and Berget, S. M. (1980) Transcription maps of adenovirus. *Meth. Enzymol.* **65,** 750–768.

12. Sambrook, J., Fritsch, E. F., and Maniatis, T. (1989) Chapters on S1 nuclease and denaturing polyacrylamide gel electrophoresis, in *Molecular Cloning: A Laboratory Manual,* 2nd ed., Cold Spring Harbor Laboratory, Cold Spring Harbor, NY, pp. 7.58–7.70, 11.46, 13.45–13.57.

13. Bolton, E. T. and McCarthy, B. J. (1962) A general method for the isolation of RNA complementary to DNA. *Proc. Natl. Acad. Sci. USA* **48,** 1390–1397.

14. Ausubel, F. M., Brent, R., Kingston, R. E., Moore, D. D., Seidmann, J. G., Smith, J. A., and Struhl, K. (1988) Primer extension, in *Current Protocols in Molecular Biology,* Green and Wiley-Interscience New York, pp. 4.8.1.–4.8.3.

15. Chirgwin, J. J., Przbla, A. E., MacDonald, R. J., and Rutter, W. J. (1979) Isolation of biologically active ribonucleic acid from sources enriched in ribonuclease. *Biochemistry* **18,** 5294–5299.

16. Puissant, C. and Houdebine, L.-M. (1990) An improvement of the single-step method of RNA isolation by acid guanidinium thiocyanate-phenol-chloroform extraction. *Biotechniques* **8,** 148–149.

17. Ausubel, F. M., Brent, R., Kingston, R. E., Moore, D. D., Seidmann, J. G., Smith, J. A., and Struhl, K. (1988) Preparation and analysis of RNA, in *Current Protocols in Molecular Biology,* Green and Wiley-Interscience, New York, pp. 4.0.1–4.5.5.

CHAPTER 9

Mapping of *Trans*-Acting Factor Binding in the Nervous System

Gary Weisinger and Edmund F. La Gamma

1. Introduction

The complexity of neural gene expression results from an interplay of a multitude of signal transduction pathways and *trans*-acting transcription factors interacting at *cis*-acting DNA elements *(1)*. Clearly, a major challenge is to define the pathways and molecular mechanism(s) interposed between protein kinase activation (action from the cell surface) and the *cis*-acting DNA elements (controlling target gene expression, in the nucleus), via intermediary protein factors. In general, investigators performing similar studies on various genes use large amounts of nuclear extracts made from cultured established cell lines. Simplicity of preparation is the main reason for this, as opposed to the difficulty of handling small quantities of primary neural tissues. Unfortunately, if only cell line extracts are used, not only is the information passed in vivo between cells in their natural context lost, but also information gained from these related established lines may have little physiological relevance to normal cells growing in their normal tissues. The methods presented in this chapter have been used *(2,3)* and may further be used to permit resolution of complex patterns of DNA–protein interactions generated from behavioral changes occurring at the level of the whole animal in healthy or diseased states.

As a first step in the study and mapping of neural *trans*-acting factors, appropriate nuclear extracts of the tissue of interest must be made. In the first instance, we suggest that the preparation of nuclear

From: *Methods in Molecular Biology, Vol. 13: Protocols in Molecular Neurobiology*
Edited by: A. Longstaff and P. Revest Copyright © 1992 The Humana Press, Totowa, NJ

extracts as the *trans*-acting factors that control gene expression most likely will be present in the nuclei themselves. These extracts are prepared from isolated nuclei (from the tissue of interest) using a protocol adapted from Dignam and colleagues *(4)*, and modified by us *(2)*. These extracts, together with a fragment from the gene of interest, can then be used for either gel retardation analysis or methylation interference footprinting. The gel retardation assay resolves DNA-bound factors, which are higher in mol wt than the free DNA, by polyacrylamide gel electrophoresis (PAGE). This same principle, together with Maxam and Gilbert sequencing chemistry *(5)* and stereochemical binding restraints, is used in the methylation interference footprinting (*see* Fig. 1).

More specifically, after the end-labeled DNA probe is randomly "G" methylated, such that any particular DNA molecule contains between a half to a quarter of its deoxyguanosine nucleotides unmethylated, the gel retardation assay is performed. In this way, not only should the DNA-binding factors recognize specific sequences, but they will also only recognize those that remain unmethylated. In this type of assay, the PAGE will indirectly separate methylated DNA molecules from those that have unmethylated factor recognition sites. These two populations of DNA molecules are analyzed in the rest of the assay (*see* Fig. 1). Hence, using this physiologically relevent approach, in vivo information maybe obtained using a minimal amount of tissue from single animals.

2. Materials

2.1. Nuclear Extract Preparation

Mold an acrylic pestle by allowing epoxy resin to harden with a plastic handle in it, within a 1.5-mL microfuge tube. Any reference to H_2O used in these procedures or buffers is with the understanding that it (the H_2O) has been double-distilled as well as autoclaved (15 psi for 20–30 min) before use.

1. Homogenization buffer: 50 mM Tris-HCl, pH 7.5, 1.5 mM MgCl$_2$, 2 mM β-mercaptoethanol (BME), 10 mM NaF, 200 mM sucrose, and 0.1% Triton X-100.
2. Wash buffer: homogenization buffer minus Triton X-100.
3. Extraction buffer: 50 mM Tris-HCl, pH 7.5, 420 mM KCl, 5 mM MgCl$_2$, 1 mM EDTA, 2 mM dithiothreitol (DTT), 0.1 mM phenylmethylsulfonyl fluride (PMSF), 20 mM NaF, 10% sucrose, and 20% glycerol.

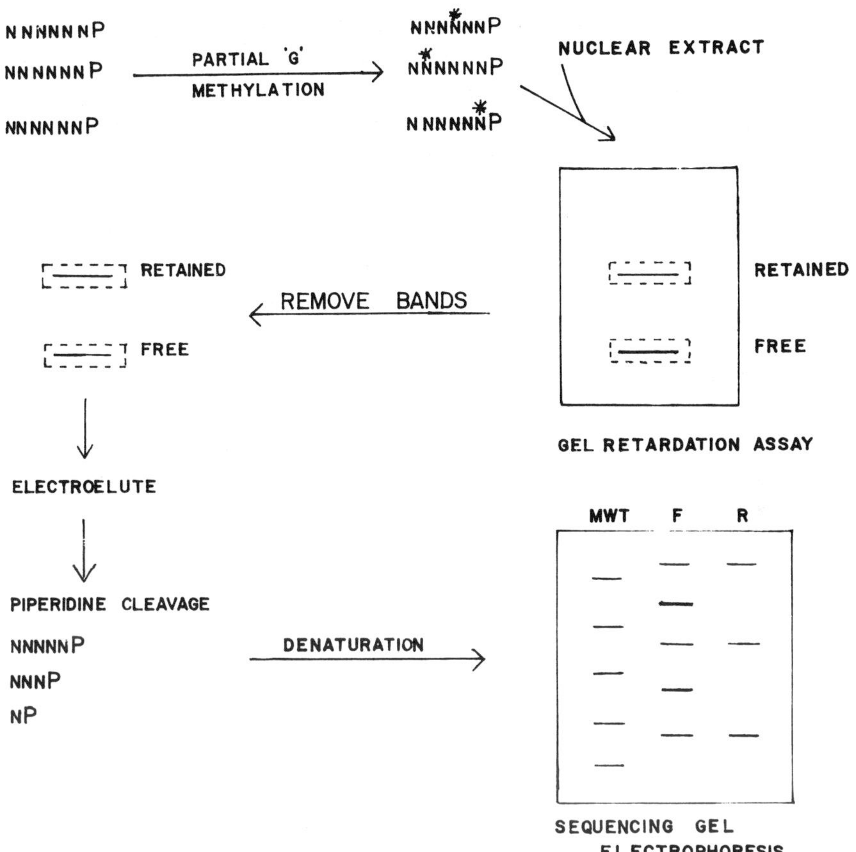

Fig. 1. Flow diagram of a methyl interference assay. N represents any nucleotide, P following a string of Ns represents a ^{32}P end-labeled phosphate, and an asterix above an N represents a methylated G.

4. Dialysis buffer: 20 mM Hepes, pH 7.4, 0.2 mM EDTA, 0.5 mM PMSF, and 0.2 mM DTT.

2.2. Gel Retardation Assay

1. Preparative vertical and horizontal gel apparatus, gel drier, electroeluter, and autoradiography equipment will be necessary.
2. Absolute ethanol and 70% aqueous ethanol cooled to –20°C are used throughout these procedures.
3. Appropriate restriction enzymes with their reaction cocktails and high-

quality plasmid DNAs will be necessary for preparation of probe and unlabeled competitor DNA fragments.

4. Bacterial Alkaline Phosphatase (BAP): Store at –20°C.
5. TE: 10 mM Tris-HCl, pH 8.0 and 1 mM EDTA.
6. 1M Tris-HCl, pH 8.0: Autoclaved 15 psi 25–30 min.
7. Proteinase K: 5 mg/mL H_2O. Store at –20°C.
8. 3M and 0.3M sodium acetate: Autoclaved at 15 psi for 25–30 min.
9. Denaturing (DN) buffer: 0.2M Tris-HCl, pH 9.5, 10 mM spermidine, and 1 mM EDTA.
10. Linker kinase buffer (10X): 700 mM Tris-HCl, pH 7.4, 100 mM $MgCl_2$, and 50 mM DTT.
11. [γ-^{32}P] ATP: (10 mCi/mL, 3000 Ci/mmol). Use only with the appropriate precautions, including training, shielding, monitoring, personal protection and institutional supervision.
12. T4 polynucleotide kinase: Store at –20°C.
13. Ficoll gel loading buffer (10X): 25% Ficoll (Type 400) in H_2O and 25% of 1% dye mix; can be stored indefinitely in a sealed container at ambient temperature.
14. 1% Dye mix: A 1% mixture of bromophenol blue and xylene cyanol in H_2O.
15. Siliconized glass wool: Loosely fill a pyrex bottle with glass wool and cover with 10% dimethyldichlorosilane in CCl_4 for 1 min. Pour off (the siliconization solution is reusable), and wash glass wool briefly with acetone and then ddH_2O which is followed by autoclaving at 15 psi for 30 min.
16. 10% Dimethyldichlorosilane in CCl_4: (HAZARD: extremely flammable, corrosive, causes burns, and is mutagenic.) Handle in a fume hood with heavy rubber gloves and appropriate protective clothing.
17. Sephadex G-50 slurry: Hydrate 1 g/15 mL H_2O over 2 d at room temperature, and replace water with TE. This material should be stored at 4°C. Some people add 0.02% sodium azide to the TE to stop bacterial growth.
18. 40% Acrylamide-1.3% *bis*-acrylamide and 30% acrylamide-1% *bis*-acrylamide (w/v): Use molecular biology grades and store at 4°C. (HAZARD: neurotoxin.)
19. TBE (10X): 0.89M Tris-borate, 0.89M boric acid, and 20 mM EDTA.
20. Ammonium persulfate.
21. TEMED: *N,N,N',N'*-tetramethyl-ethylenediamine.
22. X-ray film: Both 8 × 10" as well as 35 × 43 cm formats. We use Kodak XAR-5 film.
23. 7.5M ammonium acetate: This reagent should be filtered through a 0.45 µm filter and stored at room temperature.

24. Chloropane: Made by mixing 300 mL of freshly melted (65°C) recrystalized phenol with 300 mL CHCl$_3$, 150 mg 8-hydroxy-quinoline, and 250 mL ANE (1X). After allowing the phases to separate, the top aqueous phase is discarded, and the bottom organic phase is transferred into a brown bottle and stored at 4°C.
25. ANE (10X): 100 mM sodium acetate, pH 6.0, 100 mM NaCl, and 1 mM EDTA.
26. CHCl$_3$: isoamylalcohol (24:1 v/v).
27. DNA-binding buffer: 10 mM Tris-HCl, pH 7.5, 1 mM DTT, 1 mM EDTA, 5% glycerol, and 1–5 µg of either poly dA/dT-poly dA/dT or poly dI/dC-poly dI/dC (Pharmacia, Uppsala, Sweden) in 10 mM Tris-HCl, pH 8.0.
28. Ethidium bromide (stock): (HAZARD: This powder is a mutagen, so care should be taken to avoid *any contact* with this material.) Weigh it directly into a brown bottle with a dedicated Teflon™-covered magnetic stirrer at 10 mg/mL. Stir for a few hours to allow the powder to dissolve completely.
29. TAE (10X): 67 mM Tris-HCl (pH 6.8), 33 mM sodium acetate, and 1 mM EDTA.
30. DE81 paper: Whatman DE81 paper is cut into 12 × 40 mm strips, soaked overnight at 4°C in 2.5M NaCl, rinsed three times with H$_2$O, and stored at 4°C in TBE (1X).
31. 2.5M and 5M NaCl: Autoclave at 15 psi for 25–30 min.
32. Elution buffer: 1.5M NaCl in TE.
33. Whatman 3MM paper.
34. Large fragment of DNA Polymerase I (Klenow fragment). Store at –20°C.
35. DEPC: Diethyl pyrocarbonate.
36. Yeast tRNA: 10–25 mg/mL DEPC-treated double-distilled water. Store at –20°C.
37. Dry ice.

2.3. Methylation Interference Assay: Additional Materials

1. DMS buffer: 50 mM sodium cacodylate, pH 8.0, and 1 mM EDTA. It is usually unnecessary to adjust the pH (HAZARD: irritant and toxic).
2. Dimethyl sulfoxide (DMS): (HAZARD: very toxic and mutagenic) Use only in fume hood.
3. DMS stop buffer: 1.5M sodium acetate, pH 7.0, 1.0M BME, and 0.1 mg/mL poly(dA-dT) • poly(dA-dT).
4. Piperidine: The material sold by Sigma Chemical Co. (St. Louis, MO) is 11M. Make a 1:10 dilution with H$_2$O just before use.
5. Formamide denaturing gel loading buffer (formamide dye mix): 80%

formamide, 10% of 1% dye mix, and 10% TBE (10X). Ideally, it should be made up fresh for each day of use, but can be frozen as aliquots for later use. We use Fluka brand formamide without deionizing.
6. Urea: Molecular biology grade.

3. Methods

3.1. Nuclear Extract Preparation

The following protocol was based on the method of Dignam and colleagues *(4)* as modified by us *(2)*. We have used this protocol to make nuclear extracts from a number of rat tissues, including adrenal medulla, liver, or brain striatum *(2,6)*. After the tissue of interest has been dissected from the animal, it is very important that further manipulations are performed at 4°C (on ice). Additionally, it is required that all the following procedures utilize autoclaved micropipet tips and microcentrifuge tubes.

1. Sacrifice adult or neonate animal, microdissect the appropriate tissues, and place them into 1.5-mL microcentrifuge tubes on ice.
2. Homogenize the tissue in 0.5 mL of homogenization buffer in a 1.5-mL microcentrifuge tube with 15 strokes of a hand-held, molded acrylic pestle. We use 10 medullae (10 mg) or 50 mg striatum or liver *(2,3,6)*.
3. Allow the cellular ghosts to swell on ice for 5 min, followed by five further pestle strokes.
4. Microcentrifuge the nuclei (1600*g*, 5 min, 4°C), and save the supernatant as the "cytosolic" fraction (if required).
5. Wash the remaining pellet twice in 0.5 mL of wash buffer by vortex resuspension (10 s) of the pellet in the wash buffer, followed by recentrifugation (1600*g*, 5 min, 4°C), twice.
6. Extract the nuclear protein by resuspending the pellet in 0.1 mL of extraction buffer (vortex 10 s), and then allow the mixture to incubate for 1 h at room temperature (21°C).
7. Centrifuge these nuclei (12,700*g*, 15 min, 4°C). This resultant supernatant is designated as the "Nuclear extract."
8. In some circumstances, the salt concentration of the cytoplasmic or nuclear extracts may require modification (*see* Note 1). This can be undertaken by dialyzing 0.25 mL of each cytoplasmic extract and/or 60 µL of each nuclear extract against 50–150 vol of dialysis buffer at various salt concentrations in a 12,000–14,000 dalton mol wt cut off dialysis membrane for 1 h at 4°C. If this step is included, it is very likely that a significant amount of protein will be lost.

9. Determine the protein concentration, in our case by the method of Bradford *(7)*.

10. Aliquot the sample in 10–20 µL aliquots into microcentrifuge tubes, snap freeze in liquid nitrogen (or a dry-ice/ethanol bath), and store preferably at –80°C. Some extracts change on multiple freeze thawings, so ideally, each aliquot should be thawed only once.

3.2. Gel Retardation Assay

There are a number of component procedures involved in performing these assays. In the previous chapter in this volume, we described the preparation of T4 kinase end-labeled DNA probes, which is the first step in this assay. In the following section, we develop the assay itself.

1. Incubate the 5' end-labeled probe (5–10 ng) (*see* Note 3) at 21°C for 30 min with different amounts (initially 0.5–5 µg) of cytosolic or nuclear extracts in DNA-binding buffer, with or without unlabeled competitor DNA (we use a 20 µL total reaction vol).

2. If competitor DNA is used, a 30–150 molar excess is usually enough to see competition (*see* Section 4.2., Note 4). We recommend using an unlabeled fragment equivalent to the labeled probe. This will differentiate between probe-specific and nonspecific DNA-binding factors. We prepare the competitor DNA by resolving the cut fragment on a polyacrylamide gel stained with ethidium bromide (stain gel by gentle agitation in 1X TBE with ethidium bromide [0.5 µg/mL] for 5 min followed by visualization on a UV table). *See* Section 3.2.1. of the previous chapter and Notes 5–10.

3. Isolate the fragment, and melt it into a 1% agarose gel, place salt-treated TAE-equilibrated DE81 paper between the plug with the fragment and the positive terminal, and electrophorese the DNA onto the DE81.

4. Remove the paper and place into a 1.5-mL microcentrifuge tube. Add 0.8 mL of elution buffer (1.5*M* NaCl in TE), mix by vortexing for 1 min (to fragment paper), incubate at 37°C for 2 h, and centrifuge (10,000–14,000*g*, 15 min at room temperature) the mixture.

5. Keep the supernatant on ice while the paper is retreated with elution buffer and incubated for a second time.

6. The two aqueous supernatants are treated with an equal vol of *n*-butanol, the upper organic phase is removed, and the bottom aqueous phase retreated in the same way at least two more times.

7. To the aqueous phase add 2.5 vol of cold ethanol and leave overnight at –20°C. Only in this last step is salt (final concentration of 0.3*M* sodium acetate) added, since much NaCl is carried over in the previous precipitations.

8. Centrifuge (10,000–14,000g, 15 min, 4°C) the solution, wash the resulting pellet with 1 mL of cold 70% ethanol, vacuum dry the pellet, and resuspend it in a small vol (20–50 µL) of TE.

9. Electrophorese the entire mixture in recirculating 1X TAE at 150 V, constant voltage, for 90 min (for fragments larger than 150 bp or less for smaller fragments) on a 4% polyacrylamide/0.2% *bis*-acrylamide gel *(8,9)*. This type of gel is necessary, since weaker factor–DNA complexes will not survive electrophoresis in a regular 5% gel. Briefly, to make the appropriate 4% gel, mix the following:

    8 mL 30% acrylamide-1% *bis*-acrylamide (w/v)
    6 mL TAE (10X)
    45.9 mL ddH$_2$O
    10 µg ammonium persulfate

    Once it is clear, filter and degas the mixture, add 50 µL TEMED, pour the solution into a 1-mm thick preparative vertical gel glass assembly, and insert the well former. Allow 10 min for polymerization to be completed. The gel is usually prepared before the DNA-binding reaction is begun, so it will be ready when the binding reaction needs to be resolved. It is useful to recycle the buffer, either continuously or manually every 15 min, otherwise the pH difference that develops between the upper and lower electrode tanks may lead to poorly resolved bands.

10. At the conclusion of the electrophoresis, remove one plate and press a piece of Whatman 3MM paper onto the gel. Then, as one corner of the paper is lifted gently, the gel should remain on the paper.

11. Dry the gel and paper in a gel drier, and expose for autoradiography at –80°C, with an enhancing screen. The gel should not be wrapped in plastic wrap for the autoradiography, since static electricity will cause artifactual lines to form on the X-ray.

3.3. Methylation Interference Assay

This protocol can be divided into four sections. Initially, we partially "G" methylate the end-labeled fragment, then perform a gel retardation assay, and expose the gel undried to film at 4°C. The bands are then isolated, and the methylated DNA cleaved with piperidine. They are then separated and identified by high-resolution denaturing polyacrylamide gel electrophoresis followed by autoradiography (*see* Section 4.3.).

3.3.1. Partial Methylation of End-Labeled Probe by DMS (5,8)

1. Add 7.5 µL of end-labeled fragment (a few hundred or more cps) to 0.3 mL of DMS buffer in a 1.5-mL Eppendorf tube. Chill in ice water for 5 min (0°C).

2. Add 1.5 µL of stock DMS, and incubate at 20°C (in a water bath) for 3 min.
3. Stop the reaction by adding 75 µL of DMS stop buffer and 1.12 mL cold ethanol (this should just fit into a single microfuge tube).
4. Place in a dry-ice/ethanol bath for 10 min, and then microcentrifuge (10,000–14,000g, 10–15 min, 4°C) (*see* Note 13).
5. Resuspend the DNA in 0.2 mL of 0.3M sodium acetate (pH 7.0, 4°C), and then add 0.6 mL of cold ethanol. Vortex mix.
6. Repeat step 4 above then wash the pellet with 1 mL of cold 70% ethanol and vacuum dry. Resuspend the pellet in TE (100–400 cps/ 5 µL).

3.3.2. Gel Retardation Assay (See Section 3.2.1.)

1. Perform the reaction as in 3.2.1. steps 1–11, except that it should be scaled up 6–10-fold in a 20–30 µL reaction vol using 1/4 of the poly(dA-dT) • poly(dA-dT). Take into account the poly(dA-dT) • poly(dA-dT) that is already with the methylated probe, from the DMS stop buffer. More specifically, if 5 ng of labeled probe was used in 3.2.1. step 1, then use 30–50 ng at this step. In the same way, if 1 µg of protein was optimal in 3.2.1. step 1, then use between 6 and 10 µg in this step. We reduce the poly(dA-dT), because methylating the probe in this protocol effectively reduces the probe's concentration fourfold.
2. After the low-ionic-strength gel has run, expose it to film under plastic wrap overnight at 4°C (the low temperature reduces the background) with the glogos mentioned in the previous chapter (Section 3.2.1., step 13). You must wrap the gel in plastic wrap with one side still on the glass, since otherwise the water and salt will ruin the film.
3. Excise the different complex and free bands, and electroelute at 140 V for 30 min to 1 h, depending on the size of the fragments. For example, DNA fragments below 150 bp are electroeluted at 140 V for 30 min; fragments between 150 and 500 bp are electroeluted at 140 V for 60 min. (We use an "IBI" electroeluter; *see* the previous chapter, Section 3.2.1., steps 15 and 16.)
4. Extract the DNA-containing supernatant (salt bridge from electroeluter) with chloropane 1:1, and discard the organic phase. Extract the resulting DNA-containing supernatant (plus 10 µg tRNA carrier) with an equal vol of $CHCl_3$:isoamylalcohol (24:1), and discard the organic phase. The ^{32}P-DNA can be monitored with a handheld Geiger counter. Precipitate the remaining DNA-containing supernatant in the 7.5M ammonium acetate, i.e., the salt bridge/ DNA fragment of the IBI electroeluter, with 2.5 vol of cold ethanol in a dry ice/ethanol bath for 10 min.

5. Microcentrifuge (10,000–14,000*g*, 10–15 min, 4°C).
6. Rinse the DNA pellet with 1 mL of 70% cold ethanol, and vacuum dry for 5 min.
7. Resuspend the resulting pellet in 0.1 mL of freshly prepared 1.1*M* aqueous piperidine.

3.3.3. Methylated-G Cleavage, High-Resolution Denaturing Polyacrylamide Gel Electrophoresis, and Final Autoradiography

1. Place the piperidine/DNA mixture into a 95°C water bath for 45–60 min (fully submerged), and then quench the tubes on ice for 30 s.
2. Add 25 µg of tRNA carrier with 10 µL 3*M* sodium acetate and 2.5 vol cold absolute ethanol.
3. Vortex the tube and place in a dry-ice/ethanol bath for 10–15 min, followed by microcentrifugation (10,000–14,000*g*, 10 min, 4°C).
4. Resuspend the pellet in 0.2 mL of 0.3*M* sodium acetate, and add 0.5 mL of absolute ethanol. Repeat step 3 (second precipitation).
5. Wash the resulting pellet with 1 mL of cold 70% ethanol, and microcentrifuge again (10,000–14,000*g*, 5 min, 4°C).
6. Vacuum dry the pellet for 5–10 min and resuspend in formamide gel loading buffer (1–10 cps/3 µL). The details of urea sequencing gel preparation are given in Section 3.3.4. When the urea sequencing gel has completed the prerun, which takes at least 30 min, denature the sample at 90°C for 3 min and place on ice for at least 3 min. Load the sample onto the gel, and run in 1X TBE at 60 W constant power. We usually run a 10% gel until the xylene cyanol has run 15–25 cm into the gel. The samples are loaded with a few different concentrations of the "G" ladder (this is the probe processed without ever being bound to protein). If there are enough counts, a few different concentrations of the free band (this is the nonprotein bound band eluted from gel retention assay) and all the cps from each retained band can be in single lanes.
7. After the gel has run its course, separate the plates gently. The gel should have adhered to the nonsiliconized plate. Transfer the gel to an oversized piece of previously developed scrap X-ray film (this is done preferably while the plates are still hot) by massaging the film over all the gel contact area with tissue paper. Cover the gel on the film backing with plastic wrap. Gently massage away any air bubbles out of the sandwich, and seal the edges of the plastic completely with cellotape to the backing film. (This prevents freeze-drying of the gel during exposure, hence preventing urea from coming out of solution and quenching the counts from the sample.)

8. The film is now subject to autoradiography at −80°C with one intensifying screen (we find the Cronex Lightning Plus™ from Dupont (Wilmington, DE) quite satisfactory) for the necessary period of time. The length of the exposure is dependent on the number of counts loaded onto the gel (usually between 2 wk and 2 mo for footprints) (*see* Notes 14–16).

3.3.4. Gel Preparation for High-Resolution Denaturing Polyacrylamide Gel Electrophoresis

The high degree of resolution necessary for reproducible footprint analysis can only be attained if care is taken in the preparation and handling of the gel and gel plates. The choice of the acrylamide concentration used for the gel depends on the size of the DNA fragments to be analyzed. To determine the size of fragments less than 50 bp long accurately, gels should be cast with 10 or 12% acrylamide *(9,12)*. Fragments between 50–400 bp should be resolved on 6 or 8% gels *(9–11)*. The methods and apparatus used in our laboratory are as in the previous chapter described (Chapter 8, Section 3.2.4.).

4. Notes

4.1. Nuclear Extract Preparation

1. Different patterns of DNA-binding proteins from similar nuclear extracts were noted when the dominant cations were varied in the extracts (Figs. 3 and 5 in ref. 2). This was only accomplished reproducibly if the extracts underwent dialysis (step 8 in Section 3.1.). Since this step usually led to a large protein loss as well, one must decide whether the dialysis step is really necessary for the particular experiment. Alternatively, we found that a 1/10 to 1/20 dilution of the undialyzed extracts directly into the DNA-binding experiment (i.e., 1–2 µL extract into 20 µL reaction vol) was enough of a salt dilution and, hence, adequate for most of our needs.
2. Most of the labor-intensive time will be required initially on microdissection of the animal tissue. This time is difficult to predict, since it will be unique to each experiment and tissue. If dialysis is excluded from the procedure, allow 2 h from tissue to extract. With dialysis, allow another 1.5 h.

4.2. Gel Retardation Assay

3. The procedure for preparing the end-labeled DNA probe discussed in Chapter 8, Section 3.2.1. only details labeling 5' ends with T4 kinase. If 3' end labeling is required, the Klenow fragment of *E. coli*

polymerase (2–6 U) plus [γ- ^{32}P] dATP (40–80 µCi) can be added directly to the restriction enzyme reaction mixture after the digestion is complete (Chapter 8, Section 3.2.1., step 1). This labeling reaction is allowed to proceed for 30 min at 37°C, within appropriate shielding, followed by an equal vol of chloropane extraction (i.e., add 1:1 chloropane, vortex mix, microfuge 3 min, and take aqueous supernatant for the next step). The resulting supernatant is then loaded onto a G-50 Sephadex minicolumn and the procedure completed.

4. To test the nuclear factor specificity of retained bands, add unlabeled DNA fragment (same as the probe) to the DNA-binding reaction in at least 30-fold (molar) excess, before nuclear extract addition. If the retained band is no longer seen, then the nuclear factor(s) that originally gave rise to the shifted band is (are) specific to some sequence within the probe DNA fragment. These competition experiments can also be performed with smaller oligomers representing known nuclear factor recognition sequences as well as with antibodies to known transcription or other DNA-binding factors. In the latter case, either the retained band will disappear, as just described, or it will appear higher in the gel. This last observation is characteristic when antibodies are used and represents a higher mol-wt complex comprising the original components plus the antibody.

5. To check whether any retained bands seen in the gel retardation assay are protein factors, add 1 µL of proteinase K (5 mg/mL) to the binding reaction after binding is completed, and incubate at ambient temperature for 5–10 min. Then the low-ionic-strength PAGE is continued as usual. If in the proteinase-K-treated sample no retained bands are detected on the autoradiogram, then the binding factors probably have an essential protein component.

6. A similar assay (as in the preceding step) can be performed to determined the role of factor phosphorylation on DNA binding. In this case, add BAP enzyme, as in the preceding step and proceed accordingly.

The most common problems that arise in this procedure are:

7. Weak binding, which may be owing to steric hindrance of many factors associated with a very active *cis*-acting DNA region. This can be overcome by reducing the size of the DNA probe being used. The other reason for this problem may simply be a lack of strong DNA binding factors. Steric hindrance may also manifest itself by hiding subtle differences in gel retardation between different treatments (*see* Fig. 2). This can be overcome by adding specific factor antibodies or DNA competitors in the initial binding reaction. An example of this is demonstrated in Fig. 2. (Please *see* ref. *3* for antibody competition.)

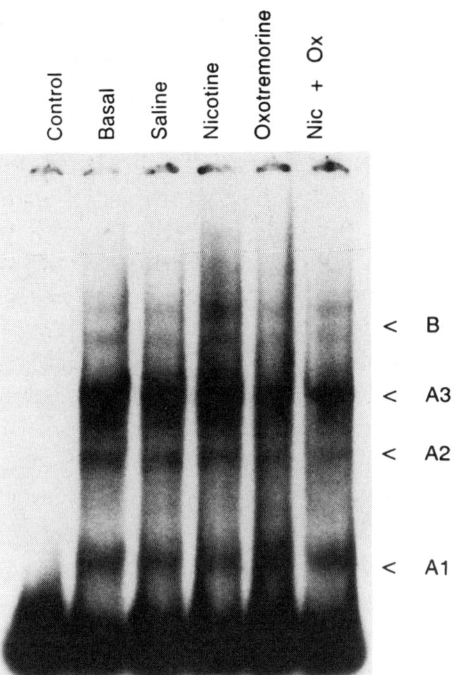

Fig. 2. Striatum-cholinergic induction. Gel shift analysis of nuclear extracts from cholinergic-agonist-treated rats using an end-labeled preproenkephalin probe, from −248 to −82 bp 5' of the transcriptional start site. Retained bands do not differ between lanes (Nicotine 5 mg/kg/d × 4 d, oxotremorine 1 mg/kg/d × 4 d). *See refs. 2,3,6.*

8. Fuzzy bands, which maybe mainly owing to the gel being run too hot or to a significant pH difference generated between the electrode baths. We do not run the gel above 150 V without cooling, and we recycle the buffer to reduce the pH difference.
9. Nonspecific DNA-binding proteins, which can in the simplest state be seen as retained bands that do not appear to enter the gel. Through increased competition with nonspecific DNA, this problem can be overcome (i.e., increase the poly[dA-dT] • poly[dA-dT] concentration in the binding reaction).
10. The preparation of a simple end-labeled probe requires a full day's work. Allow another 30 min to prepare the gel, 30 min reaction time, 90 min PAGE, and up to 60 min to dry the gel. Depending on the strength of binding and how many cps are loaded, exposures can be as short as an hour or as long as a few days.

4.3. Methylation Interference Footprinting

11. Methylation interference was used initially by Siebenlist and Gilbert *(10)* to study the interaction of RNA polymerase and the T7 promoter. It allows the indirect determination of specific nucleotides that are in relatively close contact with the DNA-binding proteins. As a result, much more information is obtained about the binding site (Fig. 3) than from the comparatively large protected region produced by another common footprinting method, DNaseI footprinting *(8)*. DMS methylates guanine residues at the N-7 position that protrude into the major groove and also at the N-3 position of adenines that protrude into the minor groove. Hence, adenines can be detected by this assay, but are much weaker than are the guanines *(5,11)*.

12. Before attempting this assay, one must optimize the conditions for the simpler gel retardation assay (Section 3.2.). This is because maximizing the amount of specific retained complexes is critical if low-abundance DNA-binding complexes are expected. This was found to be an issue by us for striatal extracts *(3)*. After optimization, if low abundance was still a serious problem, we found that the wet gel exposure to film (Section 3.3.2., step 2) could be done at –80°C with an intensification screen, greatly speeding the process of footprinting.

13. A very large DNA pellet observed after the first ethanol precipitation (Section 3.3.1., steps 4 and 5) indicates that the DMS stop buffer should be made up again.

14. If no uncleaved probe is observed at the top of the final footprint autoradiogram and a nonhomogeneous ladder of bands is noted, reduce the time of the DMS reaction.

15. Sometimes, no contacts are detected on one strand of the probe. This may mean that no guanines or adenines are in the DNA-binding site. This problem usually can be rectified by using the other strand.

16. Probe preparation is at least a full day's work. The methylation procedure requires 1–2 h. The gel retardation assay takes 3–4 h plus the overnight exposure of the film (16–24 h). Electroelution, piperidine cleavage, and high-resolution urea PAGE require another 6–8 h. The exposure of the final footprint autoradiogram can take from a few days to a few months depending on the amount of cps loaded per lane (e.g., 1 cps determined by a hand-held Geiger of a 166-bp fragment loaded onto one lane may take a month of exposure).

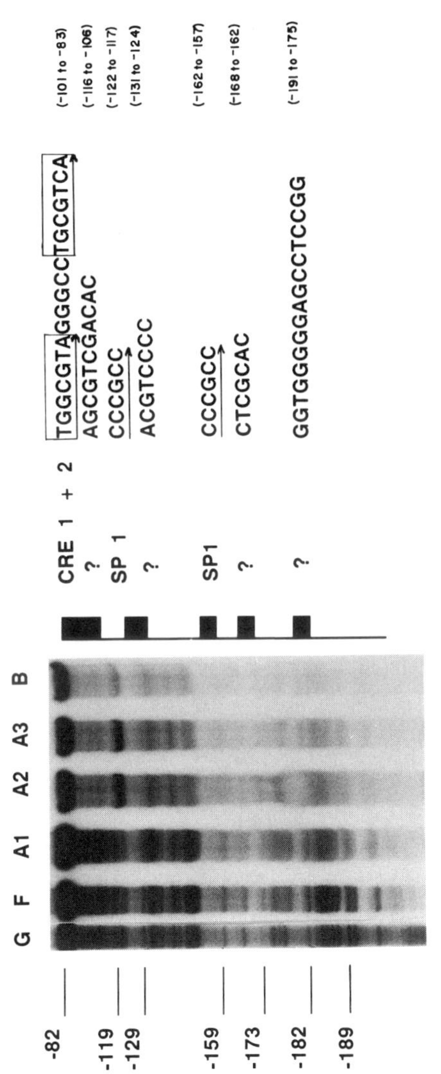
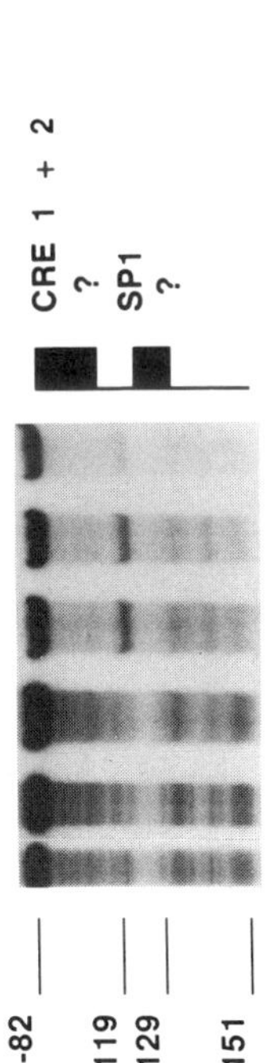

Fig. 3. Methylation interference footprint of bands labeled A1, A2, A3 and the duplex B seen in the preceding figure. "G" represents the Maxam and Gilbert G-reaction-treated fragment (*11*); "F" represents the "free" (unbound) probe band, which runs the fastest into the gel. Panels represent different exposures of the same footprint autoradiogram.

Acknowledgments

This work was supported by grants from the National Science Foundation (#BNS8719872) and the March of Dimes Foundation.

References

1. Johnson, P. F. and McKnight, S. L. (1989) Eukaryotic transcription regulatory proteins. *Annu. Rev. Biochem.* **58,** 799–839.
2. LaGamma, E. F., Goldstein, N. K., Snyder, J. B., and Weisinger, G. (1989) Preproenkephalin DNA-binding proteins in the rat: 5' flanking region. *Mol. Brain Res.* **5,** 131-140.
3. LaGamma, E. F. and Weisinger, G. (1991) Inducible preproenkephalin DNA binding proteins in the rat straitum. *Mol. Cell. Neurosci.* **2,** 427–439.
4. Dignam, J. D., Lebovitz, R. M., and Roeder, R. G. (1983) Accurate transcription initiation by RNA polymerase II in a soluble extract from isolated mammalian nuclei. *Nucleic Acids Res.* **11,** 1475–1489.
5. Maxam, A. M. and Gilbert, W. (1977) A new method for sequencing DNA. *Proc. Natl. Acad. Sci. USA* **74,** 560–564.
6. LaGamma, E. F., DeCristofaro, J. D., Agarwal, B. L., and Weisinger, G. (1989). Ontogeny of the opiate phenotype: An approach to defining transsynaptic mechanisms at the molecular level in the rat adrenal medulla. *Int. J. Dev. Neurosci.* **7,** 499–511.
7. Bradford, M. M. (1976) A rapid method for the quantitation of microgram quantities of protein utilizing the principle of protein dye binding. *Anal. Biochem.* **72,** 248–254.
8. Weisinger, G., Remmers, E. F., Hearing, P., and Marcu, K. B. (1988) Multiple negative elements upstream of the murine *c-myc* gene share nuclear factor binding sites with the SV40 and polyoma enhancers. *Oncogene* **3,** 635–646.
9. Singh, H., Sen, R., Baltimore, D., and Sharp, P. (1986) A nuclear factor that binds to a conserved sequence motif in transcriptional control elements of immunoglobulin genes. *Nature* **319,** 154–158.
10. Siebenlist, U. and Gilbert, W. (1980) Contacts between E. coli RNA polymerase and an early promoter of phage T7. *Proc. Natl. Acad. Sci. USA* **77,** 122–126.
11. Maxam, A. and Gilbert, W. (1980) Sequencing end-labeled DNA with base-specific chemical cleavages. *Methods Enzymol.* **65,** 499–560.
12. Sambrook. J., Fritsch, E. F., and Maniatis, T. (1989) Denaturing polyacrylamide gel electrophoresis, in *Molecular Cloning, A Laboratory Manual* (2nd ed.), Cold Spring Harbor Laboratory, Cold Spring Harbor, NY, pp., 1345–1357.

CHAPTER 10

In Situ Hybridization to Brain Tissue Sections Using Labeled Single-Strand Complementary RNA Probes

Giulio M. Pasinetti

1. Introduction

Because of the vast organizational complexity of the brain, it is crucial to study gene regulation at the level of individual cells. Most brain regions or individual neurons lying adjacent to each other may contain different neurotransmitters or different receptors, which may differentially regulate the gene of interest. Moreover, the brain expresses a much higher proportion of genes than other tissues, such that there are >50,000 distinct mRNA, whose cellular distribution is unknown. This extreme complexity in biochemical composition means that any study of changes in neuronal gene expression in response to pharmacological manipulations *(1)* or in response to lesions *(2)* should be analyzed at the single-cell level for maximal sensitivity to detect changes in heterogeneous cell populations.

In situ hybridization is a method originally developed for localizing specific nucleotide sequences on chromosomes *(3)* and viral genome in host cells *(4)*. Present technology allows convenient localization and assay of several mRNA on brain histological tissue sections *(5)* using a specific labeled nucleic acid probe. By analogy, since immunocytochemical techniques have proved invaluable to the elucidation of the chemical and functional nature of specific proteins in the brain, *in situ* hybridization (hybridization histochemistry) allows the identification and quantification of specific nucleic acid molecules, rather than proteins or other antigens.

Two major areas of expertise are involved in the *in situ* hybridization technique. First, understanding of basic molecular biology techniques involved in the preparation of labeled probes and consequent hybridization, and second, basic concepts in histochemistry involved in the tissue preparation and data interpretation. This chapter describes a protocol for *in situ* hybridization on either paraffin or frozen rat brain tissue sections. Alternate protocols are described for each case. Hybridization conditions are described for [^{35}S]-labeled single-strand RNA (cRNA) probes, antisense to the sense-strand complementary target molecule of interest. Although not described here, this protocol is optimized for combined *in situ* hybridization and immunocytochemical studies on the same tissue sections *(5)*. This enables quantification of a specific nucleic acid sequence in a cell containing a specific protein or neurotransmitter.

2. Materials

2.1. Glassware, Microscope Slides, and Solutions

1. Gloves are worn when handling microscope slides and glassware to protect against ribonuclease (RNAse) contamination of the tissue sections. Glassware is baked (250°C, 4 h), and plasticware is autoclaved (30 min at 15 psi on liquid cycle).
2. Solutions are autoclaved (where possible), and distilled water is diethylpyrocarbonate (DEP) treated. To prepare DEP-treated water, add DEP to a final concentration of 0.1%, allow it to sit overnight, and then autoclave before using.
3. For preparation of microscope glass slides, a sulfuric acid/potassium dichromate solution is prepared as follows: potassium dichromate (200 g) diluted in concentrated sulfuric acid (750 mL), final vol 1.75 L (with distilled water). The solution is dark red-brown, and the mixture may be reused for soaking until it becomes dark green. Use this solution with extreme care: It is highly corrosive.
4. Poly-L-lysine hydrobromide for pretreatment of histology slides. 50 μg/mL poly-L-lysine in 10 mM Tris-HCl, pH 8.0.

2.2. Tissue Processing for Paraffin Sections Using an Automatic Tissue Processor

1. We use an automatic tissue processor, which consists of a master shift carriage, to transfer tissues from one fluid to the next automatically at time intervals predetermined by a clock. This processing allows the dehydration and infiltration of tissue with embedding medium for sectioning.
2. Glass or plastic vials (20 mL).
3. Paraformaldeyde (PFA) (4%) fixative, freshly prepared in 0.1 M phosphate buffered saline (PBS), pH 7.4 (heat to 50°C to solubilize), filtered (0.45-μm filter), and then kept at 4°C.
4. Paraffin or an appropriate commercial tissue-embedding medium.
5. Ethanol solutions (50, 70, 95, and 100%).
6. Histological clearing agent, i.e., toluene or xylene.
7. Embedding molds.

2.3. Tissue Processing for Frozen Sections

1. Ice-cold 0.01 M PBS, pH 7.4.
2. Cooled 2-methylbutane (–25°C).
3. Aluminum foil.

2.4. Single-Strand cRNA Probe

DNA plasmid vectors constructed to include RNA synthesis promoters are commercially available (e.g., pBluescript Sk+/– from Stratagene [La Jolla, CA] or pGEM from Promega Biotech [Madison, WI]). Virtually any DNA (cloned cDNA, copied from the original mRNA sequence) can be inserted into these vectors *(6)*. Labeled antisense single-strand cRNA (complementary to the tissue target mRNA) probes are synthesized according to the transcription vector manufacturer's directions (e.g., Promega Biotech; *Protocols and Applications Manual)*.

2.5. In Situ *Hybridization and Autoradiography*

1. [^{35}S]-UTP, specific activity 1300 Ci/mmol.
2. 10X cRNA transcription buffer: 400 mM Tris-HCl, 60 mM MgCl$_2$, 20 mM spermidine, and 100 mM NaCl, pH 7.5.
3. Dithreothreitol (DTT) 100 mM.
4. Cold ribonucleotides for single-strand cRNA transcription, i.e., ATP, GTP, CTP. Stock solutions (10 mM).

5. RNA polymerase (20–50 U).
6. Ribonuclease (RNAse) inhibitor.
7. cRNA hydrolysis solution: 40 mM NaHCO$_3$, and 60 mM Na$_2$CO$_3$, pH 10.2.
8. Clearing agents: toluene or xylene.
9. Ethanol solutions: 100, 95, 70, 50, and 30%.
10. Stock 10X Tris-EDTA (TE) buffer: 100 mM Tris-HCl, and 10 mM EDTA, pH 7.4; 1X and 5X can be prepared easily by dilution.
11. Proteinase K (PK) stock solution: Dissolve PK at a concentration of 20 mg/mL in distilled water. Dispense into aliquots (100 µL) and store at –20°C.
12. Acetic anhydride (AA) solution.
13. Triethanolamine (TEA) solution, 0.1M, pH 8.0.
14. Stock 20X sodium salt citrate (SSC) buffer: 3M NaCl and 300 mM Na Citrate, pH. 7.4.
15. Stock 50X Denhardt's solution: 1% ficoll, 1% poly-vinyl-pyrrolidone (PVP) (Pharmaceutical Grade, M_r 40,000), and 1% bovine serum albumin (BSA). This should be filtered (0.45 µm), dispensed in aliquots, and stored at –20°C.
16. tRNA stock solution: Prepare a stock of 25 mg/mL in DEP-treated water, and store at –20°C.
17. RNA homopolymers poly(A) and poly(C): Prepare stock of 25 mg/mL in DEP-treated water, and store at –20°C.
18. Hybridization mix composition (10 mL final vol):

50% Distilled formamide	5 mL of stock (100%)
4X SSC	2 mL of 20X stock
5X Denhardt's	1 mL of 50X stock
1% Sodium dodecyl sulfate (SDS)	0.1 g
250 µg/mL tRNA	0.1 mL of 25 mg/mL
25 µg/mL poly(A)	0.01 mL of 25 mg/mL
25 µg/mL poly(C)	0.01 mL of 25 mg/mL
10% Dextran sulfate	1 g
0.1M DTT	0.08 g

Heat at 60°C to obtain the mix in solution, if necessary (this method of preparation leaves 10% of the vol for probe addition). Dispense into aliquots and store at –20°C.
19. Ribonuclease A (RNAse A) stock solution: Dissolve RNAse A at a concentration of 10 mg/mL in 10 mM Tris-HCl, 15 mM NaCl, pH 7.5. Heat to 100°C for 15 min (to inactivate DNAse), and allow to cool slowly to room temperature. Dispense into aliquots and store at –20°C.
20. Preparation of Kodak NTB-2 autoradiographic emulsion: Melt the

emulsion in a 45°C waterbath in darkness or using a safelight equipped with a Wratten red filter #1. Dilute the emulsion into 1:1 with 600 mM ammonium acetate, also warmed to 45°C. Aliquot the emulsion into 10 mL batches in scintillation vials, and double-wrap the vials with aluminum foil. Store the vials in a light-proof container at 4°C.
21. Emulsion coating chamber.
22. Kodak D-19 developer and fixer.
23. Cresyl violet for Nissl staining, 0.5% aqueous solution.

3. Methods

3.1. Microscope Slide Preparation

Slides are stored in potassium dichromate at least overnight. Rinse slides for at least 3 h in running tap water and dry. Immerse slides in poly-L-lysine for 10 min at room temperature. Air-dry in a dust-free place.

3.2. Fixed Tissue Processing: Paraffin Embedding

1. Brain preparation (rat brain): Brains are immersion fixed in approx 20 mL of PFA (use 20-mL vials) for an appropriate time (for an entire rat brain, this would be 48 h) at 4°C, and, subsequently, in 70% ethanol for an equal time (*see* Note 1).
2. Brains are then dehydrated through a graded ethanol series, impregnated with paraffin (by Automatic Tissue Processor), and cast into paraffin blocks. Thus, automatic tissue processing includes: dehydration with two changes each of 1 h in 70, 85, 95, and 100% ethanol; clearing agent treatment with two changes of 2 h each; paraffin infiltration with two changes of 2 h at 58–59°C followed by 1 h under vacuum. Brains are then transferred to embedding molds.
3. Paraffin tissue block preparation: The use of tissue molds and embedding rings is a convenient and simple method for preparing tissue for cutting on a rotatory microtome. The bottom of the mold is covered with paraffin, and the brain tissue is oriented. Paraffin for embedding should be 2–4°C above the melting point. The mold is then covered by the properly identified embedding ring. The combined mold and ring are filled with paraffin and immediately cooled (15 min) by placing the mold on ice. Slow hardening or too rapid hardening of paraffin may cause crystallization and consequent artifacts of brain tissue. On the other hand, overheating of the paraffin may cause hardening of the brain tissue. The stainless steel base

molds are reusable. The paraffin tissue block is now ready for microtome sectioning.

4. Paraffin tissue sectioning: The block must be trimmed so that the edges parallel to the knife are not crooked with respect to each other. A fine brush and small forceps may be used to handle the ribbon of sections as it comes off the microtome knife. Place the section ribbon in a warm bath (distilled, autoclaved, DEP-treated water) that should be 5–10°C below the melting point of the embedding medium. The bath is usually kept between 47–51°C for tissue sections embedded in paraffin. Sections floating on the water are picked up on treated glass slides. Drain the water from the slides, and dry them at room temperature.

3.3. Unfixed Tissue Processing: Frozen Brain

1. Brain tissue preparation: Holding the brain by the spinal cord quickly wash the brain in cooled (4–10°C) PBS to remove excess blood and hair. Then blot it on a paper towel to remove excess liquid. Dip in 2-methylbutane cooled to –25°C, holding the tissue by the spinal cord to prevent brain compression. Once the brain is sufficiently hardened (20–30 s) immerse it in 2-methylbutane for 2–3 min. Place the brain in a labeled aluminum foil wrapper, and store at –70°C.

2. Cutting frozen sections (cryostat sections): Using a cryostat is a rapid and easy means of preparing large, thin sections of single or multiple pieces of fresh frozen tissue. To begin, set the knife angle and the thickness scale (we routinely cut 10 µm thick sections). Place sufficient embedding medium for the frozen tissue block on the tissue holder, and freeze it slightly with dry ice. Place the brain in the slightly frozen medium, add more medium around the tissue, and place into the cryostat to freeze solid (the optimal cryostat temperature is –18°C). Transfer the frozen tissue mounted on the tissue holder to the microtome head, and start sectioning. Use antiroll device (cryostat is normally equipped with antiroll device) to stroke the sections onto the microtome knife as the tissue moves down over the knife. A glass slide (stored at room temperature) is placed at a close distance over the frozen section, and the section will be attracted directly to the warm slide. Do not use pressure, since this will cause distortion on the tissue sections. The tissue slides are stored in slide boxes at –70°C.

3.4. Single-Strand cRNA Probe Synthesis

cRNA synthesis kits are commercially available and include reagents for RNA transcription from a linearized DNA plasmid vector containing the cDNA of choice. Briefly, the vector is linearized downstream of the cDNA insert by a restriction cut. The linearized vector/cDNA is combined with labeled ribonucleotide and RNA polymerase in the proper buffer (the example given below uses the pGEM transcription vector, Promega Biotech). Single-strand RNA synthesis starts at the specific RNA polymerase promoter just upstream of the cDNA. Thus, discrete single-stranded RNA transcripts are produced that represent only the gene of interest.

1. RNA transcription: The protocol described below is a modification of the procedure described by Melton et al. (1984) *(7)* for the in vitro synthesis of high specific activity (2×10^9 dpm/µg) cRNA probe. This is achieved by omitting dilution of the labeled nucleotide (in this case [^{35}S]-UTP) with unlabeled nucleotide and maintaining a final concentration of UTP of about 20 µ*M*. We use labeled UTP for generating RNA probes, since higher incorporation is achieved using this nucleotide. To keep a final reaction vol from exceeding 10 µL, it is necessary to dry down the labeled nucleotide using a concentrator before adding the remaining components. Start with 250 µCi of labeled [^{35}S]-UTP (generally 20 µL vol). This will give a final concentration of labeled UTP of 20 µ*M* in 10 µL reaction vol.
   To the tube containing the dry [^{35}S]-UTP, add:
   a. 1.0 µL of 10X cRNA transcription buffer;
   b. 1.0 µL of DTT;
   c. 2.0 µL of cold ribonucleotides mix, 2.5 m*M* each of ATP, CTP, and GTP (mix 1 vol water with 1 vol each of 10 m*M* ATP, CTP, and GTP stock solutions).
   d. 1.0 µL of linearized DNA template in water (use 50–100 ng).
   e. 1.0 µL of RNA polymerase: Use appropriate polymerase for sense or antisense strand, depending on the cDNA orientation (insert) into the transcription vector.
   f. 1.0 µL RNAse (ribonuclease) inhibitor.
   g. Add water, if necessary, to 10 µL final vol, and incubate at 37°C for 1 h.
2. Removal of DNA template: After performing the in vitro transcription reaction, add RQ1 RNAase-free DNAse to a final concentration of 1 U/µg DNA (vector), and incubate for 15 min at 37°C *(see* Note 2).
3. Removal of unincorporated labeled ribonucleotides. Unincorporated ribonucleotides can be easily removed from the labeled probe by size-exclusion chromatography, for example, with small G-100 or G-50 Sephadex columns in 10 m*M* Tris-HCl, pH 7.5, containing 0.1% sodium dodecyl sulfate (SDS) *(6)*.
4. Quantification of cRNA mass transcribed: The percentage of incor-

poration of labeled nucleotide determines the mass of newly synthesized cRNA probe. Incorporation is determined by the estimation of the percent of acid precipitable counts (cpm) in the probe *(8)*. With an average 50–60% incorporation, the described reaction conditions allow the synthesis of approx 150–200 ng of cRNA probe with a specific activity of 2×10^9 dpm/µg cRNA. Store the probe at –20°C in a shielded container.

5. Probe hydrolysis. This step is described because shorter probe segments give higher *in situ* hybridization signal and lower background (higher signal-to-noise ratio) *(9)*. When necessary, we hydrolyze long cRNA probes (1–2 kb) to segments of about 300–400 bp. (a) Concentrate the cRNA probe using a vacuum concentrator. (b) Add to the tube 60 µL of cRNA hydrolysis solution, and incubate in a sealed test tube at 60°C for an appropriate time (*see* Note 3).

3.5. In Situ *Hybridization (on Paraffin Sections)*

1. Deparaffinization and hydration: Deparaffinize sections in clearing agent (two 5-min periods). Hydrate the sections by stepwise exposure to a descending series of ethanol concentrations (100, 95, 70, 50, and 30, water, 3 min each) (*see* Note 4).

2. PK digestion (*see* Note 5): 5 µg/mL PK at 37°C for 30 min in 5X TE buffer.

3. Acetic anhydride (AA) treatment: Following the PK digestion rinse the slides in water and then in 0.1 *M* triethanolamine (TEA), pH 8.0, at room temperature (5 min each). Blot the slide carrier on a paper towel to remove excess buffer. Add undiluted AA to an empty, dry staining dish, and place the slide carrier in the dish and add sufficient TEA, to cover slides and give final AA concentration of 0.25% (v/v). Incubate at room temperature for 10 min, and rinse in TEA and then in water (*see* Note 6).

4. Dehydrate sections by stepwise exposure to an ascending series of ethanol concentrations (from 30 to 100%, 2 min each) and air-dry for 10–15 min.

5. Hybridization: Prepare a sufficient amount of hybridization mix (10 µL/section, 2 sections/slide) (*see* Note 7). Combine the labeled probe to obtain the final concentration, and pipet the probe onto tissue sections. Then, apply one edge of the cover slip to the drop of hybridization mix, and slowly squeeze the sample between the cover slip and the slide so that bubbles will be forced out. The cover slips are sealed to the sides with rubber cement and the slides are then incubated at 50°C for about 5 h.

6. After removing the rubber cement, rinse slides in 4X SSC containing 10 m*M* DTT until the cover slips are completely removed, rinse slides in a fresh solution of 4X SSC (no DTT) (*see* Note 8), and then incubate in RNAse A solution, 10 µg/mL, 37°C, for 30 min. RNAse A digestion is done in 0.5*M* NaCl, 5X TE buffer.
7. Stringency washes: Wash in 2X SSC for 12 h at room temperature and then for 1 h in 0.1X SSC at 60°C. These washes should be done in a gently agitated 2–3 L vol (*see* Note 9).

3.6. **In Situ** *Hybridization (on Frozen Sections)*

1. Tissue section processing: Frozen sections are removed from the slides boxes (stored at –70°C), and immediately postfixed by immersion in fresh 4% PFA (30 min, room temperature) and washed in PBS (three washes of 10 min each at room temperature) (*see* Section 2.2., Step 3).
2. From this point on, *in situ* hybridization on frozen section or paraffin sections is identical, except that the PK digestion is omitted on frozen sections. Go to step 3 of paraffin section *in situ* hybridization (*see* Section 3.5.).

3.7. Identification of Hybridization Signal and Cellular Localization by Emulsion Autoradiography

1. Coat the slides in autoradiographic emulsion. Melt 10 mL of prediluted emulsion in a 45°C waterbath (in dark or using safe-light). Pour the emulsion slowly into the coating chamber, so as to avoid the formation of bubbles. The chamber is kept at 45°C in the waterbath. Allow the emulsion to sit for 10–15 min, and then dip some blank slides into the emulsion until surface bubbles have been removed. Dip the slides by immersing them slowly twice and as smoothly as possible. However, do not remove the slides too slowly from the emulsion, since this may lead to uneven coating. Place the slides vertically in a test tube rack, and allow them to dry for about 1 h. Transfer the slides to a light tight box, and store at 4°C for an appropriate time.
2. Development and fixation: Place the light tight boxes at room temperature, and allow them to come to approx 15°C. This will take about 15 min. Transfer the slides to a glass-slide carrier, and immediately develop for 2.5 min in a D-19 Kodak developer at 15°C. Longer developing times and higher temperature may produce grains in the emulsion background. Rinse in water also at 15°C, and fix in Kodak fixer for 5 min at 15°C. After a rinse in distilled water, the slides are washed under running tap water for 20–30 min.

3. For visualization of neurons (or other brain cell types), stain the tissue sections in cresyl violet (Nissl staining) for an appropriate time, and dehydrate the sections in 90 and 100% ethanol (1 min each). Clear the sections for 5 min and mount using cover slips with a mounting medium for subsequent handling and microscopic examination.

4. Notes

1. Optimal fixation and processing time gives good morphology as well as good signal-to-noise ratio after *in situ* hybridization.
2. After transcription, the DNA template (DNA vector) should be removed by digestion with DNAse. The RQ1 RNAse-free preparation of DNAse (Promega Biotech) has been tested for its ability to degrade DNA while maintaining the integrity of newly synthesized cRNA-labeled probe.
3. Hydrolysis time can be estimated by the following equation:

$$t = L_o - L_f / k L_o L_f$$

   where: t = time in minutes to incubate at 60°C; L_o = initial length (kb); L_f = desired length in (kb); and k is approx 0.11 cuts/kb/min. Use a denaturing 2.2M formaldehyde, 2% agarose gel electrophoresis for probe length analysis *(6)*.
4. Toluene or xylene can be used for deparaffinization, and either may be substituted with no alteration in protocol. A descending series of ethanol is used for tissue-section hydration. Use baked histology dishes and glass-slide carriers to transfer tissue sections (mounted on slides) from one fluid to the next.
5. The extent of PK digestion may be varied by adjusting the proteinase concentration. Different degrees of digestion may be required for different tissue and/or fixation conditions. The idea is to use sufficient proteinase to maximize accessibility to target RNA, without causing loss of RNA from sections or deteriorating morphology.
6. The purpose of this step is to neutralize positive charges on the sections and slides, and thus reduce electrostatic binding of the probe.
7. The best signal-to-noise ratios are achieved at probe concentrations just sufficient to saturate target RNA strands. For abundant and moderate RNA prevalence, this can be achieved using 0.3 µg/mL probe/kb of probe length *(5)* using a [^{35}S]-cRNA probe with a specific activity of 2×10^9 dpm/µg cRNA. The kinetics of the hybridization *in situ* depends on both hybridization time and probe concentration. However, these variables do not appear to be interchangeable *in situ (9)*.

8. DTT may inhibit RNAse activity.
9. For a first trial, it may be advisable to hybridize at a slightly reduced criterion, and then try several wash temperatures comparing signals with the test probe and background binding with sense-strand probe (1X SSC, 150 mM NaCl, 15 mM Na citrate, pH 7.4).

References

1. Pasinetti, G. M., Morgan, D. G., Johnson, S. A., Millar, S. L., and Finch, C. E. (1990) Tyrosine hydroxylase mRNA concentration in midbrain dopaminergic neurons is differentially regulated by reserpine. *J. Neurochem.* **55,** 1793–1799.
2. Pasinetti, G. M., Lerner, S. A., Johnson, S. A., Morgan, D. G., Telford, N. A., and Finch C. E. (1989) Chronic lesions differentially decrease tyrosine hydroxylase messenger RNA in dopaminergic neurons of the substantia nigra. *Mol. Brain Res.* **5,** 203–209.
3. Henderson, A. S. (1982) Cytological hybridization to mammalian chromosomes. *Int. Rev. Cytol.* **76,** 1–46.
4. Brahic, M., Haase, A. T., and Cash, E. (1984) Simultaneous in situ detection of viral RNA and antigens. *Proc. Natl. Acad. Sci. USA* **81,** 5445–5448.
5. Pasinetti, G. M., Morgan, D. G., Johnson, S. A., Lerner, S. P., Myers, M. A., Poirier, J., and Finch, C. E. (1989) Combined in situ hybridization and immunocytochemistry in the assay of pharmacological effects on tyrosine hydroxylase mRNA concentration. *Pharmacol. Res.* **21,** 299–311.
6. Sambrook, J., Fritsch, E. F., and Maniatis, T. (1989) *Molecular Cloning: A Laboratory Manual,* vol. 3, 2nd ed., Cold Spring Harbor Laboratory, Cold Spring Harbor, NY.
7. Melton, D., Krieg, P. A., Rebagliati, M. R., Maniatis, T., Zinn, K., and Green, M. R. (1984) Efficient *in vitro* synthesis of biologically active RNA and RNA hybridization probes from plasmids containing a bacteriophage SP6 promoter. *Nucleic Acids Res.* **12,** 7035–7056.
8. Berger, S. L. (1987) Quantifying ^{32}P-labeled and unlabeled nucleic acid. *Meth. Enzymology* **158,** 49–54.
9. Cox, K. H., DeLeon, D. V., Angerer, L. M., and Angerer, R. C. (1984) Detection of mRNA in sea urchin embrios by *in situ* hybridization using asymmetric RNA probes. *Develop. Biol.* **101,** 485–502.

CHAPTER 11

In Situ Hybridization Histochemistry Using Alkaline Phosphatase-Labeled Oligodeoxynucleotide Probe

Hiroshi Kiyama, Piers C. Emson, and Masaya Tohyama

1. Introduction

Recent progress in the field of molecular biology has provided a new method to localize an mRNA signal within a cell or tissue section. This technique, called "*in situ* hybridization histochemistry," gives us dynamic information about gene expression at individual cellular levels. For the detection of a particular target mRNA, a specific probe that has complementary sequence to the target mRNA is used. This probe has to include a reporter molecules in order to visualize the position of the probe-mRNA hybrid on a tissue section *(1)*. Conventional *in situ* hybridization procedures use radioisotopes (e.g., ^{3}H, ^{35}S, ^{32}P) as reporter molecules that can be visualized by autoradiography after hybridization *(2–4)*. The use of autoradiography, both film autoradiography and emulsion autoradiography, restricts the wide application of this method because of the following reasons.

1. A long exposure time is required;
2. Only a low resolution can be achieved;
3. There are difficulties in coexpression studies (in the simultaneous visualization of two mRNAs);

From: *Methods in Molecular Biology, Vol. 13: Protocols in Molecular Neurobiology*
Edited by: A. Longstaff and P. Revest Copyright © 1992 The Humana Press, Totowa, NJ

4. There are difficulties in an application of electron microscopic study; and
5. Special equipment is needed for radioactive work and the disposal of the radioactive waste produced.

It is particularly impossible to overcome both points 1 and 2 at the same time. In order to resolve such problems simultaneously, alternative methods in which some other reporter molecule is used have been sought. In order to be a valid reporter molecule, it must have a small size and have a specific detector, such as an antibody. So far, acetylaminofluorene *(5,6)*, biotin *(7–10)*, digoxigenin *(11)*, dinitrophenol *(12–14)*, mercury *(15,16)*, and sulfonation *(17,18)* have all been used as a reporter molecules (Table 1, column 1). In these cases, specific detectors (Table 1, column 2) were used to label these molecules, and finally they were labeled with an enzyme, such as alkaline phosphatase, to precipitate color reaction products. Such chemical amplification can increase the detectability of these probes. However, in some cases, the background is also increased. This kind of indirect detection requires a longer protocol and gives a low reliability for semi-quantification, because such chemical amplification can also increase the error range. However, the greatest problem in the use of these nonradioactive methods is the low sensitivity. The detectable amounts of mRNA in a tissue are too low to allow reliable quantitative comparisons with a conventional method using a radioactive probe.

The alkaline phosphatase-linked probe has recently been developed, has a higher sensitivity than other nonradioactive methods, and is comparable to the sensitivity of conventional radioactive probes *(25–27)*. The major advantage of this probe is the result of its low background, which gives a high contrast between positive and negative cells. Furthermore, since this method is a so-called "direct" method, which does not require any chemical amplification, the protocol becomes simpler and quicker. In this chapter, we introduce a method of nonradioactive *in situ* hybridization histochemistry using an alkaline phosphatase-linked probe.

2. Materials and Solutions

1. 0.2M phosphate buffer (PB)(1 L): 5.9 g $NaH_2PO_4 \cdot 2H_2O$, 58 g $Na_2HPO_4 \cdot 12H_2O$.
2. 0.1M phosphate buffered saline (PBS) (1 L): 2.95 g $NaH_2PO_4 \cdot 2H_2O$, 29 g $Na_2HPO_4 \cdot 12H_2O$, 9 g NaCl.
3. 20X saline sodium citrate (SSC): 175.3 g NaCl, 88.2 g Na_3 Citrate •

Table 1
Nonradioactive Reporter Molecules for Labeling of Probes

Indirect method Reporter molecule	Detector	Reference
Acetylaminofluorene (AAF)	Antibody	Tchen et al., 1984 *(5)* Landegent et al., 1984 *(6)*
Biotin		
Biotin-dUTP	Streptavidin (antibody)	Langer et al., 1981 *(7)* Leary et al., 1983 *(8)*
5'-Biotin	Streptavidin (antibody)	Agrawal et al., 1986 *(9)*
Photobiotin	Streptavidin (antibody)	Foster et al., 1985 *(10)*
Digoxigenin (-dUTP)	Antibody	Heiles et al., 1988 *(11)*
Dinitrophenol (DNP)	Antibody	Shroyer et al., 1988 *(12)*
Photo-DNP	Antibody	Keller et al., 1988 *(13)*
Br-guanine (C8)-DNP	Antibody	Keller et al., 1989 *(14)*
Mercury	HS-hapten and antibody	Dale and Ward, 1975 *(15)* Hopman et al., 1986 *(16)*
Sulfonation	Antibody	Sverdlov et al., 1974 *(17)* Poverenny et al., 1979 *(18)*

Direct method Reporter molecule	Modification of probe	Reference
Fluorescent primer	5'Amino	Smith et al., 1985 *(19)*
Enzyme (alkaline phosphatase)	Amino-thymidine (C-5)	Ruth et al., 1985 *(20)* Jablonsky et al., 1986 *(21)*
	5'-Amino	Li et al., 1987 *(22)*
	3'-Amino-SH	Chu et al., 1988 *(23)*
	Amino-cytosine (C-5)	Urdea et al., 1988 *(24)*

$2H_2O$, 800 mL H_2O. Adjust pH to 7.0–6.9 with $1N$ HCl. Add H_2O to make 1 L. The 20-fold dilution of this solution is 1X SSC.

4. Hybridization buffer (10 mL): 5 mL Deionized formamide, 2 mL 20X SSC, 200 μL 50X Denhardt's solution, 400 μL salmon sperm DNA(10 mg/mL), 1 g dextran sulfate, 2.4 mL H_2O.

5. Pretreatment buffer (100 mL): 100 mL H_2O, 2.4 mL triethanolamine, 250 μL aceto-anhydride, 0.9 g NaCl.

6. Color reaction buffers: Buffer A (500 mL): 25 mL $2M$ Tris base, 4.5 g NaCl, 475 mL H_2O. (Adjust pH to 7.2–7.5 with conc. HCl.) Buffer B (500 mL): 25 mL $2M$ Tris base, 2.92 g NaCl, 5.07 g $MgCl_2$, 475 mL

H_2O. (Adjust pH to 9.0–9.5 with conc. HCl.) Buffer C (500 mL): 2.5 mL 2M Tris base, 4.5 g NaCl, 475 mL H_2O, 186 mg EDTA. (Adjust pH to 7.2–7.5 with HCl.)

7. Substrate stock solutions: 75 mg/mL nitro blue tetrazolium (NBT). Dilute with 70% dimethyl formamide. 50 mg/mL 5-bromo-4-chloro-3-indolyl phosphate (BCIP). Dilute with absolute dimethyl formamide. Store these stock solutions at –20°C. These are usually diluted to: 340 g/mL NBT, 170 g/mL BCIP in buffer B just before use.

3. Methods

3.1. Structure and Preparation of Probe

The procedure of an enzyme labeling of oligodeoxynucleotides is described by Ruth and his colleagues *(20,21)*. The modified base that carries a linker arm with a primary amine group on the top is included during the synthesis of the oligodeoxynucleotide (Fig. 1A). This modified base is incorporated directly into the automated synthesis carried out on a DNA synthesizer. An alkaline phosphatase is crosslinked to a primary amine group on the linker arm using a primary amine reactive bifunctional crosslinker, such as disuccinimidyl suberate (DSS)(Figs. 1B,2). Briefly the modified oligodeoxynucleotide and DSS are reacted and the DSS-linked oligodeoxynucleotide separated from free DSS by gel filtration. The DSS-linked oligodeoxynucleotide is then linked with an alkaline phosphatase. The alkaline phosphatase–oligodeoxynucleotide conjugate is finally separated by anion-exchange chromatography. During the gradual change of NaCl concentration (0–1M) the free alkaline phosphatase elutes first, and then the conjugate is removed. The free oligodeoxynucleotide (DSS-oligo) can be eluted at higher concentrations of NaCl(0.3M). The fractions that contain conjugate or free alkaline phosphatase can be identified by the absorbance ratio at A_{260}/A_{280}, and the activity of alkaline phosphatase can be assayed spectophotometrically by the hydrolysis of p-nitrophenyl phosphate at 405 nm.

3.2. Hybridization Histochemistry

3.2.1. Tissue Preparation

Both perfused and freshly frozen tissue can be used for this method. However, in order to minimize the time before hybridization, we recommend the use of freshly frozen preparation. The freshly frozen tissue is cut on a cryostat (5–20 µm thickness), thaw-mounted on a gelatin

A Modified Base

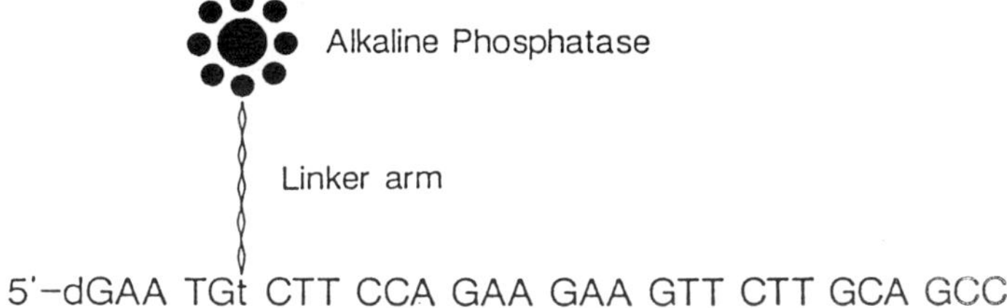

B Alkaline Phosphatase Labeled Probe

Fig. 1. Structure of a modified dTTP (**A**) and alkaline phosphatase-labeled oligodeoxynucleotide probe (**B**). An NH_2 is attached at the position of C-5 in dTTP (**A**). This modified nucleic acid can be incorporated into a probe using automated DNA synthesizer. **B** shows an example of an alkaline phosphatase-labeled probe.

coated slide, and dried quickly using a hair drier. Chrome-alum gelatin (0.5% gelatin, 0.05% $CrK[SO_4]2\text{-}12H_2O$), poly-L-lysine (1 mg/mL), or 3-aminopropyl triethoxy silane (2% in acetone) is used for the slide coating. Before coating with these solutions, slide glasses are defatted with acetone for 5–10 min and then washed with absolute ethanol for 5 min. The rinsed and dried slides are then dipped with one of the solutions for a few seconds. In case of silane coating, the dipped slides have to be rinsed with acetone once or twice. All coated slides are kept in a cold place (4°C).

3.2.2. Hybridization

A variety of methods for hybridization can be used for this type of novel probe. For us, the following procedure gave good results (*25–29*).

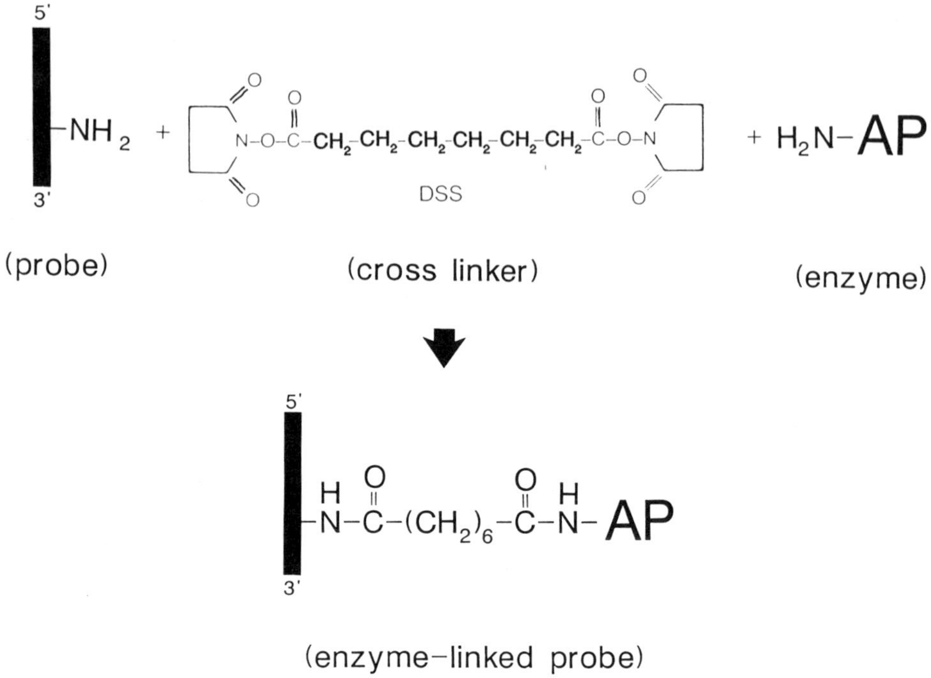

Fig. 2. The crosslinking of a probe and an alkaline phosphatase molecule by DSS.

1. Fix the sections for 30 min in 4% paraformaldehyde in 0.1 *M* PB or PBS, pH 7.5.
2. Wash the sections with 0.1 *M* PBS for 10 min to remove the fixative.
3. Pretreat the sections with pretreatment buffer for 10 min (*see* Note 1).
4. After the pretreatment, dehydrate the sections with ethanol (70% up to 100%).
5. Defat the sections with absolute chloroform for 10 min.
6. Wash again with absolute ethanol.
7. Put 50 µL of hybridization buffer on a section, and leave the sections to hybridize in the hybridization buffer at 37°C overnight.
8. After hybridization, wash the sections with 1X SSC at room temperature to remove the hybridization solution. Then put the sections through four washes with 1X SSC or 0.5X SSC at 40–55°C each for 15–20 min.
9. Following the washes, the sections are incubated in Buffer A for 15–30 min.
10. Incubate the sections in buffer B for 10 min.

11. Incubate the sections in diluted substrate solution overnight or longer (up to 6 d) to visualize the sites of hybridization. This should be done at room temperature in the dark (*see* Note 2).
12. Stop the color reaction with Buffer C for 30 min after a suitable time, which can be judged by the microscopic observation (*see* Note 3).
13. Finally, mount the sections under cover slips with buffer C containing 50% glycerol (*see* Notes 4–8).

4. Notes

1. Blocking endogenous alkaline phosphatase activity: To use this procedure in tissue that contains an amount of endogenous alkaline phosphatase, treat with $0.2M$ HCl for 10 min to block the activity. This step can be used between steps 2 and 3 *(28)*.
2. Polyvinyl alcohol: The addition of polyvinyl alcohol (PVA, 15–17%) in the substrate solution (step 11) could be useful for incubations of more than two nights. The increase of background in the substrate solution with polyvinyl alcohol is slower than in the solution without PVA; therefore the contrast between signal and background could be increased *(28)*.
3. Color of the reaction product: One of the advantages of using alkaline phosphatase-linked probes is that the color of the final product can be selected from red, blue, black, or purple-brown depending on which substrate of alkaline phosphatase is used. Moreover, the recent development of chemiluminescence substrates allows us to use chemicals, such as adamantyl-1,2-dioxatane phosphate (PPD), for the visualization of alkaline phosphatase. The localization of alkaline phosphatase can be detected by the chemiluminescence of the reacted substrate with photographic film or other photon detectors.
4. Reaction products: An mRNA positive signal by the alkaline phosphatase-labeled probe demonstrates quite high resolution and contrast (Fig. 3A). These are fairly better than ordinary radioactive method visualized by autoradiography (Fig. 3B). The hybridization signal is identified as the alkaline phosphatase reaction product whose localization is normally observed in the somatic cytoplasm (Figs. 3,4). In some cases, the reaction products show deposit patterns in the cytoplasm, and these positions seem to correspond to the localization of rough endoplasmic reticulum (Fig. 4). The primary dendrites and initial part of axon sometimes have alkaline phosphatase reaction products. This might suggest the existence of mRNA in a dendrite or axon. Such precise localization of mRNA has not been obtainable with the ordinary radioactive methods.

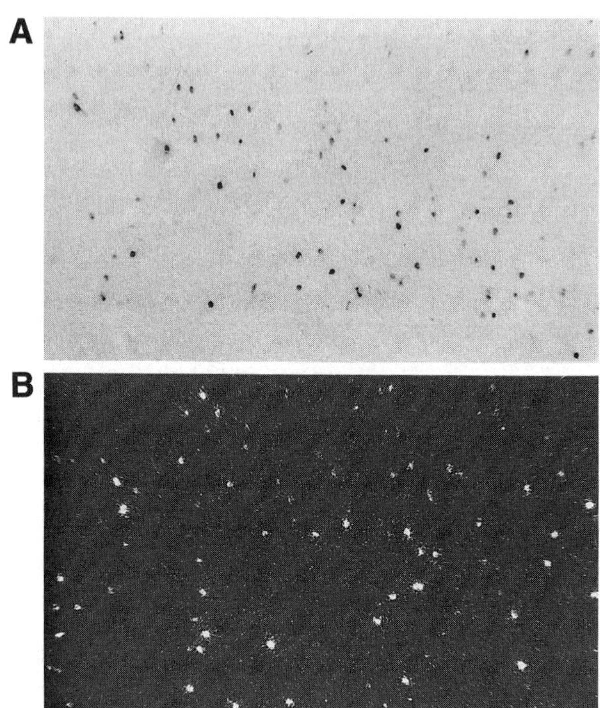

Fig. 3. Comparison between radioactive and nonradioactive probes. Somatostatin mRNA gene expression in rat cerebral cortex is demonstrated by both alkaline phosphatase-labeled probe (**A**) and ^{35}S labeled probe (**B**). **A** shows alkaline phosphatase positive cells obtained after one night of color development, whereas **B** shows emulsion-autoradiogram by dark-field illumination, which took 2 wk for exposure.

5. Control experiments: We do not have an optimum control experiment of *in situ* hybridization at the moment, but a number of methods can be used as controls, such as ribonuclease pretreatment, competition with excess nonlabeled probe, and a hybridization with sense-sequence probe. The ribonuclease pretreatment control tests the possibility of nonspecific binding of the probe to an unexpected protein or something other than RNAs. Generally, a section is treated with ribonuclease A (10 µg/mL) for 30 min before hybridization. This treatment digests most ribonucleic acids on the tissue section, and the specific signal is expected to disappear because of the lack of

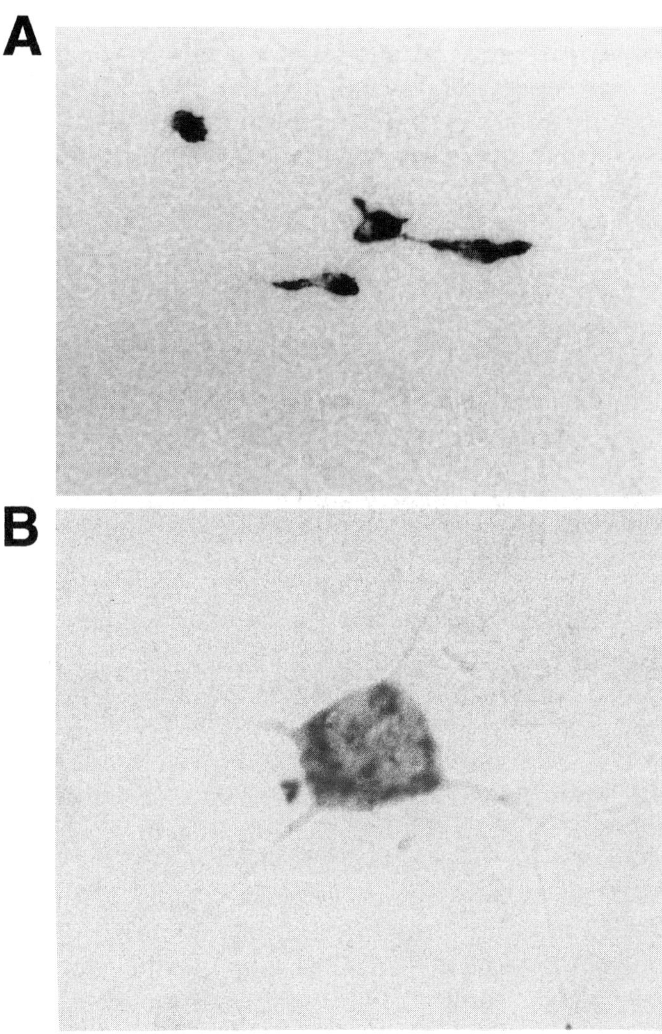

Fig. 4. The localization of alkaline phosphatase reaction products in a cell. **A** demonstrates oxytocin mRNA positive cells by the alkaline phosphatase-labeled probe in the accessory nucleus of hypothalamus. Some stained processes can be seen with high resolution, and these cells lack the stain at the position of nuclei. **B** is a high-power magnification of cortical primary cultured cell that has been stained with alkaline phosphatase-labeled probe for somatostatin mRNA. Several deposits by the reaction of alkaline phosphatase can be observed in the somatic cytoplasm.

target mRNA. The competition control is suitable for use with an oligodeoxynucleotide probe. Usually a >100-fold higher concentration of unlabeled probe is added to the hybridization buffer, which contains the usual working concentration of labeled probe. A sense-probe hybridization is sometimes used as a control experiment for a complementary RNA probe. However, in the case of the present alkaline phosphatase-labeled probe, an alkaline phosphatase-labeled sense-probe has to be newly synthesized just for the control experiment, so we do not recommend this type of control for the present experiment. In addition to these histological control experiments, it is particularly important to know if the probe can recognize mRNA of the correct size by Northern blotting. Finally, to be certain a signal is a true positive, more than one of the control experiments described earlier should be carried out. If probes that code different sites of a single mRNA molecule show the same distribution, this would also be a good control.

6. Simultaneous demonstration of two mRNAs: One of the advantages of using the present nonradioactive probe is that the simultaneous visualization of two mRNAs is possible. In our hands, the combination of alkaline phosphatase-labeled probe and ^{35}S-labeled probe shows good results *(25,29)*. Both radioactive and nonradioactive probes are added to the hybridization buffer and follow the same protocol as shown earlier. When the color reaction has finished as usual, the sections are then processed with emulsion autoradiography. A few points we should emphasize are that the section should be washed very carefully after the color reaction, and the incubation time for emulsion development in the developer should be shorter than ordinary development. Since a developer solution is generally rather high alkaline, the substrates remaining on the section react nonspecifically in the high-alkaline solution and exhibit intense background staining. Another point to remember for the coexpression study is that the developed section has to be mounted on the cover slip with buffer C, for example, containing 50% glycerol. Do not use a mountant using organic solvent.

7. Combination with tracer dyes: In our hands, both fast blue and fluoro gold can be used for this type of *in situ* hybridization. These tracer dyes are injected beforehand and follow the ordinary hybridization protocol. These dyes can survive in some cell bodies even after coexpression procedure, although slight decreases of the dye fluorescein could be observed.

8. Cultured cells: This alkaline phosphatase-labeled probe shows good results when used with cultured tissue. Since chloroform is used during

hybridization, glass chamber slides that have wells on the glass slide that can be removed after cell culture or glass cover slips are recommended as culture plates, rather than tissue culture plastic for a culture dish. Also, care must be taken to ensure that the cultured slide is dried completely with a hair drier before fixation. This makes the cells stick tightly to the glass slide.

Acknowledgments

We are grateful to Dr. J. Ruth (Molecular Biosystem, San Diego, CA, USA) and Dr. M. Gait (MRC Laboratory of Molecular Biology, Cambridge, UK) for their help in developing the described *in situ* hybridization method. H. K. acknowledges the support of J. S. P. S. (Japan) and Royal Society (UK).

References

1. Haralambidis, J., Chai, M., and Treger, G. W. (1987) Preparation of base-modified nucleotides suitable for non-radioactive label attachment and their incorporation into synthetic oligodeoxy ribonucleic acids. *Nucleic Acids Res.* **15,** 4875–4876.
2. Keller, G. H. and Manak, M. M. (1989) *DNA Probes.* Macmillan, New York.
3. Polak, J. M. and Mcgee, J. O'D. (1990) *In Situ Hybridization Principles and Practice.* Oxford University Press, Oxford, UK.
4. Uhl, G. R. (1986) *In Situ Hybridization in Brain,* Plenum, New York.
5. Tchen, P., Fuchs, R. P. P., Sage, E., and Leng, M. (1984) Chemically modified nucleic acids immunodetectable probes in hybridization experiments. *Proc. Natl. Acad. Sci. USA* **81,** 3466–3470.
6. Landegent, J. E., Jansen in de Wal, N., Baan, R. A., Hoeijmakers, J. H. J., and Van der Ploeg, M. (1984) 2-Acetylamino fluoren modified probes for the indirect hybridcytochemical detection of specific nucleic acids sequence. *Exp. Cell Res.* **153,** 61–72.
7. Langer, P. R., Waldrop, A. A., and Ward D. C. (1981) Enzymatic synthesis of biotin labelled polynucleotides: novel nucleic acid affinity probes. *Proc. Natl. Acad. Sci. USA* **78,** 6633–6637.
8. Leary, J. L., Brigati, D. J., and Ward, D. C. (1983) Rapid and sensitive colorimetric method for visualizing biotin-labelled DNA probes hybridization to DNA or RNA immobilized on nitrocellulose: bio-blots. *Proc. Natl. Acad. Sci. USA* **80,** 4045–4049.
9. Agrawal, S., Christodoulou, C., and Gait, M. J. (1986) Efficient methods for attaching non-radioactive labels to 5' ends of synthetic oligodeoxyribonucleotides. *Nucleic Acids Res.* **14,** 6227–6245.
10. Foster, A. C., McInnes, J. L., Skingle, D. C., and Symons, R. H. (1985) Non-radioactive hybridization probes prepared by chemical labelling of DNA and RNA with a novel reagent photobiotin. *Nucleic Acids Res.* **13,** 745–761.

11. Heiles, B. J., Genersch, E., Kessler, C., Neumann, R., and Eggers, H. J. (1988) In situ hybridization with digoxigenin-labelled DNA of human papillomaviruses (HPV 16/18) on Hela and Sila cells. *Biotechnique* **6,** 978–981.

12. Shroyer, K. R. and Nakane, P. K. (1983) Use of DNP-labelled cDNA in in situ hybridization. *J. Cell Biol.* **97,** 377a.

13. Keller, G. H., Huang, D. P., and Manak, M. M. (1989) Labelling of DNA probes with a photoactivatable hapten. *Anal. Biochem.* **177,** 392–395.

14. Keller, G. H., Cumming, C. U., Huang, D. P., Mannak, M. M., and Ting, R. (1988) A chemical method introducing hapten onto DNA probes. *Anal. Biochem.* **170,** 441–450.

15. Dale, R. K. M. and Ward, D. C. (1975) Mercurated polynucleotides: new probes for hybridization and selective selective polymer fractionation. *Biochemistry* **14,** 2458–2469.

16. Hopman, A. H. N., Wiegant, J., Tesser, G. I., and Van Duijn, P. (1986) A non-radioactive in situ hybridization method based on mercurated nucleic acids probes and sulphydryl-hapten ligands. *Nucleic Acids Res.* **14,** 6471–6488.

17. Sverdlov, E. D., Monastyrskaya, G. S., Guskova, L. I., Levitan, T. L., Sheishenko, V. J., and Bukowsky, E. J.(1974) Modification of the cytidine residues with a bisulfite-*O*-methyl hydroxylamine mixture. *Biochem. Biophys. Acta* **340,** 153.

18. Poverenny, A. M., Podgorodnichenko, V. K., Bryksina, L. E., Monastyrskaya, G. S., and Sverdolov, E. D. (1979) Immunochemical approaches to DNA structure investigation I. *Mol. Immunol.* **16,** 313–316.

19. Smith, L. M., Fung, S., Hunkapiller, M. W., Hunkapiller, T. J. and Hood, L. E. (1985) The synthesis of oligonucleotides containing an aliphatic amino group at 5' terminus: synthesis of fluorescent DNA primers for use in DNA sequence analysis. *Nucleic Acids Res.* **13,** 2399–2412.

20. Ruth, J., Morgan, C., and Paska, A. (1985) Linker arm nucleotide analogs useful in oligonucleotide synthesis. *DNA* **4,** 93.

21. Jablonski, E., Moomaw, E. W., Tullis, R. H., and Ruth, J. (1986) Preparation of oligodeoxynucleotide alkaline phosphatase conjugates and their use as hybridization probe. *Nucleic Acids Res.* **14,** 6115–6128.

22. Li, P., Moden, P., Skingle, D. C., Lanser, J. A., and Symons, R. H. (1987) Enzyme linked synthetic oligonucleotide probes nonradioactive detection of Escherichia coli in fecal specimens. *Nucleic Acids Res.* **15,** 5275–5287.

23. Chu, E. C. F. and Orgel, L. E. (1988) Ligation of oligonucleotides to nucleic acids or protein via disulfide bonds. *Nucleic Acids Res.* **16,** 3671–3691.

24. Urdea, M. S., Warnner, B. D., Running, J. A., Stempien, M., Clyne, J., and Horn, T. (1988) A comparison of non-radioisotopic hybridization assay methods using fluorescent, chemiluminescent and enzyme-labelled synthetic oligodeoxynucleotide probes. *Nucleic Acids Res.* **16,** 4937–4956.

25. Kiyama H. and Emson P. C. (1990) Evidence for the coexpression of oxytocin and vasopressin mRNAs in magnocellular neurosecretory cells: simultaneous demonstration of two neurohypophysin mRNAs by hybridization histochemistry. *J. Neuroendocrinol.* **2,** 257–259.

26. Kiyama, H., Emson, P. C., Ruth, J., and Morgan, C. (1990) Sensitive non-radioactive in situ hybridization histochemistry: demonstration of tyrosine hydroxylase gene expression in rat brain and adrenal. *Mol. Brain Res.* **7**, 213–219.
27. Kiyama, H., Emson, P. C., and Tohyama M. (1990) Recent progress in the use of the technique of non-radioactive in situ hybridization histochemistry: new tools for molecular neurobiology. *Neurosci. Res.* **9**, 1–21.
28. Kiyama, H. and Emson, P. C. (1991) An in situ hybridization histochemistry method for the use of alkaline phosphatase-labeled oligonucleotide probes in the small intestine. *J. Histochem. Cytochem.* **39**, 1377–1384.
29. Kiyama, H., McGowan, E. M., and Emson, P. C. (1991) Co-expression of cholecystokinin mRNA and tyrosine hydroxylase mRNA in population of rat substantia nigra cells: a study using a combined radioactive and non-radioactive in situ hybridization histochemistry. *Mol. Brain Res.* **9**, 87–93.

The Identification of Neuropeptide Gene Regulatory Elements in Transgenic Mice

Malcolm J. Low

1. Introduction

The typical mammalian genome contains 3×10^9 bp of DNA. This amount of DNA is believed to include $3–4 \times 10^4$ unique protein-encoding genes. Even complex, highly specialized cells, such as neurons, express a small fraction, perhaps 20%, of all the possible genes, however. A major problem in neurobiology, therefore, is to characterize the mechanisms involved in neural-specific gene expression.

Significant progress has been made in our understanding of the factors that restrict the expression of genes to particular cell lineages. A particularly good example is the growth hormone (GH) gene. Growth hormone belongs to a gene family that includes prolactin and chorionic somatomammotropin. Despite a high degree of homology between these genes derived from a single common ancestral gene, GH is exclusively expressed in one cell type in the mammal, the pituitary somatotroph. The first step in defining the molecular basis of somatotroph-specific gene expression was the identification of the minimal nucleotide sequences within the GH gene that confer strict pituitary-specific expression. Utilizing pituitary-derived cell lines transfected with reporter genes that contained putative cell-specific human or rat GH gene regulatory sequences, several groups of investigators delineated two binding sites in the promoter region that are essential

From: *Methods in Molecular Biology, Vol. 13: Protocols in Molecular Neurobiology*
Edited by: A. Longstaff and P. Revest Copyright © 1992 The Humana Press, Totowa, NJ

for cell-specific expression *(1–3)*. Contained within these sequences is a common motif 5' A(A/T)TTATNCAT 3' that is the core binding site for a nuclear protein. This site has been shown to be functionally essential for GH gene expression in both cell-free in vitro transcription reactions *(4)* and in the pituitary of transgenic mice *(5)*. Nucleotide sequences in addition to the 5' promoter region of GH gene may also be important for less well understood aspects of tissue-specific expression. For example, many fusion genes containing the structural portion of GH are expressed ectopically in the CNS, presumably because of some signals in the introns or protein-coding sequence *(6,7)*. Our laboratory demonstrated that the 3' flanking region of the GH gene contains a cryptic signal that directs the unexpected expression of fusion genes to the gonadotroph, and not somatotroph cells, in the pituitary of transgenic mice *(8,9)*.

The nuclear factor that specifically binds to the GH promoter sites has been named GHF-1 *(1,4)* or Pit-1 *(10)*, and belongs to a superfamily of transcriptional factors containing a homeo-domain (named for the earliest described factors that interact with *Drosophila* genes involved in homeotic mutations) *(11,12)* and a POU domain *(13)*. Expression of the GHF-1 gene is, in turn, highly restricted to the pituitary gland during development, suggesting that GHF-1 is the final common mediator leading to both induction of GH gene expression in development and maintenance of the somatotroph phenotype in adult pituitaries *(14,15)*. This one piece of the puzzle of cell-specific gene expression in mammals is understood to a fair degree in molecular terms. In general, we expect neural systems to contain similar combinations of cell-specific DNA elements and their cognate protein–DNA binding factors.

The experimental approach to understanding cell-specific expression of neuropeptides is more complicated than that for the GH gene for several reasons. First, unlike GH, which is produced in a single homogenous cell population, most neuropeptides are expressed in a large number of neurons, neuroendocrine cells, and possibly glia that have little obviously in common, except for the particular neuropeptide phenotype. Second, many neuropeptides have peculiar developmental-stage specific patterns of expression often characterized by transiently high expression in certain groups of embryonic or early postnatal neurons, and the later absence of expression in these areas in the adult brain. Thus, a molecular explanation of cell-specific neuropeptide gene expression must account for this temporal sequence in addition to the spatial information. Finally, there are insufficient or inadequate cell lines available to serve as experimental models for the study of many neuropeptides in distinction to the well-characterized somatotroph cell lines.

Because of these complexities, our laboratory is investigating the regulation of neuropeptide genes in transgenic mice. Our general approach is to use reporter fusion genes whose products can be detected in individual cells within the mice. The particular strength of our approach is that it allows a screening of reporter gene expression in all possible cell types and at every stage of development.

Transgenic mice can be produced by a variety of methods, including direct oocyte injection, retroviral infection of preimplantation embryos, and blastocyst injection of in vitro manipulated pluripotential embryonic stem cells *(16,17)*. Some of these methods have also been applied successfully to a number of other mammalian and nonmammalian species. The most widely applicable method has been the microinjection of cloned DNA into fertilized oocytes *(18)*.

There are three major approaches to the design of reporter genes for use in the production of transgenic mice (Fig. 1). Commonly, an intact nonmurine gene containing substantial lengths of both 5' and 3' flanking sequences is purified for egg injection. The resulting mRNA or encoded protein can then be distinguished from endogenous mouse products by RNA hybridization, or with antibodies directed against nonhomologous regions of the transgene and its murine counterpart *(18a)*. Extensive experience has demonstrated that transgenes from a variety of mammalian and avian species are appropriately expressed in transgenic mice *(19)*. Subtle differences have occurred, such as the stage-specific developmental regulation of human globin genes *(20)*.

A second approach is to "mark" a murine gene so that its products can be distinguished from the endogenous homologous gene. The deletion of internal exons results in a minigene whose mRNA is detectably shorter than the normal mRNA by Northern blot analysis *(21)*. Alternatively, a short oligonucleotide sequence (20–30 bp) can be inserted into a presumably neutral site in regard to affecting gene expression, such as the 5' untranslated region found between the mRNA start and the translation initiation codon, and the resulting mRNA in transgenic mice can be detected by a solution RNA assay. A further refinement of this strategy that is important for the detection of cell-specific elements in neural genes is the anatomic localization of

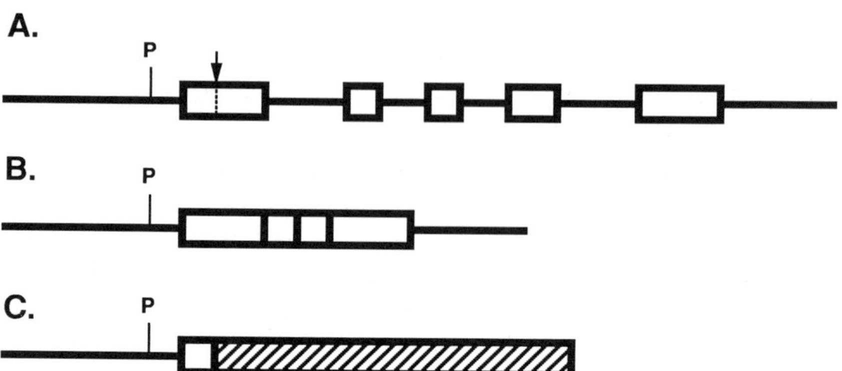

Fig. 1. Gene constructions used for gene transfer experiments. **(A)** A typical eukaryotic gene is represented schematically with five exons (open boxes). The exons are separated by intervening sequences (solid line) and the promoter (P) is indicated in the 5' flanking DNA. The gene can be "marked" by the insertion of an oligonucleotide sequence (arrow) in the 5' untranslated sequence of exon 1. **(B)** A minigene is constructed by a deletion of all the intervening sequences and exon 4. **(C)** A reporter gene contains the 5' flanking sequence, including the promoter, and the 5' untranslated region of exon 1 fused to the coding sequences (hatched box) of another gene, typically an enzyme, e.g., bacterial chloramphenicol acetyl transferase or β-galactosidase.

the oligo-marked mRNA product by *in situ* hybridization using an oligonucleotide probe *(21a)*.

A third approach to marking transgenes is to clone a chimeric gene containing putative regulatory sequences, a promoter, and the initial segment of 5' untranslated exon 1 sequences from the gene under study, ligated to the coding sequences for a reporter protein. The reporter can be a hormone, such as human growth hormone, or a nuclear protein, such as SV40 T antigen, both of which can be identified immunohistochemically in individual cells *(22,23)*. The reporter molecule may produce an easily detectable phenotype, such as transformed foci from an oncogene or cell death from a toxin *(24,25)*. A variety of genes encoding enzymes also have been used as reporters. Bacterial chloramphenicol acetyl transferase (CAT), β-galactosidase, and firefly luciferase activities have all been detected in extracts of transgenic mouse tissues *(26–28)*. In addition, many enzymes can be detected immunohistochemically. β-Galactosidase has the further advantage of being analyzed *in situ* by a simple histochemical procedure, allowing the identification of individual cells in tissue sections without the use of antisera *(29,29a)*.

In all likelihood, a combination of approaches to study gene expression in transgenic mice will yield the correct information regarding the localization of important DNA regulatory elements. Large genomic fragments have the highest probability of being expressed at physiologic levels (i.e., regarding transcript number per cell, and normal spatial and temporal regulation). These large genomic sequences are most likely to contain the multiple elements that act coordinately to produce a normal expression phenotype. Chimeric genes are invaluable for more narrowly delimiting individual DNA elements, however. To be effective for this purpose, reporter genes should satisfy several criteria. Importantly, the sequences encoding the reporter protein should not themselves influence gene expression from the putative regulatory sequences under study. Additionally, the reporter should display a high penetrance of expression between cells, be suitably stable for sensitive detection, and have a low endogenous background by whatever detection method is used.

The remainder of this chapter describes my laboratory's detailed method for the production of transgenic mice by nuclear microinjection and the analysis of tissue-specific expression of two useful reporter genes. The first is the *E. coli* enzyme β-galactosidase. All our fusion genes have used a 3.8 kb *Hind* III-*Bam H*I fragment from the clone pCH126 containing the *LacZ* gene *(30)*. We have detected cell-specific expression of this reporter by the methods described here using the regulatory regions from the rat proopiomelanocortin (POMC), rat somatostatin, human vasoactive intestinal peptide, and mouse metallothionein-I genes. The second reporter is the K1 (nontransforming) mutant of the SV40 virus *(31)*. We have used a 2.7-kb *Stu* I-*BamH* I fragment encoding the T antigen (Tag) early region ligated to the promoter/enhancer regions of the rat POMC, rat somatostatin, and human thyrotropin-releasing hormone genes. In all cases, we have obtained neuroendocrine cell-specific expression from these neuropeptide-K1Tag fusion genes.

2. Materials

2.1. Essential Equipment

1. Kopf model 720 vertical pipet puller or equivalent (David Kopf Instrs., Tujunga, CA).
2. Defonbrunne microforge (Kramer Scientific, Yonkers, NY) or equivalent.
3. Fixed-stage (inverted or upright) microscope equipped with a plain

4X objective and a long working distance 40X objective with differential interference contrast (Nomarski) optics.
4. Left- and right-handed manual micromanipulators.
5. Stereo dissecting microscopes with zoom objective, diascopic base for specimen transillumination, and a fiberoptic source for epillumination.
6. Microinjection and holding systems composed of 100-µL Hamilton syringes mounted in Stoelting syringe holders (Wood Dale, IL) and fitted with 20-g luer stub needles connected to PE 100 (Clay Adams) tubing. The hydraulic system is back-filled with light paraffin oil and must be completely free of all air bubbles.

2.2. Special Supplies

1. Hanging drop center depression slides medium thick (Fischer Scientific, Springfield, NJ).
2. Pyrex glass capillary tubes, 100 mm length, 1 mm od, 0.75 mm id (Drummond Scientific, Broomall, PA).
3. Pregnant mares' serum gonadotropins (PMS) (Sigma Chemical Corp., St. Louis, MO), 50 IU/mL: Dissolve in sterile-filtered 0.9% NaCl, and store in 1-mL aliquots at –20°C until thawed once for use. Solution is stable for at least 1 mo if frozen.
4. Human chorionic gonadotropin (HCG) (Sigma Chemical Corp.), 500 IU/mL. Dissolve in sterile filtered water, and lyophilize in 0.1-mL aliquots. May be stored at –20°C indefinitely. Prior to use, dissolve one aliquot in 1 mL sterile filtered 0.9% NaCl.
5. Elutip-d DNA purification and concentration columns (Schleicher and Schuell, Keene, NH).
6. Vectastain ABC Kit (Vector Labs, Burlingame, CA) for immunohistochemistry.

2.3. Solutions

1. Whitten's medium (bicarbonate buffered embryo culture medium): *See* Section 2.3.1. for preparation method:

Compound	g/100 mL	mM
NaCl	0.5140	88.0
KCl	0.0350	4.7
KH_2PO_4	0.0170	1.2
$NaHCO_3$	0.1910	23.0
Glucose	0.1000	5.5
L(+)-Lactic acid hemicalcium salt	0.0525	4.8
Pyruvic acid, sodium salt	0.0025	0.2

DL lactic acid, sodium salt	0.37 mL of	
	60% syrup	19.8
$MgSO_4 \cdot 7H_2O$	0.0290	1.2
Bovine serum albumin (BSA)		
Fraction V	0.3000	
EDTA (free acid)	0.0037	0.1
Penicillin G, potassium salt		
(final conc., 120 U/mL)	0.0075	
Streptomycin sulfate		
(final conc., 50 µg/mL)	0.0050	
Phenol Red	0.0010	
Milli-Q H_2O	Up to 100 mL	

2. TBE (20X concentrated stock):

Compound	g/1 L	M
Tris base	121	1
Boric acid	61.7	1
EDTA—$Na_2 \cdot 2H_2O$	7.44	0.02
H_2O to make	1 L	

Store at room temperature.

3. Low-salt solution (for Elutip columns):

Stock solution	Amount/50 mL	mM
5M NaCl	2 mL	200
1M Tris-HCl, pH 7.4	1.0 mL	20
0.5M EDTA—$Na_2 \cdot 2H_2O$	0.1 mL	1
Milli-Q H_2O	46.9 mL	

4. High-salt solution (for Elutip columns):

Stock solution	Amount/50 mL	mM
5M NaCl	10.0 mL	1000
1M Tris-HCl, pH 7.4	1.0 mL	20
0.5M EDTA—$Na_2 \cdot 2H_2O$	0.1 mL	1
Milli-Q H_2O	38.9 mL	

5. 5 mM Tris/ 0.1 mM EDTA solution for DNA microinjection:

Stock solution	Amount/50 mL	mM
1M Tris-HCl, pH 7.6	250 µL	5
0.5M EDTA—$Na_2 \cdot 2H_2O$	10 µL	0.1

Filter sterilize solution with a 0.22-µm syringe filter.

6. HEPES-buffered embryo medium (HBM, used for embryo manipulation outside incubator; *see* Section 2.3.1. for preparation method):

Compound	g/100 mL	mM
NaCl	0.5200	89.0
KCl	0.0360	4.8

$CaCl_2 \cdot 2H_2O$	0.0250	1.7
KH_2PO_4	0.0160	1.2
$MgSO_4 \cdot 7H_2O$	0.0290	1.2
HEPES (free acid)	0.5950	25.0
Pyruvic acid, sodium salt	0.0028	0.3
DL lactic acid, sodium salt	0.42 mL of 60% syrup	22.5
Glucose	0.1000	5.5
EDTA (free acid)	0.0037	0.1
Streptomycin sulfate	0.0133	
Penicillin G, potassium salt	0.0061	
Phenol Red	0.0010	
BSA	0.5000	
Milli-Q H_2O	Up to 100 mL	

7. Avertin anesthetic (2% solution): Dissolve 1 g 2,2,2-tribromoethanol (Aldrich Chemical Co., Milwaukee, WI) in 50 mL distilled water over very low heat just until crystals disappear. Remove from heat, and store in dark container at 4°C for up to 1 mo. Each batch should be tested for potency by intraperitoneal (ip) injection at the standard dose of 0.15 mL/10 g body wt. Surgical anesthesia should be reached within 5 min and last up to 30 min.

8. Genomic DNA extraction buffer:

Compound	Amount/50 mL	mM
0.5M EDTA—$Na_2 \cdot 2H_2O$	10.0 mL	100
1M Tris-HCl, pH 8.0	2.5 mL	50
10% SDS	2.5 mL	
Proteinase K	25.0 mg	
Milli-Q H_2O	35.0 mL	

Combine EDTA, Tris, and H_2O first, and then add SDS to prevent its precipitation. Add Proteinase K fresh just before use.

9. TE solution for genomic DNA:

Stock solution	Amount/500 mL	mM
1M Tris-HCl, pH 7.6	5 mL	10
0.5M EDTA—$Na_2 \cdot 2H_2O$	1 mL	1

10. Preperfusion flush solution:

Compound	g/100 mL	mM
NaCl	0.90	154
Glucose	5.00	277
Procaine hydrochloride	0.06	
Heparin	1500 U	
Milli-Q H_2O	Up to 100 mL	

Combine all ingredients, and filter-sterilize with a 0.22-μm membrane. Store at 4°C.

11. Phosphate-buffered paraformaldehyde: Dissolve 40.0 g of paraformaldehyde into 400 mL of distilled water. Heat to 80°C (do not boil, use a fume hood) with stirring, and add several drops $1M$ NaOH, until the solution clears. Bring the final vol to 500 mL. Add 500 mL $0.2M$ $NaPO_4$ buffer, pH 7.4. Filter through Whatman paper. Store at 4°C. Best when used fresh.

12. Potassium phosphate-buffered saline (KPBS, 10X concentrated stock):

Compound	Amount/L
NaCl	81.81 g
K_2HPO_4	34.83 g
Milli-Q H_2O	Up to 1 L

Dilute one part 10X concentrate with nine parts Milli-Q H_2O to obtain working solution (140 mM NaCl, 20 mM K_2HPO_4, pH 7.4). Stable at room temperature.

13. X-gal (5-Bromo-4-chloro-3-indolyl-β-D-galactopyranoside) solution:

Compound	Amount/30 mL	mM
30 mM potassium ferricyanide	1 mL	1
30 mM potassium ferrocyanide	1 mL	1
200 mM $NaPO_4$ buffer, pH 7.4	2.5 mL	17
$5M$ NaCl	1 mL	166
$1M$ $MgCl_2$	60 μL	2
Nonidet P-40	3 μL	
Sodium deoxycholate	3 mg	
Milli-Q H_2O	Up to 28.5 mL	
X-gal (2% solution in dimethylformamide, DMF)	1.5 mL	

Prepare stock solutions, and combine all ingredients except the X-gal. Filter-sterilize through a 0.22-μm membrane. Store the filtered aqueous solution and the 2% X-gal solution separately in the dark at 4°C. Just before use, combine the two solutions in the ratio of 19:1, and vortex immediately to prevent precipitation of the highly water-insoluble X-gal.

14. Fluorometric β-galactosidase assay buffer:

Compound	Amount/10 mL	mM
200 mM $NaPO_4$ buffer, pH 7.4	4.5 mL	90
$1M$ $MgCl_2$	90.0 μL	9
$14.4M$ β-mercaptoethanol	0.31 mL	446
Milli-Q H_2O	Up to 10 mL	

15. 4-Methylumbelliferyl-β-D-galactopyranoside (4MUFG) stock solution:

Dissolve 3.1 mg 4MUFG into 20 mL DMF (455 μM).

16. β-galactosidase standards: Make a concentrated stock in distilled water from grade VI lyophilized powder in buffered salts from *E. coli* (Sigma Chemical Co., St. Louis, MO). Stable at –20°C.

17. Tris-buffered saline (TBS, 10X concentrated stock):

Compound	Amount/L
$1 M$ Tris-HCl, pH 7.6	500 mL
NaCl	90 g
Milli-Q H_2O	Up to 1 L

Dilute one part 10X concentrate with nine parts Milli-Q H_2O to obtain working solution (50 mM Tris-HCl, 154 mM NaCl).

2.3.1. Preparation of Embryo Culture Media

1. Soak all glassware in phosphate detergent overnight.
2. Rinse 10 times in tap H_2O to remove all detergent residue.
3. Acid clean glassware in chromic sulfuric acid (Chromerge) solution.
4. Rinse 10 times in tap H_2O, five times in distilled H_2O, and five times in Milli-Q H_2O.
5. Sterilize for 2 h at 110°C in dry baking oven.
6. Measure dry ingredients, except BSA, into an acid-cleaned and sterile bottle.
7. Add 50 mL Milli-Q H_2O, mix, then add sodium lactate syrup, and mix again.
8. Sprinkle BSA onto surface of media, and mix by gentle inversion. Do not shake, since the BSA will foam and denature.
9. Adjust pH of the HEPES-buffered media to 7.4 with $1 M$ NaOH.
10. Adjust vol to 100 mL with Milli-Q H_2O.
11. Sterile-filter through a 0.22-μm low-protein binding filter.
12. In a laminar flow hood, pour the media into new acid-cleaned, sterile bottles and cap tightly.
13. Store at 4°C for up to 2 wk. Always open the bottles under aseptic conditions.

    Note: All ingredients should be tissue-culture grade.

3. Methods

3.1. Production of Transgenic Mice

3.1.1. Animal Colony

Mice are housed in a facility free from any contact with other animals that might introduce infectious disease into the colony. Lights are on from 5 :00 AM to 7 :00 PM. Breeding females are given a high-fat, high-protein diet, whereas other mice are fed maintenance rodent chow. A breeding colony consists of individually caged mating pairs of a DBA/2 male × C57B1/6 female to produce B6D2F1 hybrids. Mice are weaned at 18–20 d of age. New breeding stock, male and female Swiss Webster mice are bought periodically from suppliers of virus-free animals. Sexually experienced stud and vasectomized males are housed in individual cages. Stud males are generally used for mating only once per week to maintain high fertilization rates.

3.1.2. Collection and Culture of Mouse Embryos

1. Three days prior to oocyte collection, inject three to seven 4-wk-old B6D2F1 female mice with 5 IU PMS ip at 2:00 PM. Each mouse will yield 20–40 eggs.
2. One day prior to oocyte collection, inject the PMS-primed mice with 5 IU HCG ip at 1:00 PM, and place each female with a separate stud B6D2F1 male overnight. At the same time, 10 8-wk-old females are selected for vaginal signs of estrous and housed overnight with 5–10 vasectomized SW males to obtain pseudopregnant females for embryo transfer.
3. On the experimental day, select the females that have a copulation plug consisting of coagulated semen. Return any females that did not mate to the colony for use at a later date.
4. Fill five 35-mm culture dishes with 3 mL of Whitten's medium, and place in the incubator to warm and equilibrate with the CO_2. On an additional dish, place two 30-μL drops of Whitten's medium containing 1% hyaluronidase (HA) and four separate 30-μL drops of plain Whitten's medium, cover with light mineral oil, and store in the incubator.
5. Prepare six embryo-transferring pipets from borosilicate or soda lime glass Pasteur pipets using an alcohol lamp. Pipets should be pulled over the flame to yield a range of final interior diameter ranging from 100–500 μm. The pipets should have a clean, square end.
6. Kill the donor B6D2F1 mice by cervical dislocation, and expose the reproductive tract through a horizontal lower abdominal incision. Rapidly remove each coiled oviduct with a minimum of residual adipose tissue, ovary, or uterus, and place in one of the dishes of prewarmed Whitten's medium.
7. Use a dissecting microscope and transillumination to identify the distended, translucent loop of each oviduct containing the cumulus mass, and free it into the media with watchmaker forceps and a 25-g needle. Discard the empty oviducts.
8. After teasing all the cumulus masses free, use the largest transfer pipet (all pipeting is done by mouth) to place the eggs in the drops

of HA. After approx 5 min, gently triturate the oocytes to free them of all adherent cumulus cells, and wash the oocytes four times in successive drops of Whitten's medium. Transfer the washed eggs in groups of 20–30 into fresh oil-covered drops of Whitten's medium. The eggs can be injected and transferred to oviducts the same day, or incubated up to 5 d, at which time the majority will develop into blastocysts and "hatch" from their confining zona pellucidas.

3.1.3. Preparation of Plasmid DNA for Microinjection

1. Digest up to 5 μg of the reporter gene-containing clone with restriction endonucleases, according to standard protocols *(32)*, to remove all plasmid or phage vector sequences.
2. Size-separate the resulting DNA fragments on a 0.8% agarose gel containing ethidium bromide (add 5 μL of a 10 mg/mL stock to each 100 mL of gel) in TBE.
3. Visualize the gel under UV light, and cut out the band containing the fragment for microinjection (position of the band should be determined by the inclusion of appropriately sized mol-wt markers).
4. Add the agarose slab to 350 μL TBE in a short segment of rinsed 3/4-in dialysis tubing (mol-wt cutoff 12,000–14,000 daltons) clipped at both ends. Be sure to remove all air bubbles. Electroelute the DNA at 250 V for 15–30 min by laying the tubing perpendicular to the electric field in a standard horizontal electrophoresis apparatus. Reverse the current for 1 min to remove adherent DNA from the dialysis membrane.
5. Collect the buffer, and rinse the dialysis tubing with an additional 150 μL of TBE. Extract once with a half-vol of saturated phenol, once with an equal vol of phenol/chloroform/isoamyl alcohol (20/19/1), and once with an equal vol of chloroform/isoamyl alcohol (19/1). For each extraction, vortex for 30 s and centrifuge (12,000g, 2 min, 25°C). Collect the top aqueous phase, leaving behind all the interface, into a new microfuge tube.
6. Precipitate the DNA by adding 0.1 vol 3*M* sodium acetate, pH 5.4, and 2.5 vol ethanol. Freeze on dry ice for 30 min and centrifuge (12,000g, 30 min, 4°C). Gently wash the tear-shaped pellet, so as not to dislodge it from the walls of the tube, with 200 μL 70% ethanol (4°C). Dry the pellet under vacuum in a rotary (Speed-Vac) evaporator for 10 min.
7. Resuspend the DNA pellet in 1 mL low-salt solution.
8. Prepare an Elutip-d column as described in the manufacturer's instructions. Load the DNA solution, and reapply the eluent (1X).

Wash with 3 mL of low salt solution. Elute the DNA from the matrix with 0.4 mL high-salt solution.

9. Reprecipitate the DNA with 2.5 vol cold ethanol (do not add any extra salts or carrier transfer RNA), and resuspend in an appropriate vol of sterile-filtered 5 mM Tris-HCl, 0.1 mM EDTA solution to obtain an estimated concentration of 10–100 ng/μL.
10. Determine the actual concentration of DNA using the fluorescent dye *bis*-benzimidazole (Hoechst 33258) as described by Labarca and Paigen *(33)*.
11. Store the DNA at 4°C. Avoid repeated freezing and thawing.

3.1.4. Microinjection of DNA into the Pronucleus

1. First, prepare several holding and injection pipets. Pull the holding pipets from the capillary tubing (Section 2.2., Step 2) using a filament setting of 13 and initial solenoid setting of 3.0 (these initial settings were determined empirically for the Kopf pipet puller). Just as the glass begins to melt, turn the solenoid dial to 0. This maneuver produces long tips of a uniform 100–50 μm taper. Place the pipet into the microforge, and rest the tubing at a point where it is 50–75 μm in outer diameter (use a calibrated reticle) on the filament. Apply heat with a filament setting of 5.5 until the glass starts to bend. Immediately turn off the filament, and simultaneously turn on the blower. As the filament cools, it will retract and leave a clean square break in the pipet. Reheat the filament at a setting of 7.0 (the filament should be red-hot), and heat-polish the flush end of the pipet leaving a small unfused aperture (*see* Fig. 2A).
2. Pull injection pipets using a filament setting of 12.5 and solenoid setting of 4.0 (settings will have to be adjusted periodically for each filament). The resulting pipets should have a narrow shank and a final stinger-like shape consisting of a point 50–75 μm long and 1–2 μm in diameter (Fig. 2A). These pipets are fused and have to be chipped to produce a patent tip.
3. Insert the large end of a holding pipet into the oil-filled PE tubing connected to the holding Hamilton syringe (*see* Section 2.1., step 6), and insert the instrument holder assembly into one of the micromanipulators. Fill one of the injecting pipets with oil, using a long 25-g needle and syringe, leaving as small a dead air space as possible in the final taper of the pipet. Attach the oil-filled injecting needle to a second Hamilton syringe and micromanipulator.
4. Under low-power magnification, align the tips of the two pipets in the center of the field in the same focal plane. The pipets should be

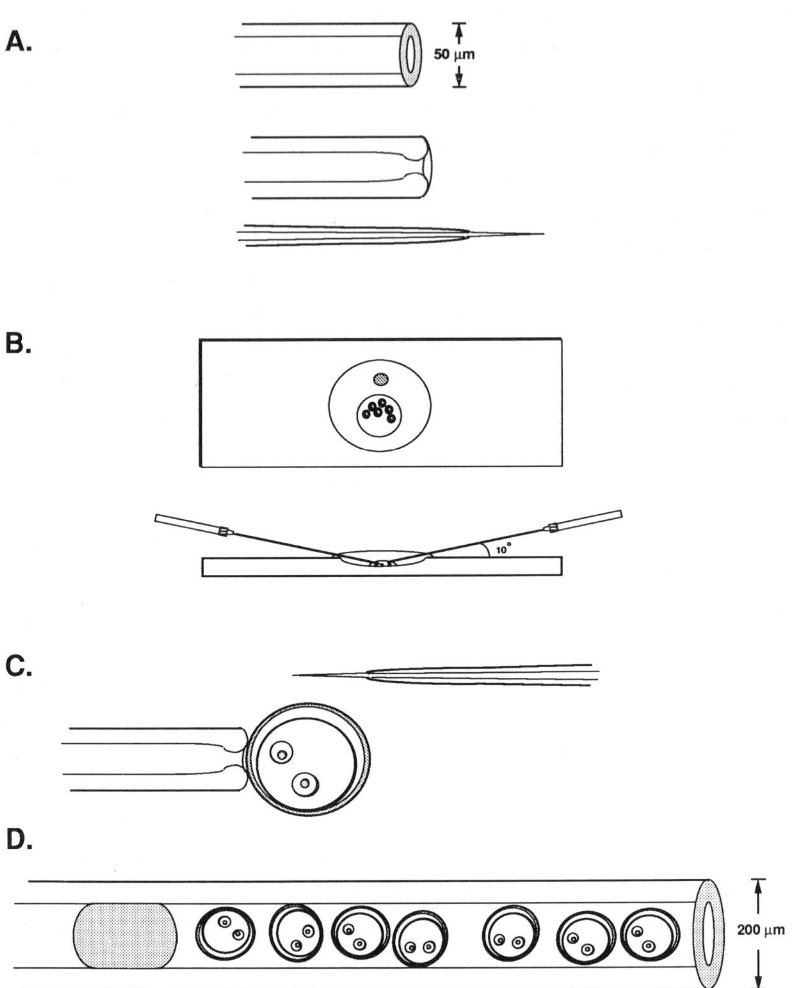

Fig. 2. Diagrams of pipets used for manipulating mouse oocytes. **(A)** A typical holding pipet before (upper) and after (middle) heat polishing with the microforge. The injection pipet (lower) is drawn to the same scale. **(B)** Top view of a hanging drop slide showing the relative size and position of the media drop containing several oocytes and the DNA drop (stippled). Side view of the slide shows the oil drop flooding the central depression covering the media drop and the two pipets entering the drops at a shallow angle. **(C)** The relative size and position of the holding pipet, fertilized oocyte containing two pronuclei and surrounded by its zona pellucida, and the injecting pipet. **(D)** A typical oviduct transfer pipet filled with media (clear), an air bubble (stippled), and seven microinjected oocytes.

angled approx 10° from the microscope stage, and there should be sufficient distance between the stage and micromanipulators, on each side, so that the instrument holders do not touch the stage when the tips of the pipets are moved across the stage's midpoint. The proper height of the stage and focal plane for the pipet tips needs to be set only once, and depends on the thickness of the hanging drop slide at its midpoint.

5. To chip the tip of the injecting pipet, first apply back pressure with four complete clockwise turns of the Hamilton-syringe micrometer. Under high-power magnification, briskly tap the injecting needle against the end of the oil-free holding pipet. A successful chip is not immediately visible and leaves a long point (*see* Note 1). After several seconds, a stream of fine oil drops will emerge from the newly opened tip. Allow the pressure to equilibrate until the oil flow stops.

6. Prepare a slide for microinjection by placing a 5-μL drop of HBM into the center of the well on a hanging drop slide. Cover the media drop completely with mineral oil, and then place a 1-μL drop of DNA solution adjacent to the larger media drop (Fig. 2B). Place 20–30 eggs in the media drop at the edge nearest to the DNA drop.

7. Place the loaded slide onto the microscope stage, and align both pipets so that they enter the drop containing the eggs and are freely moveable without dragging on the bottom of the well. Before the holding pipet is used, it should be purged of all air with oil from the reservoir in the holding syringe system. Insert the tip of the injection pipet into the DNA drop, and fill the needle with DNA by slowly turning the injecting micrometer counterclockwise. The easily visible meniscus formed at the DNA solution/oil interface should always be kept within the low-power objective field of view. Fill the holding pipet similarly with a small amount of media from the media drop.

8. Hold each egg for microinjection in position on the end of the holding pipet by applying gentle negative pressure with the micrometer syringe. Rotate the egg until the larger pronucleus is positioned slightly off the midline towards the pipet tip and both the edge of the nuclear membrane and the surrounding zona pellucida are sharply in focus (use the microscope fine focus) (Fig. 2C).

9. Next, align the tip of the injection needle in the same focal plane as the pronucleus using the vertical adjustment on the micromanipulator. Draw back the injecting needle, and puncture the zona pellucida and cytoplasmic membrane using a rapid motion of the micromanipulator joystick (most other movements are done best with the coarse control dials), thereby impaling the desired pronucleus. Gently turn the micrometer clockwise to produce a steady flow of

DNA through the injection pipet. A successful injection is apparent by a rapid swelling of the pronucleus, and the needle should be immediately withdrawn (*see* Note 2). Discard and replace any needles that immediately lyse the eggs, or become obstructed at their tip and prevent DNA flow.

10. After all the eggs on the slide are injected, wash them in fresh Whitten's medium and return to the incubator. A total of 50–70% of injected eggs survive, and are ready for immediate oviduct transfer or in vitro embryo culture.

3.1.5. Oviduct Transfer

1. Anesthetize a 0.5-d pseudopregnant (*see* Section 3.1.2.) SW female mouse with Avertin 0.15 mL/10 g body wt ip.
2. Place the mouse on her right side, and make a 1-cm flank incision at a point level with the knee. Make a small puncture in the muscle layers, and enlarge the opening by blunt dissection. Find the opalescent white ovarian fat pad, and use it to pull the ovary out of the peritoneal cavity. Maintain gentle traction on the ovary and attached uterus by clipping the fat pad to the body wall with a fine bull-dog clamp. Under the stereo dissecting microscope, tear a small opening in the bursa surrounding the oviduct using a pair of superfine watchmaker's forceps, being careful not to break any blood vessels.
3. Identify the ampulla of the oviduct whose opening will be pointed horizontally. Gently insert the tip of a fire-polished embryo-transfer pipet that has been loaded with 10–15 microinjected eggs as pictured in Fig. 2D. Using positive air pressure, expel the eggs into the lumen of the oviduct until the air bubble next to the eggs is visible in an oviduct loop. The air bubble will prevent retrograde movement of the eggs back out into the bursal sac.
4. Place the uterine horn, oviduct, and ovary back into the peritoneal cavity, and suture the abdominal wall muscles with 000 silk. Close the skin incision with autoclips, and repeat the procedure on the contralateral side if suficient numbers of injected embryos are available. No special postoperative care is needed. Injected embryos can be maintained in tissue culture overnight and transferred on d 1.5 to d 0.5 pseudopregnant recipients, if necessary.

3.1.6. Identification of Transgenic Mice

1. Mice that develop from the injected eggs are weaned from their foster mothers at 18–20 d of age and separated by sex. Lightly anesthetize the mice with Avertin 0.15 mL/10 g, number the mice by ear punches, and cut off the terminal 1–2 cm of tail with a single-edge

razor blade. Cauterize the tail stump, and keep the mice on a 37°C slide warming tray until they recover from the anesthetic.

2. Score the tail segment with a razor blade, remove the bone, and place the skin in a 1.5-mL microcentrifuge tube that contains 0.6 mL DNA extraction buffer. Continually mix the tubes overnight using an orbital shaker placed inside a 55°C incubator.

3. Centrifuge (12,000g, 5 min, 25°C) and transfer the supernatants without the lipid layer to fresh tubes. Extract the supernatants once with an equal vol of phenol:chloroform:isoamyl alcohol (20:19:1) followed by a second extraction with 1 vol of chloroform:isoamyl alcohol.

4. Add 0.1 vol of 3M sodium acetate, pH 5.4, and 2 vol 95% ethanol to the aqueous phase, and vortex. The DNA strands will precipitate immediately. Pellet the high-mol-wt DNA by centrifugation (12,000g, 10 min, 25°C). Decant the supernatant, rinse the pellet with 200 μL 70% ethanol, vacuum dry the pellet, and resuspend in 100 μL of Tris/EDTA buffer.

5. After the DNA is completely in solution (you may have to warm tubes to 55°C in a waterbath), quantitate by fluorimetry as described in Section 3.1.3.

6. Use 5–10 μg of each DNA sample for either Southern blot or dot-blot analysis *(17–19,32)* to determine the presence, number, and organization of integrated transgenes. Ideally, the radioactive hybridization probe should be complementary only to sequences contained in the transgene to reduce background and easily permit the detection of single-copy integrants.

3.2. Analysis of Reporter Protein Expression in Transgenic Mice

3.2.1. β-Galactosidase Enzyme Histochemistry

1. Fixation: Perfuse anesthetized adult mice through the left ventricle using a 23-g butterfly needle and 10 mL preperfusion buffer immediately followed by 30 mL of ice-cold phosphate-buffered paraformaldehyde. Individual organs may then be postfixed overnight in 4% paraformaldehyde containing 30% sucrose w/v. For developmental studies of gene expression, embryos and fetuses are fixed by immersion in 4% paraformaldehyde for times ranging from 5 min to overnight for two cell embryos to E18 fetuses, respectively. Rinse tissue twice in KPBS after fixation.

2. Cut sections of the desired thickness (range of 10–50 μm) with either a cryostat or vibratome. Do not embed tissues in paraffin, because the high temperatures required destroy β-galactosidase enzyme activity.

3. Incubate the sections, either free-floating or attached to gelatin-coated slides, in the X-gal histochemical solution at 37°C. Maximal azure color development will occur in 5 min to several hours, depending on the concentration of enzyme. Background staining varies in mammalian tissues, is time dependent, lowest in nervous tissues and pituitary, and highest in testis, kidney, and spleen (*see* Note 3).

4. Sections can be counterstained (methyl green, hematoxylin, eosin, and cresyl violet), dehydrated through alcohols and xylenes, and permanently mounted under cover slips *(34)*.

3.2.2. Simultaneous Dual Immunofluorescent Localization of β-Galactosidase and a Second Endogenous Antigen

1. Prepare tissue exactly as described in Section 3.2.1.

2. Wash sections twice for 20 min each in KPBS on a rotating table.

3. Preincubate free-floating sections in KPBS plus 0.3% Triton X-100 for 30 min at room temperature.

4. Incubate sections in KPBS containing 0.3% Triton X-100, 2% normal goat serum, rabbit anti-β-galactosidase IgG 10 µg/mL (Cappel, Organon Teknika Corporation, West Chester, PA), and an appropriate dilution of a mouse monoclonal IgG directed against an endogenous antigen, overnight at 4°C.

5. Wash sections twice for 20 min each in KPBS, 0.3% Triton X-100 on a rotating table.

6. Incubate in KPBS containing 0.3% Triton X-100, goat antirabbit IgG (H + L) fluorescein isothiocyanate, 20 µg/mL (Tago, Inc., Burlingame, CA), and goat antimouse IgG (H + L) rhodamine isothiocyanate, absorbed to rat, human, bovine, and horse serum, 20 µg/mL (American Qualex, La Mirada, CA).

7. Repeat step 5.

8. Mount sections on plain glass slides and cover slip with 50% glycerol, 50 m*M* sodium carbonate, pH 8.

9. Observe and photograph using an epifluorescent microscope equipped with appropriate filters for fluorescein and rhodamine excitation/emission.

3.2.3. Quantitation of β-Galactosidase Gene Expression by a Fluorometric Enzyme Assay

1. Place fresh or flash-frozen tissues into 250 m*M* Tris, pH 7.4 (use at least 10 µL/1 mg tissue).

2. Homogenize small tissues in microcentrifuge tubes using freeze-thaw

and sonication or Teflon™ pestles, and larger tissues in 50 mL polypropylene tubes using a polytron.

3. Pellet the insoluble debris, and transfer the supernatants to fresh tubes.
4. Combine 200 µL of tissue extract (or smaller vol diluted to a final vol of 200 µL with additional 250 mM Tris-HCl) with 33 µL fluorometric assay buffer and 66 µL of 455 µM 4MUFG solution dissolved in dimethylformamide. Mix well.
5. Incubate at 37°C for up to 16 h, then add 0.5 mL 800 mM NaOH to each tube to inhibit further enzyme activity.
6. Measure fluorescence at excitation 365 nm and emission at 458 nm. Construct a standard curve using purified β-galactosidase over the concentration range 5.5×10^{-6} to 4×10^{-3} U/reaction tube. Compare serial dilutions of unknowns to the β-galactosidase standard curve. Negative controls from wild-type mice must be included to correct for background lysosomal galactosidases.

3.2.4. Immunohistochemical Localization of SV40 T Antigen in Transgenic Tissues

1. Prepare tissue sections as described in Section 3.2.1.
2. Wash the sections twice for 20 min each in KPBS on a rotating table.
3. Wash the sections twice for 20 min each in KPBS, 0.3% Triton X-100.
4. Incubate the sections for 30 min at room temperature in KPBS, 0.3% Triton X-100, and 1% H_2O_2 (or until all bubbling stops).
5. Repeat step 3.
6. Incubate the sections overnight at 4°C in KPBS, 0.3% Triton X-100, 2% normal goat serum, and 1:10,000 dilution of rabbit antiTag IgG (polyclonal antiserum kindly provided by D. Hanahan, University of California, San Francisco).
7. Repeat step 3.
8. Incubate the sections for 2 h at room temperature in KPBS, 0.3% Triton X-100, and biotinylated goat antirabbit IgG 7.5 µg/mL.
9. Repeat step 3.
10. Incubate sections for 1 h at room temperature in avidin–biotin–peroxidase complex (supplied with the Vectastain kit) diluted in KPBS, Triton X-100.
11. Repeat step 2.
12. Incubate sections in TBS containing 0.25 mg/mL diaminobenzidine (CAUTION: suspected carcinogen) and 0.1% H_2O_2 until the desired balance between specific staining and background is obtained (time varies from 1–10 min).
13. Wash twice in TBS, and mount sections in gelatin onto microscope slides.

14. Counterstain, dehydrate, and apply a cover slip as described in Section 3.2.1., step 4.

4. Notes

1. We have found the most temperamental step in the microinjection process to be the proper manufacture of the injecting pipets. If the final point is too long (>100 μm) and narrow, chipping the tip will be difficult because of excess flexibility in the glass. After a successful break, DNA injection may still be impaired because of the excessive injection pressure that must be generated to overcome the resistance of a long, narrow lumen. Conversely, stubby or blunt needles will puncture too large a hole in the plasma membrane and lead to rapid lysis of the eggs. Needles should be replaced if either of these situations occurs. Some investigators prefer using a Flaming-Brown-type horizontal micropipet puller and micropipet beveler. More complete discussions of the manufacture and use of microinjection pipets can be found in refs. *18* and *35*.

2. We routinely use a metallothionein *E. coli lac Z* gene as a test gene for inexperienced researchers. The ease with which β-galactosidase enzyme can be detected histochemically in newly injected eggs and E12 embryos makes this gene ideal for monitoring the success rate of microinjections, embryo implantation, and transgene integration/expression. If embryos are to be examined at the two-cell stage, inject a 5 ng/μL solution of supercoiled MT-*lac Z* plasmid. Incubate the nonlysed eggs in Whitten's medium containing 50 μM CdSO$_4$ overnight. The embryos' growth will be blocked at the two-cell stage because of metal toxicity. Fix the embryos as described above (Section 3.2.1.) in drops of 4% paraformaldehyde under mineral oil. After incubation in the standard x-gal solution, 50–80% of embryos should turn blue. There is a high rate of mosaicism in the transgene expression (our unpublished data and ref. *36*).

3. We have found the X-gal solution listed above to produce an optimal color reaction. Final concentrations of the ferro- and ferricyanide components between 0.5 and 0 mM cause increased color intensity, but unacceptable diffusion of the reaction product. Concentrations between 3 and 40 mM cause a dramatic decline in color intensity. pH 7.4 is optimal for the bacterial β-galactosidase. pH 4.0 will result in a blue color in most tissues owing to the presence of endogenous lysosomal galactosidases with an acidic pH optimum. pH 9.0 results in a loss of all activity, except in the mouse testis because of overwhelm-

ing concentrations of endogenous enzyme. Fixation with both 4% paraformaldehyde and 2% paraformaldehyde/0.2% glutaraldehyde have given comparable results in our hands. We have found neuronal expression of peptide promoter-*lacZ* genes to be very stable with preservation of β-galactosidase activity in fixed tissue sections stored for at least 3 mo in KPBS at 4°C. For unknown reasons, the activity in the pituitary glands of transgenic mice expressing POMC-*lacZ* fusion genes declines rapidly under the same conditions. For these reasons, it is recommended that tissue be processed as rapidly as possible from mice with new *lacZ* fusion genes until the product stability is individually determined.

Acknowledgments

The author thanks K. Ebert, G. Hammer, R. Ventimiglia, and V. Fairchild-Huntress for their contributions to the work described in this chapter, and S. Byars for preparation of the manuscript. This work was supported by NIH grant DK 40457 and a Scholars Award for New Faculty from Pfizer, Inc.

References

1. Lefevre, C., Imagawa, M., Dana, S., Grindlay, J., Bodner, M., and Karin, M. (1987) Tissue-specific expression of the human growth hormone gene is conferred in part by binding of a specific *trans*-acting factor. *Embo. J.* **6,** 971–981.
2. West, B. L., Catanzaro, D. F., Mellon, H., Cattini, P. A., Baxter, J. D., and Reudelhuber, T. L. (1987) Interaction of a tissue-specific factor with an essential rat growth hormone gene promoter element. *Mol. Cell. Biol.* **7,** 1193–1197.
3. Nelson, C., Crenshaw, E. B., III, Franco, R., Lira, S. A., Albert, V. R., Evans, R. M., and Rosenfeld, M. G. (1986) Discrete *cis*-active genomic sequences dictate the pituitary cell type-specific expression of rat prolactin and growth hormone genes. *Nature* **322,** 557–562.
4. Bodner, M. and Karin, M. (1987) Activation of cell-specific expression of rat growth hormone promoter in extracts of nonexpressing cells. *Cell* **50,** 267–275.
5. Lira, S. A., Crenshaw, E. B., III, Glass, C. K., Swanson, L. W., and Rosenfeld, M. G. (1988) Identification of rat growth hormone genomic sequences targeting pituitary expression in transgenic mice. *Proc. Natl. Acad. Sci. USA* **85,** 4755–4759.
6. Swanson, L. W., Simmons, D. M., Arriza, J., Hammer, R., Brinster, R., Rosenfeld, M. G., and Evans, R. M. (1985) Novel developmental specificity

in the nervous system of transgenic animals expressing growth hormone fusion genes. *Nature* **317**, 363–366.

7. Russo, A. F., Crenshaw, E. B., III, Lira, S. A., Simmons, D. M., Swanson, L. W., and Rosenfeld, M. G. (1988) Neuronal expression of chimeric genes in transgenic mice. *Neuron* **1**, 311–320.

8. Low, M. J., Lechan, R. M., Hammer, R. E., Brinster, R. L., Habener, J. F., Mandel, G., and Goodman, R. H. (1986) Gonadotroph-specific expression of metallothionein fusion genes in pituitaries of transgenic mice. *Science* **231**, 1002–1004.

9. Low, M. J., Goodman, R. H., and Ebert, K. M. (1989) Cryptic human growth hormone gene sequences direct gonadotroph-specific expression in transgenic mice. *Mol. Endocrinol.* **3**, 2028–2033.

10. Nelson, C., Albert, V. R., Elsholtz, H. P., Lu, L. I. W., and Rosenfeld, M. G. (1988) Activation of cell-specific expression of rat growth hormone and prolactin genes by a common transcription factor. *Science* **239**, 1400–1405.

11. Bodner, M., Castrillo, J. L., Theill, L. E., Deerinck, T., Ellisman, M., and Karin, M. (1988) The pituitary-specific transcription factor GHF-1 is a homeobox-containing protein. *Cell,* **55**, 519–529.

12. Ingraham, H. A., Chen, R., Mangalam, H. J., Elsholtz, H. P., Flynn, S. E., Lin, C. R., Simmons, D. M., Swanson, L., and Rosenfeld, M. G. (1988) A tissue-specific transcription factor containing a homeodomain specifies a pituitary phenotype. *Cell* **55**, 519–529.

13. Herr, W., Sturm, R. A., Clerc, R. G., Corcosan, L. M., Baltimore, D., Sharp, P. A., Ingraham, H. A., Rosenfeld, M. G., Finney, M., Ruvkun, G., and Horvitz, R. (1988) The POU domain: a large conserved region in the mammalian *pit*-1, *oct*-1, *oct*-2, and *Caenorhabditis elegans unc*-86 gene products. *Genes Dev.* **2**, 1513–1516.

14. He, X., Treacy, M. N., Simmons, D. M., Ingraham, H. A., Swanson, L. W., and Rosenfeld, M. G. (1989) Expression of a large family of POU-domain regulatory genes in mammalian brain development. *Nature* **40**, 35–42.

15. Dollé, P., Castrillo, J. L., Theill, L. E., Deernick, T., Ellisman, M., and Karin, M. (1990) Expression of GHF-1 protein in mouse pituitaries correlates both temporally and spatially with the onset of growth hormone gene activity. *Cell* **60**, 809–820.

16. Jaenisch, R. (1988) Transgenic animals. *Science* **240**, 1468–1473.

17. Brinster, R. L. and Palmiter, R. D. (1986) Introduction of genes into the germ line of animals. *Harvey Lectures* **80**, 1–38.

18. Hogan, B., Costantini, F., and Lacy, E. (1986) *Manipulating the Mouse Embryo.* Cold Spring Harbor Laboratory, Cold Spring Harbor, NY.

18a. Kumar, R., Fairchild-Huntress, V., and Low, M. J. (1992) Gonadotrope-specific expression of the human follicle stimulating hormone beta subunit gene in pituitaries of transgenic mice. *Mol. Endocrinol.* **6**, 81–90.

19. Palmiter, R. D. and Brinster, R. L. (1986) Germ-line transformation of mice. *Ann. Rev. Genet.* **20**, 465–499.

20. Chada, K., Magram, J., and Costantini, F. (1986) An embryonic pattern of expression of a human fetal globin gene in transgenic mice. *Nature* **319**, 685–689.
21. Hammer, R. E., Krumlauf, R., Camper, S. A., Brinster, R. L., and Tilghman, S. M. (1987) Diversity of alpha-fetoprotein gene expression in mice is generated by a combination of separate enhancer elements. *Science* **2**, 53–58.
21a. Rubinstein, M., Goodman, R. H., and Low, M. J. (1992) Targeted expression of somatostatin in vasopressinergic magnocellular hypothalamic neurons of transgenic mice. *Mol. Cell. Neurosci.* (in press).
22. Gordon, J. I. (1989) Intestinal epithelial differentiation: new insights from chimeric and transgenic mice. *J. Cell Biol.* **108**, 1187–1194.
23. Efrat, S., Teitelman, G., Anwar, M., Ruggerio, D., and Hanahan, D. (1988) Glucagon gene regulatory region directs oncoprotein expression to neurons and pancreatic alpha cells. *Neuron* **1**, 605–613.
24. Hanahan, D. (1985) Heritable formation of pancreatic β-cell tumors in transgenic mice expressing recombinant insulin/simian virus 40 oncogenes. *Nature* **315**, 115–122.
25. Behringer, R. A., Mathews, L. S., Palmiter, R. D., and Brinster, R. L. (1988) Dwarf mice produced by genetic ablation of growth hormone-expressing cells. *Genes Dev.* **2**, 453–461.
26. Khillan, J. S., Schmidt, A., Overbeek, P. A., de Crombrugghe, B., and Westphal, H. (1986) Developmental and tissue-specific expression directed by the alpha2 type 1 collagen promoter in transgenic mice. *Proc. Natl. Acad. Sci. USA* **83**, 725–729.
27. Goring, D. R., Rossant, J., Clapoff, S., Breitman, M. L., and Tsui, L-C. (1987) *In situ* detection of β-galactosidase in lenses of transgenic mice with a gamma-crystallin/*lac Z* gene. *Science* **235**, 456–458.
28. DiLella, A. G., Hope, D. A., Chen, H., Trumbauer, M., Schwartz, R. J., and Smith, R. G. (1988) Utility of firefly luciferase as a reporter gene for promoter activity in transgenic mice. *Nucleic Acids Res.* **16**, 4159.
29. Zakany, J., Tuggle, C. K., Patel, M. D., and Nguyen-Huu, M. C. (1988) Spatial regulation of homeobox gene fusions in the embryonic central nervous system of transgenic mice. *Neuron* **1**, 679–691.
29a. Hammer, G. D., Fairchild-Huntress, V., and Low, M. J. (1990) Pituitary-specific and hormonally regulated gene expression directed by the rat proopiomelanocortin promoter in transgenic mice. *Mol. Endocrinol.* **4**, 1689–1697.
30. Hall, C. V., Jacob, P. E., Ringold, G. M., and Lee, F. (1983) Expression and regulation of *Escherichia coli. lac Z* gene fusions in mammalian cells. *J. Mol. Appl. Gen.* **2**, 101–109.
31. Pipas, J. M. (1988) SV40 Large T antigen mutant data base. *In Vitro Cell. Devel. Biol.* **24**, 1147.
32. Sambrook, J., Fritsch, E. F., and Maniatis, T. (1989) *Molecular Cloning: A Laboratory Manual,* 2nd ed. Cold Spring Harbor Laboratory, Cold Spring Harbor, NY.

33. Labarca, C. and Paigen, K. (1980) A simple, rapid, and sensitive DNA assay procedure. *Anal. Biochem.* **102,** 344–352.
34. James, J. (1976) *Light Microscopic Techniques in Biology and Medicine.* Martinus Nijhoff, The Hague.
35. DePamphilis, M. L., Herman, S. A., Martinez-Salas, E., Chalifour, L. E., Wirak, D. O., Cupo, D. Y., and Miranda, M. (1988) Microinjecting DNA into mouse ova to study DNA replication and gene expression and to produce transgenic animals. *Bio. Techniques* **6,** 662–680.
36. Stevens, M. E., Meneses, J. J., and Pedersen, R. A. (1989) Expression of a mouse metallothionein-*Escherichia coli* β-galactosidase fusion gene (MT-βgal) in early mouse embryos. *Exp. Cell Res.* **183,** 319–325.

CHAPTER 13

Expression of Exogenous Ion Channels and Neurotransmitter Receptors in RNA-Injected *Xenopus* Oocytes

Nathan Dascal and Ilana Lotan

1. Introduction

1.1. Xenopus *Oocytes*

This chapter is designed to serve as a guide for neurobiologists interested in studying neurotransmitter receptors and ion channels expressed in the oocytes of the African clawed frog *Xenopus laevis.* The emphasis is on routinely employed methods, based mostly on the experience gained in our laboratories, but also taking into account the practice of many others. A short introduction provides an overview of advantages and disadvantages of the system, and of what can be done using it.

The oocyte is a well-established experimental tool for the expression of foreign proteins (reviews: *1–4).* This is usually achieved by microinjection of messenger RNA (mRNA) extracted from various tissues or the injection of RNA derived from cloned complementary DNA (cRNA). The oocytes readily synthesize many (but not all; ref. *5)* exogenous proteins *(1).* Their large size allows various manipulations, such as two-electrode voltage clamp and intracellular injections. The cells are easily singled out and stripped of the surrounding cellular and noncellular layers, allowing patch clamping to be carried out and also providing a homogeneous cellular preparation for biochemical work. Because of these features, oocytes have been used widely for such studies as screening

From: *Methods in Molecular Biology, Vol. 13: Protocols in Molecular Neurobiology*
Edited by: A. Longstaff and P. Revest Copyright © 1992 The Humana Press, Totowa, NJ

assay to clone cDNAs of membrane proteins *(6)*, as a model to study the coupling of the expressed receptors to second messenger cascades and to G proteins *(7)*, to examine the biophysics of the expressed ion channels *(8)*, the modulation of ion channels and neurotransmitter receptors by second messengers and protein kinases *(9)*, and receptor pharmacology *(10)*. Most importantly, the use of oocytes allows the most detailed study of structure–function relations of the membrane proteins. The most powerful approaches are the expression of point-mutated or chimeric cDNAs *(11)*, and coexpression or selective suppression of different subunits of multisubunit proteins *(12)*.

There are some disadvantages.

1. The expression of exogenous proteins is transient, since the cells die several days after the injection of RNA.
2. Like any other in vivo expression system, oocyte's own "native" components (regulatory proteins, membranal environment) as well as incorrect processing (e.g. glycosylation) may alter or interfere with the functioning of the expressed proteins *(13)*.
3. As in other cells, there are native ion channels and neurotransmitter and hormone receptors; most of them are well characterized and produce quite small ionic currents *(see 2,14)*, but they still can interfere with the expressed ones.

1.2. Separation of Poly (A) RNA

Most of a cell's RNA is ribosomal RNA (rRNA), whereas only about 5% is mRNA. "Total" RNA extracted from cells contains both rRNA and mRNA, and is heterologous (contains mRNAs coding for many proteins). mRNA may be selected by affinity chromatography from total RNA owing to the fact that most mRNA species contain a polyadenosine sequence; thus, the term poly (A) RNA corresponds to mRNA. The oocytes express exogenous proteins quite well when injected with total RNA; however, expression can be improved by the use of poly (A) RNA. Further enrichment can be achieved by fractionating poly (A) RNA *(15)*. The best expression is obtained with cRNA transcribed in vitro from cloned cDNA.

The quality of RNA is crucial. Expression in the oocyte is the most rigorous test of RNA quality; thus, even RNA that appears undegraded on a gel, may not be good enough for expression purposes. Therefore, the methods for preparing uncontaminated, RNase-free RNA have to be meticulously followed. RNases are RNA-degrading enzymes, and they are very resistant to protein-degrading treatments. In essence, the use of high-quality reagents and RNase-free glassware and plasticware, the right choice of the extraction procedure (for tissue-derived RNA), and the appropriate RNA storage are the clues to success *(see* Note 1). More detailed recommendations on some of the issues described here can be found in molecular biology laboratory manuals *(16–18)* and original publications *(19–23)*.

1.3. In Vitro RNA Transcription

The synthesis of RNA in vitro (cRNA) requires that the cloned cDNA is subcloned into a plasmid expression vector containing a bacteriophage promoter (recognized specifically by an RNA polymerase encoded by a bacteriophage; e.g., T3, T7, and SP6). The cDNA is inserted in the plasmid vector in a specific orientation, so that the promoter enables the appropriate polymerase to transcribe a sense RNA strand. Prior to transcription, the plasmid DNA is linearized with a restriction enzyme that has a unique cleavage site immediately downstream of the cDNA insert. The procedures for subcloning cDNA and for preparing and linearizing plasmids containing the cDNA are beyond the scope of this chapter; however, in vitro transcription of RNA from linearized plasmid, a protocol for preparation of cRNA that proved to direct expression of ion channels in the oocytes, is provided in Section 3.3. It includes capping of the transcript (attachment of a cap structure, 7 methyl guanosine attached to triphosphate, at the 5' end), which is required for the stability of cRNA in the oocytes, thereby increasing the protein synthesis severalfold; however, the capping reagent need not be used in the methylated form, since it will become methylated in the oocyte. A poly (A) tail at the 3' end is not always necessary for successful expression of the protein.

1.4. Hybrid Arrest of Expression of mRNA Coding for Channel Protein

Several laboratories have developed an approach for disrupting gene expression in living cells, based on antisense- or hybrid-arrest of translation. A complementary (antisense) RNA or DNA sequence binds specifically to the corresponding mRNA and blocks its expression. Antisense DNA has proved to be more reliable than antisense RNA in many systems *(24)*.

In *Xenopus* oocytes, short oligodeoxynucleotides (oligonucleotides), complementary to an mRNA species, specifically inactivate this mRNA by an RNase H-like degradation mechanism (RNase H is an enzyme that hydrolyzes the RNA moiety of RNA–DNA hybrids; *see* refs. *25, 26*). This method of *sequence-specific* and *irreversible* suppression of expression of a specific mRNA, in oocytes injected with heterologous RNA, can be used to test for functional roles of cloned cDNAs and for homologies among genes. We present here the method modified for hybrid arrest of rare messages coding for channel proteins *(12,27)*, by including denaturation and prolonged hybridization prior to the injection into the oocyte. Using this method, 40–90 base-long oligonucleotides proved to inhibit *selectively* the expression of ion channels coded by the target RNA, with no effect on the expression of other channels whose mRNA had sequence homology with the oligonucleotide of up to 50%. A 50 base-long oligonucleotide with 77% sequence homology to the mRNA still inhibited selectively the expression of a voltage-dependent Ca^{2+} channel.

2. Materials

2.1. Glassware and Solutions, General Considerations

All glassware should be heat-treated ("baked") at 180°C for at least 4 h. All plasticware should be RNase-free, sterile, untouched by hands. When working with small quantities of RNA, it is essential to siliconize the glassware (tubes) coming in contact with RNA. For this purpose, wash the tube briefly with a siliconizing liquid (e.g., Sigmacote, Sigma) and then with double-distilled water before baking.

All reagents coming in contact with RNA at the various stages of purification must be of high purity; molecular biology grade is recommended. All solutions must be made with double-distilled, autoclaved water. Most solutions (unless otherwise specified) should be autoclaved. If the solution should contain SDS or β-mercaptoethanol, the latter must be added after autoclaving, preferably just before use. It is also recommended to filter all solutions with sterile 0.2- or 0.45-µm filters to remove particles that may later clog the RNA injection pipet. Store the solutions at room temperature (unless otherwise specified); renew every 3–6 wk.

2.2. General Stock Solutions

1. Ethanol should be extra pure, 95% rather than absolute (to prepare absolute ethanol, redistillation with organic solvents is used, the traces of which may be harmful to RNA). We advise that stocks of 95% ethanol and 80% ethanol at –20°C should be kept readily available.

2. 10M, 4M, and 1M solutions of NaOH.
3. 2.5M Na acetate buffer, pH 5.2–5.5. Autoclaving recommended, but not critical; make in double-distilled autoclaved water.
4. 20% Solution of SDS in water. Do not autoclave. Sarcosyl can be used instead of SDS.
5. 1M Tris-HCl, pH 8.
6. 100 mM EDTA. Titrate with NaOH to pH 7.

2.3. LiCl-Urea Method

1. Lysis buffer: 10 mM sodium acetate, pH 5.2–5.5 (derived from stock buffer), 6M urea, 3M LiCl, and 0.1% SDS (add just before use, since it precipitates in the course of several minutes). This should be prepared shortly before the extraction and stored at room temperature for no more than 1 d (there is no need to autoclave, but it should be filtered).
2. Resuspension buffer: 10 mM Tris-HCl, pH 7.5, 5 mM EDTA, and 0.5% SDS (add just before use). This can be prepared using the stock solutions of Tris and EDTA. The resuspension buffer (without the SDS) should be autoclaved and filtered; it can be stored at room temperature for 2–3 wk.

2.4. Guanidine-HCl Method

1. 1M acetic acid, after filtering, can be stored for up to 2 mo at room temperature.
2. Lysis buffer A: 200 mM sodium acetate, pH 5.0–5.5, 6M guanidine hydrochloride, and 100 mM β-mercaptoethanol (add just before use). Do not autoclave, but filter; store at 4°C up to 3 wk without β-mercaptoethanol.
3. Solution B: 25 mM sodium citrate, pH 7, 7.5M guanidine hydrochloride, and 50 mM β-mercaptoethanol (add just before use). Prepare the day before extraction; store refrigerated.

2.5. Phenol-Chlorophorm Extractions

1. Phenol should be given special attention. Purchase phenol of the highest purity available, or distill it yourself at 180°C. Add approx 1 vol of 0.5M Tris-HCl buffer, pH 8/1 vol of phenol, stir well, replace the 0.5M Tris-HCl buffer with 0.1M Tris-HCl, pH 8, and titrate to pH 8 with NaOH while periodically stirring with phenol. Repeat the wash with 0.1M Tris-HCl, pH 8. Add β-mercaptoethanol to a final concentration of 0.2% and (optional) 8-hydroxyquinoline to 0.1%. Store light-protected at 4°C for no more than 4 wk.

2. Sevag mixture: 24 vol of chloroform and 1 vol of isoamylalcohol. Do not autoclave; store at 4°C for up to 3 wk. For the phenol-chlorophorm extraction, a 1:1 phenol-Sevag mixture is prepared (25:24:1 phenol:chloroform:isoamylalcohol).

2.6. Chomczinski-Sacchi Procedure

1. $2M$ sodium acetate, pH 4.
2. Denaturing solution: 25 mM sodium citrate, pH 7, 4M guanidine thiocyanate, 0.5% sarcosyl, and 0.1 M β-mercaptoethanol (add just before use). The stock solution (without β-mercaptoethanol) can be stored for up to 3 mo at room temperature.

2.7. Oligo-dT Cellulose
Separation Technique

1. Washing solution: 0.1 M NaOH and 5 mM EDTA.
2. 2X Loading buffer stock: 40 mM Tris-HCl, pH 7.6, 1M NaCl, 2 mM EDTA; and 0.2% SDS (add just before use).
3. Elution buffer: 10 mM Tris-HCl, pH 7.5, and 1 mM EDTA. We do not add SDS to this buffer.
4. 0.2% Na azide: sterile, used for the storage of oligo-dT cellulose columns.

2.8. RNA Transcription In Vivo

1. TE: 10 mM Tris-HCl, pH 7.6 and 1 mM EDTA.
2. 5X Transcription buffer:

	For T7 and T3 polymerases	For SP6 polymerase
Tris-HCl, pH 7.5	200 mM	200 mM
MgCl$_2$	30 mM	40 mM
Spermidine	10 mM	10 mM
NaCl	250 mM	50 mM

The buffers should be used with the appropriate polymerases.
3. 5X rNTPs (ribonucleotide triphosphates mixture): 2.5 mM rATP, 2.5 mM rUTP, 2.5 mM rCTP, and 0.25 mM rGTP. Keep in small aliquots, 12.5 µL each, at –20°C.
4. 25 mM rGTP. Keep in small aliquots of 1.5 µL each at –20°C.

2.9. Hybrid Arrest Experiment

1. 6X hybridization solution: 600 mM NaCl and 60 mM Tris-HCl, pH 7.6.
2. Paraffin oil, pure.
3. RNA (8–10 µg/µL).
4. Oligonucleotide (40–80 bases long) solution in water, in several concentrations: 1 mg/mL, 300 µg/mL, and 30 µg/mL.

2.10. Oocyte Culture

1. ND-96 solution: 5 mM Hepes-NaOH, pH 7.6, 96 mM NaCl, 2 mM KCl, 1.8 mM CaCl$_2$, 1 mM MgCl$_2$.
2. NDE-96: Enriched ND-96 solution, used for long-term culture of oocytes. To prepare NDE-96, add to ND-96 solution: 2.5 mM sodium pyruvate, 100 U/mL penicillin, and 100 µg/mL streptomycin.
3. Collagenase solution for defolliculation of the oocytes: Collagenase is dissolved at 1.5–2 mg/mL in sterile Ca^{2+}-free ND-96 solution (omit CaCl$_2$ when preparing the solution in which collagenase is dissolved. The Ca-free ND-96 solution is also used to wash the oocytes before placing them in collagenase; *see* Section 3.5.3.).

3. Methods

3.1. Extraction of Total RNA

In Sections 3.1.1. (LiCl-urea) and 3.1.2. (guanidine HCl), the tissue is lysed, the proteins denatured and RNA is separated from DNA and protein by precipitation in a lysis buffer, in one or two steps. The next step is resuspension of RNA and removal of the remaining protein by phenol-chloroform extractions (Section 3.1.3.). Procedure 3.1.4. (Chomczynski-Sacchi method) does not require such extractions. *In all procedures. it is important to homogenize the tissue very thoroughly in the lysis buffer.* All procedures should be carried out on ice, unless otherwise indicated, and as fast as possible (*see* Note 1).

3.1.1. LiCl-Urea Method

This method is applicable to soft tissues (brain) of embryos and young animals (modified from *19*).

1. Add about 15 mL of lysis buffer/1 g of tissue or 0.5 mL of packed cultured cells.
2. Homogenize the tissue immediately in the lysis buffer. Put the homogenate on ice at once. Store overnight at 4°C.
3. On the next day, if there is any foam on the surface of the homogenate, aspirate it away. Centrifuge the homogenate for 20 min at 10,000g.
4. Discard the supernatant and dissolve the pellet in the resuspension buffer (1–2 mL/1 g of the original tissue) by pipeting and light vortexing. If part of the material does not dissolve, it is not RNA; centrifuge (10,000g, 5 min, 4°C) and discard the pellet.
5. The RNA solution (in resuspension buffer) is subjected to phenol-chloroform extractions: proceed to Section 3.1.3.

3.1.2. Guanidine Hydrochloride Method

This method is modified from *(20)* (*see* ref. *21*), applicable to both fibrous and soft tissues, and gives high-mol-wt RNA; however, it is not efficient for small-mol-wt RNA (<120 bases). Note that RNA is very sensitive to heat in guanidine hydrochloride; perform all steps on ice or at –20°C as indicated.

1. Add about 20–25 mL of buffer A/g tissue or 0.5 mL packed cells.
2. Homogenize the tissue on ice. Keep on ice for 10–15 min.
3. Add 0.65 vol of cold 95% ethanol, mix well, and keep at –20°C for at least 6 h, or overnight.
4. If there is any foam on the surface of the homogenate, aspirate it away. Centrifuge (10,000g, 20 min, 4°C).
5. Discard the supernatant, and dissolve the pellet in ice-cold solution B by pipeting and light vortexing (the amount of buffer B is about one-tenth of A, or 2 mL/g tissue). If a part of the material does not dissolve, it is not RNA; centrifuge (10,000g, 5 min, 4°C) and discard the pellet. Transfer the supernatant to a clean tube.
6. Add 0.025 vol of 1 M acetic acid, stir, and then add 0.5 vol of cold 95% ethanol, mixing well. Incubate at –20°C for at least 3 h or overnight.
7. Centrifuge for 10–15 min at 10,000g, 4°C, and discard the supernatant. Using 0.5–1 mL water/1 g of the original tissue or 0.5 mL of packed cells, dissolve the pellet in water (on ice) by triturating and vortexing. The resuspension buffer from LiCl-urea procedure can be used instead of water (*see* Section 2.3.). Do not discard anything that does not dissolve in any case, but try triturating for a longer time or adding more water.
8. The RNA solution is subjected to phenol-chloroform extractions: proceed to Section 3.1.3.

3.1.3. Phenol-Chloroform Extraction

1. Prepare phenol-Sevag mixture (25:24:1 phenol:chloroform isoamylalcohol). Add about 1 vol phenol-Sevag mixture to 1 vol of RNA solution (use chloroform-resistant tubes) and shake for 3–5 min on ice. Centrifuge (3000–3500g, 10 min, 4°C). The bottom layer is phenol, and the aqueous phase is at the top. The interphase between phenol and water contains mostly protein; if it is very thick or the aqueous phase is not clear, transfer the mixture into glass tubes and centrifuge at 10,000g (5 min, 4°C).
2. Transfer the water phase into a new tube; discard the phenol. Do not let the interphase get into the aqueous phase.

3. Repeat the procedure described in steps 1 and 2 two more times with phenol-Sevag mixture, and then twice with Sevag alone (to remove the remaining phenol).

4. Transfer the aqueous phase (containing RNA) to a sterile tube for RNA precipitation. Add 0.1 vol of 2.5M sodium acetate, pH 5–5.5, mixing thoroughly; then add 2.5 vol of cold (–20°C) 95% ethanol while mixing. Keep at –20°C for at least 30 min. It is possible to store RNA for days or even weeks under these conditions.

5. Centrifuge (10,000g, 15–20 min, 4°C; longer for small amounts of RNA). The RNA stays in the pellet. Discard the supernatant, and wash the pellet and the tube with cold 80% ethanol without disturbing the pellet. Discard the ethanol and let the RNA dry for several minutes. Dessication is not necessary.

6. Dissolve the RNA in water (about 0.1 mL/g of original tissue). Take a 5–10 μL sample and determine the concentration of RNA as follows: Dissolve the RNA sample in 1 mL of water, or better, 10 mM Tris HCl buffer, pH 7. Place in quartz cuvets and measure absorbance (A) at 260 and 280 nm. For pure RNA, the ratio A_{260}/A_{280} is 2 at pH 7. To calculate the concentration, in mg/mL, use the equation:

$$RNA = [(\text{dilution of sample in water}) \times A_{260}]/20 \qquad (1)$$

Example: If you start with 5 μL of RNA solution and dilute it in 1 mL water to measure the A, then the dilution factor is 200; if your A_{260} was 0.25, then you have 2.5 mg/mL total RNA. To calculate the total amount of RNA you have, simply multiply this number by the vol (in mL) of your RNA solution.

7. Precipitate the rest of the RNA with sodium acetate and ethanol as in step 4. After incubation at –20°C for 0.5–1 h, centrifuge, and resuspend the pellet in 2–3 mL cold 80% ethanol (vortex and pipet) to remove sodium acetate. Centrifuge again (10,000g, 15–20 min, 4°C), discard the supernatant, and dry the pellet under vacuum for several minutes until the drops on the tube dry out and the edges of the pellet are transparent, but the rest is still white.

8. Dissolve the RNA in water to obtain a final concentration of about 10 mg/mL; measure A (*see* step 6 above), and calculate concentration and hence the total amount of RNA. Dilute with water if necessary to get the desired final concentration. Most RNAs will damage the oocytes at concentrations higher than 6–8 mg/mL.

9. Divide the RNA solution into 3–5 μL aliquots for future oocyte injections (note that repetitive freezing and thawing of RNA may severely reduce its effectiveness in the oocytes [22]). Store at –70 to –80°C for

up to 6 mo (depending on RNA batch). If selection of poly (A) RNA is desired, proceed to Section 3.2.

3.1.4. Chomczynski-Sacchi Method

Adopted from Chomczynski and Sacchi *(23)*, this method is fast, useful for small quantities of tissue or cultured cells, and preserves the low-mol-wt RNA well.

1. Homogenize the tissue on ice in denaturing buffer (10 mL buffer/ 1 g tissue). Add sequentially, mixing after each addition: 0.1 vol of 2*M* sodium acetate, pH 4, then 1 vol of phenol, and 0.2 vol of chloroform-isoamyl alcohol mixture (49:1). Shake vigorously for 10 s, and leave on ice for 15 min.
2. Centrifuge (10,000*g*, 20 min, 4°C). RNA remains in the aqueous phase (upper), whereas DNA and proteins are in the interphase and phenol phase (lower).
3. Transfer the aqueous phase (beware not to mix the interphase) to a clean tube and precipitate the RNA by adding 1 vol of isopropanol and storing at –20°C for 1 h.
4. Centrifuge (10,000*g*, 20 min, 4°C). Discard the supernatant, dissolve the pellet in the denaturing solution (about one-third of the amount used in step 2), and reprecipitate the RNA by adding 1 vol of isopropanol. Leave for 1 h at –20°C.
5. Centrifuge as in step 4. Resuspend the RNA in 80% cold ethanol (to wash out the salt remaining from the lysis buffer), centrifuge again as described earlier, discard the supernatant, and vacuum dry the pellet (as in step 7 of Section 3.1.3.).
6. Dissolve RNA in water, and determine the concentration of RNA as described in step 6 of Section 3.1.3. The RNA solution may then be diluted in water as appropriate, divided into aliquots, and stored at –70°C until required.

3.2. Separation of Poly (A) RNA on Oligo-dT Cellulose Column

This procedure is slightly modified from Sambrook et al. *(18)* to provide RNA expressible in the oocytes. All steps, unless otherwise indicated, are performed at room temperature. The yield of the column is 2–4% (i.e., 2–4% of the total RNA, which is polyadenylated, is retained by the column).

Used column: just wash out the 0.2% Na azide (in which the column was stored from the previous RNA purification; *see* step 12 below) with water and proceed to step 6 below.

New column:

1. Swell oligo dT cellulose, 0.4 g cellulose in 1 .5–2 mL autoclaved water; leave to stand for at least 2 h or overnight at 4°C.
2. Prepare the column by inserting fiberglass wad into a sterile syringe (or use commercially available column syringes). The fiberglass should be siliconized and baked before use. Add the solution of cellulose, and wash with 10 column vol of water.
3. Wash the column with 10 column vol of washing solution and then with at least 10 vol water, until the water eluted from the column has pH < 8.
4. Wash with 10 column vol of loading buffer at the final dilution (prepare by diluting the stock solution with the same vol of water and adding SDS).
5. This step is optional, but highly recommended for newly prepared columns: preloading of the column in order to saturate nonspecific binding sites. Prepare about 5 column vol of A(–) RNA *(see below)* at concentration of 400–500 µg/mL. The same vol of loading buffer (at stock concentration) is added; add SDS to 0.1%. This solution is washed through the column two or three times (washing duration about 15 min). Wash out with 10 column vol of elution buffer.
6. Rewash with 10 column vol of loading buffer.
7. Prepare the total RNA for the column by diluting it in water to a concentration of 0.4–0.8 mg/mL in water. Add equal vol of loading buffer (2X stock) with SDS. The total final vol of RNA solution should not exceed 15 column vol.
8. Pass the RNA through column two or three times over 30–40 min. (An option at this point is to heat the RNA solution for 5 min at 65°C to denaturate the RNA before passing it through the column. Note that this procedure improves the yield, but may damage RNA). The eluate contains rRNA and some mRNA, called collectively "A(–)" RNA; collect the eluate, and precipitate the A(–) RNA with sodium acetate and ethanol (as in step 4, Section 3.1.3.) for further use as control and for preloading of columns.
9. Wash the column with 10 vol of loading buffer; discard the eluate.
10. Elute poly (A) RNA with 3 column vol of elution buffer. Determine the amount of RNA using the method of step 6, Section 3.1.3. Keep RNA solution on ice.
11. Precipitate the mRNA with sodium acetate and ethanol (*see* Section 3.1.3., steps 4,5). To obtain the final concentration of RNA, proceed as in Section 3.1.3., steps 8,9. mRNA is usually injected into oocytes at concentrations of 1–3 mg/mL. Before dividing RNA into aliquots for storage, centrifuge the final RNA solution in a microfuge (3–5 min, 4°C) to pellet particles that may later clog the injection needle.

12. Wash column with 5 vol of elution buffer and then with 10 vol of water. Store the column in 0.2% solution of Na azide at 4°C. The column may be reused. It is possible to calculate the column's capacity for mRNA from the manufacturer's specifications (the amount of RNA that can be retained by the column is usually indicated in absorbance U; 1 mg mRNA corresponds to 20 absorbance U). Example: if the indicated capacity of 1 g of oligo-dT cellulose is 40 absorbance U (2 mg poly [A] RNA), and you made the column with 0.4 g of cellulose, the column may retain 0.8 mg poly (A) RNA. If you have prepared, for instance, 200 μL of poly (A) RNA at a concentration of 2 μg/μL, then the amount of poly (A) RNA that had been retained by the column was 0.4 mg, and you can use the column once more to separate a similar amount of mRNA.

3.3. RNA Transcription In Vitro

3.3.1. Preparation of DNA Template

All precautions used for the purification of RNA should be followed carefully when synthesizing in vitro RNA.

1. When linearization of the plasmid DNA is completed, it is treated with proteinase K (1 μg for every 10 μg DNA) added directly to the restriction mixture and incubated for an additional 30 min at 37°C to remove all RNases and the restriction enzyme.
2. Extract the linearized DNA twice with phenol:chloroform (1:1) followed by two chloroform extractions and ethanol precipitation (as in Section 3.1.3., except that isoamylalcohol is not added to the chloroform).
3. The linearized DNA is dissolved in RNase-free TE solution (Section 2.8.) at a concentration of 1 μg/μL.

3.3.2. In Vitro Transcription of Capped RNA

1. Set up the following reaction mixture in an autoclaved 1.5-mL microfuge tube. The amounts shown in the following are calculated for the case when 5 μg linearized DNA are used as template for RNA transcription. Scale the reaction reagents proportionally to the amount of DNA to be used as a template. Because it is possible that the DNA and spermidine precipitate at 4°C, add the components at room temperature in the order shown, and before the addition of DNA, gently mix the reagents.

5X Transcription buffer	12.5 μL
Dithiothreitol (100 m*M*)	6.2 μL
RNasine (ribonuclease inhibitor)	80 U
5X rNTPs	12.5 μL

    GpppG (5 m*M*) (diguanosine triphosphate;   6.2 μL
      cap analog)
    water (RNase-free)   to final vol of 63 μL
    Linearized DNA   5 μg
    ^{32}P-rCTP (400 Ci/mmol; 1:10 dilution in water)   1 μL
    RNA polymerase   150 U

    The total reaction vol is 63 μL.

    Incubate the reaction mixture at 37°C for 1 h, and add in the following order: 12 U RNasine; 1.3 μL 25 m*M* rGTP, and 90 U RNA polymerase. Mix gently, spin briefly for a few seconds in a microfuge. Incubate for another 2 h.

2. Add RNasine to a final concentration of 1 U/μL. Add DNase I to a final concentration of 1 U/μg DNA (*see* Note 2). Mix gently, spin briefly for a few seconds in a microfuge, and incubate for 15 min at 37°C (to digest the DNA). Place on ice.

3. Remove a 2-μL sample and add it into 48 μL water for estimation of yield of synthesized transcript (described here in steps 8–12).

4. Extract twice with phenol:chloroform followed by two chloroform extractions (*see* Section 3.3.1., step 2).

5. To collect the RNA and remove the unincorporated nucleotides, precipitate out the RNA by adding 1/9 vol of 10*M* ammonium acetate, mix, add 3 vol of cold (–20°C) 95% ethanol, mix, and place at –20°C for 30 min, centrifuge at 4°C for 30 minutes at 12,000*g* in a microfuge, and remove the supernatant.

6. Dissolve in 100 μL RNase-free water and precipitate again as in step 5.

7. Wash the pellet in cold (–20°C) 80% ethanol followed by a wash with 95% ethanol, dry the pellet briefly under vacuum, and resuspend in RNase-free water to a final concentration of 0.2–0.5 μg/μL. Store at –70°C in aliquots.

8. Spot 5 μL of the solution prepared in step 3 onto Whatman GF/C glass-fiber filter (2.4 cm diameter) for total count. Allow to dry at room temperature.

9. Add 5 μL into 500 μL of ice-cold carrier DNA solution (50 μg/mL). Vortex and then add 150 μL of ice-cold 50% trichloroacetic acid (TCA) solution. Vortex, and keep on ice for at least 20 min.

10. Filter with suction through a second glass-fiber filter. Wash the filter with 10 mL of ice-cold 10% TCA solution, then with 10 mL of cold (–20°C) 100% ethanol, and allow to dry at room temperature.

11. Place filters into scintillation vials, add 10 mL of Hydrofluor, and determine the amount of radioactivity on each filter in a scintillation counter.

12. Calculate the proportion of incorporated nucleotides:

$$\text{Proportion incorporated} = (\text{cpm in washed filter}/ \text{cpm in unwashed filter}) \tag{2}$$

Estimation of RNA yield:

$$\text{Amount of RNA (ng)} = (\text{nmol of rCTP in the reaction mixture}) \times (\text{proportion incorporated}) \times 1380, \tag{3}$$

where $1380 = 4 \times 345$ (the average mol wt of an rNMP).

3.4. Hybrid Arrest of Expression of mRNA Coding for Channel Protein

Oligonucleotides, synthesized by an automatic DNA synthesizer, are available commercially upon request. It is recommended to purify them by electrophoresis through a denaturing polyacrylamide gel (*see* ref. *18;* synthetic oligonucleotides, already gel purified, can be commercially purchased upon request). Any traces of RNase A should be then removed by a series of two phenol-Sevag and two Sevag extractions, and the oligonucleotide recovered by ethanol salt precipitation (*see* Section 3.1.3.). Heterologous RNA purified from a tissue should be used at concentration of about 10 µg/µL. This high concentration is recommended as in the hybridization mixture, which is injected into the oocyte, its actual concentration is only 2/3, and its potency to induce currents in oocytes is reduced by approx 20–50% because of the heating (depending on the quality of the purified RNA).

Hybridization reactions are carried out in autoclaved microfuge tubes (preferably of 0.5 mL vol). The minimum set of reactions in one experiment is two: (1) control reaction, which contains the RNA alone in the hybridization medium, and (2) experimental reaction, which contains RNA and oligonucleotide in the hybridization medium; both reactions are treated in the same way. It is advisable to have more reactions with different concentrations of the oligonucleotide (*see* Section 2.9). RNA and oligonucleotide are denatured by a 2-min incubation at 65°C and subsequent cooling on ice, before mixing.

1. Prepare reaction solutions: Reaction solution 1: 2 µL RNA, 0.5 µL hybridization solution, 0.5 µL water. Reaction solution 2: 2 µL RNA, 0.5 µL hybridization solution, 0.5 µL oligonucleotide solution (e.g., 300 µg/mL). Add few drops of paraffin oil to each tube, and centrifuge briefly (a few seconds).

2. Incubate the solutions for 2 min at 65° C, and then incubate for 3 h at 37°C. Place the solutions on ice before injecting into the oocytes.
3. If the oocytes injected with the reaction solution containing 1 mg/mL oligonucleotide still express the current under study, then there is definitely no significant homology between the oligonucleotide and the mRNA sequence of the channel under study. If the homology is high, the 30 µg/mL oligonucleotide should inhibit the expression of the current significantly. (*See* Note 3.)

3.5. Maintenance of Frogs and Preparation, Injection, and Culture of Oocytes

3.5.1. Maintenance of Frogs

Proper frog maintenance is essential for successful expression work. Details on frog maintenance can be found in ref. *28*. Mature female *Xenopus laevis* can be purchased from a large number of suppliers worldwide. For expression purposes, it is best to buy frogs that have been tested for productivity and have plenty of oocytes; inquire from the supplier.

1. Light and temperature are critical for the condition of the oocytes. Keep the frogs at 19 ± 2°C, and never let the temperature exceed 23°C, otherwise the survival rate of the oocytes in culture will be reduced. A constant light–dark cycle in the frog room is essential, but the duration of the light period in different laboratories varies between 9–12 h. Avoid noise in the frog room to prevent stress. Even under constant conditions of frog maintenance, seasonal variations in the viability of oocytes have been observed (with better survival in winter), although this claim has not been rigorously tested.
2. The females (no males allowed!) are kept in tanks filled with dechlorinated tap water, covered with a mesh. It is preferable to use dark-colored tanks (gray or black), spacious enough so that the animals are not crowded, e.g., 30 frogs in a 100 × 30 × 40 cm tank, and with the water at least 15 cm deep. Change the water three or four times a week, usually 3–5 h after feeding. Frogs recovering from surgery should be kept in a separate tank.
3. *Xenopus laevis* are carnivores. They should be fed twice a week with diced liver, fat-free meat or similar food, or frog brittle (8 g/meal/frog). Leave the food in the tank for at least 2 h (overnight is not recommended), and then change the water.
4. There are several diseases that may afflict the frogs, most of them bacterial, and some causing wounds and patches on the skin. Treatment is with antibiotics (e.g., 100 U/mL penicillin sulfate and 100 µg/mL

streptomycin in tank water), incubation of frogs in saline (0.9% NaCl) for several days, and/or daily 3–5 min treatment with 0.3% potassium permanganate.

3.5.2. Surgery

Work with clean, autoclaved surgical instruments. Anesthesize the frog in MS 222 (tricaine methanesulfonate; 0.15% solution in tap water. Store the MS 222 solution in the refrigerator if reused). Wait for 10–15 min until the frog is immobile and does not respond to mechanical stimuli, place it on its back, and make a 0.5-cm incision in the skin in the lower half of the belly and then in the underlying rectus abdominus muscle layer. Pull the necessary number of oocytes from the incision with forceps out and excise them; place the oocytes in sterile, Ca^{2+}-free ND-96 solution (ref. *29, see* Section 2.10. and Note 4). Suture the skin and the muscle (either together or separately) with 3/0 monofilament silk on 16-mm curved needle, and place the frog into a separate vessel with tap water, until it awakens. You can operate on the same frog again a day or several days later, through a separate incision on the second half of the abdomen. After that, let the frog recover for at least 2 mo until any further surgery is done.

3.5.3. Preparation of Oocytes

Most people use the largest oocytes of stages 5 and 6, having diameters between 0.9–1.3 mm *(30)*, but small oocytes can also be injected with RNA *(31)*. It is possible to inject oocytes while surrounding cellular and noncellular layers are still intact, i.e., as follicles *(30,32)*, or defolliculate the oocytes (i.e., remove all layers except the noncellular vitelline membrane) before RNA injection. For manual separation of follicles, use fine forceps or very fine scissors *(see* Note 5).

1. Open the clusters of oocytes removed from the ovary to allow access of collagenase to all oocytes. Wash the clusters well in the Ca^{2+}-free ND-96 solution, to remove exploded oocytes (which are a source of proteolytic enzymes and which may partially digest the membrane of the healthy oocytes, and of potassium, the elevation of which will depolarize the healthy oocytes).
2. Put clusters of about 200 oocytes/5 mL collagenase solution (Section 2.10., solution 3) in a tube, place the tube horizontally on the top of a shaker bath, and shake for 2–4 h until they are separated (the exact time will depend on the batch of collagenase and on the donor). Check for clearness of the solution periodically; if it becomes dirty and opaque, replace it with fresh solution.

3. Wash the separated oocytes with fresh NDE-96 twice, place them in a Petri dish, and choose healthy looking oocytes under a dissecting binocular microscope (20–25X magnification) for further processing. Note that at this stage many oocytes may still be surrounded by the outer layers; to remove them, pipet the oocytes up and down with a fire-polished Pasteur pipet with an inner diameter slightly smaller than the oocytes'. The remaining follicular cells can be removed by rolling the oocytes on polylysine-treated glass *(33)*, although this is not necessary for successful expression of foreign proteins. The easiest way to test whether the oocytes are defolliculated is to observe them under lateral illumination; the defolliculated oocytes have a distinct glare (owing to reflection of the light by the vitelline membrane). Let the defolliculated oocytes rest for at least 1 h or even overnight in NDE-96, and select only healthy looking cells for RNA injection.

3.5.4. RNA Injection

We inject RNA with a Drummond 510 microdispenser, but other arrangements are possible.

1. Pull the injection needle (a capillary corresponding to the type of microinjector used; inquire from the distributor of the dispenser you are using. Bake the capillaries at 180°C for 4 h). Note that the needle must be long enough to contain the plunger of the microdispenser, at least 5 mm of paraffin oil, and 2–5 mm of RNA solution.
2. Break the tip with a baked glass pipet or a sterile plastic plate to an outer diameter of 20–30 μm.
3. Inject paraffin oil into the proximal (open) end of the needle, so that the length of paraffin-oil-filled portion will be about 1 cm, and push it gently with the plunger of the injector to the tip of the needle, until the proximal end reaches its final position. Note that there should be 5–10 mm of oil between the end of the plunger and the beginning of the narrowing of the needle.
4. Place a drop of RNA solution on a sterile Petri dish, and take it up into the needle.
5. Put the oocytes into a Petri dish. To prevent the oocytes from sliding, use a dish scratched in a mesh-like fashion with a sterile needle or a baked glass pipet. To avoid accidental injection into the nucleus, insert the needle into the oocyte close to the equator or in the vegetal (light) half, and inject the desired amount of RNA solution (do not exceed 70 nL). Check the injected oocytes several hours after the injection, and discard those whose yolk is still leaking.

3.5.5. Incubation of the Oocytes

The oocytes express membrane proteins 1–5 d after the injection of RNA, depending on type of protein, quality of RNA, and batch of oocytes. The oocyte culture can be routinely maintained for 4–5 d in sterile NDE-96, and in many cases for up to 2 wk if sterility is preserved. Change the solution daily, and be sure to remove any dead or dying oocytes (which may be distinguished by the lack of glare observed in healthy denuded oocytes with lateral illumination, or by the appearance of white spots on the membrane). We incubate the oocytes at 22°C, but successful expression can be obtained with incubation at 19–20°C. Do not let the temperature exceed 24°C.

3.5.6. Preparing the Oocyte for Patch Clamp:
Removal of Vitelline Membrane

If patch clamp or internal perfusion of the oocytes is desired, the noncellular vitelline membrane still surrounding the defolliculated oocytes has to be removed. Place the oocytes for 20–30 min in hypertonic solution (e.g., normal ND-96 with the addition of 50 mM NaCl or 100 mM sucrose). The oocytes shrink, and there are areas where the vitelline membrane (seen under the dissection microscope as a clear shining sheath) is clearly separated from the plasma membrane. Clasp the vitelline membrane in one of these spots with two pairs of very sharp forceps, and tear it apart by moving the forceps in opposite directions (sometimes this procedure has to be repeated several times until the vitelline membrane is completely removed). Beware not to expose the devitellinized oocytes to air, since they would explode immediately. Devitellinized oocytes stick to almost any surface in minutes.

4. Notes

1. Wear gloves during the entire procedure of RNA preparation, either from a tissue or from a cloned cDNA, to prevent contact between RNA and your hands (a source of RNases).
2. The DNase I should be tested for its ability to degrade DNA while maintaining the integrity of RNA (choose the right manufacturer). The cap analog used is not in its methylated form, since it will become methylated in the oocyte. More polymerase should result in more RNA. An optional way to purify the transcript from unincorporated nucleotides is by chromatography on a spun column of Sephadex G-50 (ref. *18*, appendix E) that has been autoclaved in water.

3. To ensure that the inhibition is indeed specific and is not because of impurities (leading, for example, to RNA degradation), verify that the oligonucleotide does not inhibit the expression of structurally unrelated channel, preferably expressed in the same oocytes injected with the same RNA. Another control to ensure that the inhibition is not the result of mRNA to treatment with any oligonucleotide is to verify that its expression is not affected by treatment with an oligonucleotide having an irrelevant sequence.

4. All rules of sterility should be strictly followed. All culture media should be sterilized, preferably by filtering through 0.2 or 0.45 μm filters. The oocytes can be incubated in commercially available culture media (e.g., Leibowitz-15 diluted to 70%), but the best survival and synthesis are observed in simple Ringer-like solutions.

5. Defolliculation is usually performed with partially purified collagenase that contains contaminating proteolytic enzymes, which seem to help the defolliculation. Check the quality of collagenase by testing a small amount of it on oocytes before you purchase a large amount of the enzyme of the same batch.

References

1. Soreq, H. (1985) The biosynthesis of biologically active proteins in mRNA microinjected *Xenopus* oocytes. *CRC Crit Rev. Biochem.* **18,** 199–238.
2. Dascal, N. (1987) The use of *Xenopus* oocytes for the study of ion channels. *CRC Crit. Rev. Biochem.* **22,** 317–387.
3. Snutch, T. P. (1988) The use of *Xenopus* oocytes to probe synaptic communication. *Trends Neurosci.* **11,** 250–256
4. Lester, H. A. (1988) Heterologous expression of excitability proteins: route to more specific drugs? *Science* **241,** 1057–1063.
5. Thornhill, W. B. and Levinson, S.R. (1987) Biosynthesis of electroplax sodium channels in *Electrophorus* electrocytes and *Xenopus* oocytes. *Biochemistry* **26,** 4381–4388.
6. Julius, D., MacDermont, A. B., Axel, R., and Jessel, T. (1988) Molecular characterization of a functional cDNA encoding the serotonin 1c receptor. *Science* **241,** 558–564.
7. Moriarty, T., Padrell, E., Carty, D. J., Omri, G., Landau, E. M., and Iyengar, R. (1989) G_o protein as signal transducer in the pertussis toxin-sensitive phosphatidylinositol pathway. *Nature* **343,** 79–82.
8. Stuhmer, W., Methfessel, C., Sakmann, B., Noda, M., and Numa, S. (1987) Patch clamp characterization of sodium channels expressed from rat brain cDNA. *Eur. J. Biophys.* **14,** 131–138.
9. Lotan, I., Dascal, N., Naor, Z., and Boton, R. (1990) Modulation of vertebrate Na^+ and K^+ channels by subtypes of protein kinase C. *FEBS Lett.* **267,** 25–28.

10. Sugiyama, H., Ito, I., and Watanabe, M. (1989) Glutamate receptor subtypes may be classified into two major categories: A study on *Xenopus* oocytes injected with rat brain mRNA. *Neuron* **3,** 129–132.

11. Stuhmer, W., Conti, F., Suzuki, H., Wang, X., Noda, M., Kubo, H., and Numa, S. (1989) Structural parts involved in activation and inactivation of the sodium channel. *Nature* **339,** 597–603.

12. Lotan, I., Goelet, P., Gigi, A., and Dascal, N. (1989) Specific block of Ca^{2+} channel expression by a fragment of dihydropyridine receptor cDNA. *Science* **243,** 666–669.

13. Buller, A. L. and White M. M. (1990) Functional acetylcholine receptors expressed in *Xenopus* oocytes after injection of *Torpedo* β, γ, and δ subunits are a consequence of endogenous oocyte gene expression. *Mol. Pharmacol.* **37,** 243–248.

14. Parker, I. and Ivorra, I. (1990) A slowly inactivating potassium current in native oocytes of *Xenopus laevis. Proc. Roy. Soc. Lond.* B **238,** 369–381.

15. Sumikawa, K., Parker, I., Amano, T., and Miledi, R. (1984) Separate fractions of mRNA from *Torpedo* electric organ induce chloride channels and acetylcholine receptors in *Xenopus* oocytes. *EMBO J.* **3,** 2291–2294.

16. Davis, L. G., Dibner, M. D., and Battey, J. F. (1986) *Basic Methods in Molecular Biology.* Elsevier, New York.

17. Berger, S. L. and Kimmel A. R. (1987) Guide to molecular cloning techniques, in *Methods in Enzymology,* vol. 152. Academic, New York.

18. Sambrook, J., Fritsch, E. F., and Maniatis T. (1989) *Molecular Cloning. A Laboratory Manual,* 2nd ed., Cold Spring Harbor Laboratory, Cold Spring Harbor, NY.

19. Dierks, P., Van Ooyen, A., Mautei, N., and Weissmann, C. (1981) DNA sequences preceding the rabbit β-globin gene are required for formation in mouse L-cells of β-globin RNA with the correct 5'-terminus. *Proc. Natl. Acad. Sci. USA* **78,** 1411–1415.

20. Chirgwin, A. M., Przybyla A. E., MacDonald R. J., and Rutter W. J. (1979) Isolation of biologically active ribonucleic acid from sources enriched in ribonuclease. *Biochemistry* **18,** 5294–5299.

21. Snutch, T. P., Leonard, J., Nargeot, J., Lubbert, H., Davidson, N., and Lester, H. A. (1987) Characterization of voltage-gated Ca^{2+} channels in *Xenopus* oocytes after injection of RNA from electrically excitable tissues, in *Cell Calcium and the Control of Membrane Transport. Soc. Gen. Physiol. Ser.* vol. 42. (Mandel, L. J. and Eaton, D. C., eds.), Rockefeller University Press, New York.

22. Sigel, E. (1987) Properties of single sodium channels translated by *Xenopus* oocytes after injection with messenger ribonucleic acid. *J. Physiol.* **386,** 73–90.

23. Chomczynski, P. and Sacchi, N. (1987) Single-step method of RNA isolation by acid guanidinium thiocyanate-phenol-chloroform extraction. *Anal. Biochem.* **162,** 156–159.

24. Lotan, I. (1991) Hybrid arrest technique for functional roles of cloned cDNAs and to identify homologies among ion channel genes, in *Methods in Enzymology,* vol. 207. Academic, New York, in press.

25. Dash, P., Lotan, I., Knapp, M., Kandel, E. R., and Goelet, P. (1987) Selective elimination of mRNAs in vivo: complementary deoxynucleotides promote RNA degradation by an RNase H-like activity. *Proc. Natl. Acad. Sci. USA* **84,** 7896–7900.
26. Cazenave, C., Loreau, N., Thuonng, N. T., Toulmé, J. J., and Hélène, C. (1987) Enzymatic amplification of translation inhibition of rabbit β-globin mRNA mediated by anti-messenger oligodeoxy nucleotides covalently linked to intercalating agents. *Nucleic Acid Res.* **15,** 4717–4736.
27. Lotan, I., Volterra, A., Dash, P., Siegelbaum, S. A., and Goelet, P. (1988) Blockade of ion channel expression in *Xenopus* oocytes with complementary DNA probes to Na$^+$ and K$^+$ channel mRNAs. *Neuron* **1,** 963–971.
28. Brown, A. L (1970) *The African Clawed Toad Xenopus laevis.* Butterworth, London.
29. Dascal, N., Snutch, T. P., Lubbert, H., Davidson, N., and Lester, H. A. (1986) Expression and modulation of voltage operated calcium channels after RNA injection in *Xenopus* oocytes. *Science* **231,** 1147–1150.
30. Dumont, J. N. (1972) Oogenesis in *Xenopus laevis* (Daudin) 1. Stages of oocyte development in laboratory maintained animals. *J. Morphol.* **136,** 153–180.
31. Krafte, D. S. and Lester, H. A. (1989) Expression of functional sodium channels in stage 11-111 *Xenopus* oocytes. *J. Neurosci. Meth.* **26,** 211–215.
32. Dumont, J. N. and Brummett, A. R. (1978) Oogenesis in *Xenopus laevis* (Daudin) V. Relationship between developing oocytes and their investing follicular tissues. *J. Morphol.* **155,** 73–98.
33. Miledi, R. and Woodward, R. M. (1989) Effects of defolliculation on membrane current responses of *Xenopus* oocytes. *J. Physiol.* **416,** 601–621.

CHAPTER 14

Analysis of Insulin and Insulin-Like Growth Factor-I Receptors in Neural Tissues

Martin L. Adamo, Mohan K. Raizada, Joshua Shemer, Akira Ota, Colin Sumners, John Olson, and Derek LeRoith

1. Introduction

A variety of biochemical and physiological studies in recent years have made it clear that the brain possesses specific receptors and putative signal transduction pathways for insulin and the insulin-like growth factors (IGF-I and IGF-II). These observations have challenged traditional ideas concerning the physiological role of these peptides in the CNS, and have compelled us and others to consider that they are important regulators of metabolism and growth and development in the CNS (reviewed in *1*). A combination of the techniques used for culturing neural-derived cells and those involved in studying insulin and IGF-I receptors on peripheral nonneural tissue has provided a major stimulus for these studies, the results of which define insulin and the IGFs as neuropeptides. The purpose of this chapter is to present these techniques together in a comprehensive manner so as to provide the reader with an overview of how it is possible to study insulin and the IGF-I receptors in nervous tissue.

From: *Methods in Molecular Biology, Vol. 13: Protocols in Molecular Neurobiology*
Edited by: A. Longstaff and P. Revest Copyright © 1992 The Humana Press, Totowa, NJ

2. Materials

2.1. Cell Culture

1. Dulbecco's Modified Eagle's Medium (DMEM) with glutamine containing 100 U/mL penicillin G and 100 µg/mL streptomycin.
2. Plasma-derived horse serum (PDHS).
3. Fetal bovine serum.
4. Isotonic salt solution (concentrations in g/L): NaCl (8.0), KCl (0.402), Na_2HPO_4 (0.024), KH_2PO_4 (0.0299), glucose (0.99), sucrose (20.17), penicillin (100,000 U), streptomycin (0.1), fungizone (0.025), phenol red (0.4 mL of a 4% solution). Adjust the pH to 7.4, sterile filter, with 0.22-µm filter and store at –20°C. Stable for 2–4 mo.
5. Poly-L-lysine: 10 mg/L in double-distilled water, sterilize by filtration through 0.22-µm filters. Store in sterile bottles at –20°C. Stable for 2–4 mo.
6. Trypsin: 150 U/mg, 2.5 g/L in isotonic salt solution. Adjust pH to 7.4 with $2M$ NaOH, and sterilize by filtration with 0.22-µm filter. Store at –20°C in 50-mL aliquots.
7. Deoxyribonuclease I (DNase I): 16 mg/100 mL in isotonic salt solution. Sterilize by filtration, and store at –20°C.
8. Cytosine arabinoside: 28 mg/100 mL in phosphate buffered saline, pH 7.4. Sterilize by filtration, and store at –20°C in 5-mL aliquots.

2.2. Membrane-Binding Studies

1. Phosphate buffered saline (PBS; available commercially).
2. Homogenization buffer: 1 mM $NaHCO_3$, 2 mM phenylmethyl sulfonylfluoride (PMSF), 10 mg/mL leupeptin, and aprotinin (1 U trypsin inhibitor activity/mL). This buffer should be freshly prepared.
3. Calcium-free KRP buffer (KRP): 150 mM NaCl, 5 mM KCl, 1.2 mM KH_2PO_4, 16 mM K_2HPO_4, 16 mM Na_2HPO_4, and 1.2 mM $MgSO_4$ • $7H_2O$, pH 7.8. This solution may be kept at 4°C indefinitely.
4. Assay buffer A: KRP with 30 mg/mL BSA, and 3 mg/mL bacitracin, pH 8.0. Store aliquots frozen at –20°C, or prepare freshly. Do not store at 4°C.
5. Assay buffer B: KRP with 1 mg/mL bovine serum albumin (BSA), pH 7.6. Store aliquots at –20°C, or prepare freshly. Do not store at 4°C.
6. ^{125}I-insulin (monoiodo[A_{14}]) radio-receptor grade with specific activity between 250–370 µCi/µg. Store at –20°C.
7. Unlabeled insulin. 1.2 mg/mL stock solution in 10 mM HCl. Store in aliquots at –20°C.
8. ^{125}I-IGF-I. Store at –20°C.

9. Unlabeled IGF-I. Stock solution is 0.1 mg/mL in 10 mM HCl or 100 mM acetic acid. Store in aliquots at –20°C.

2.3. Receptor Solubilization and Partial Purification

1. Solubilization buffer: 50 mM Hepes, 1 mM PMSF, 1% Triton X-100, pH 7.6. This buffer, without the PMSF, may be stored indefinitely at 4°C.
2. Wheat germ agglutinin (WGA) agarose (2 mL gel, from ICN Biomedicals, Costa Mesa, CA).
3. WGA column buffer I: 50 mM Hepes, 150 mM NaCl, 0.1% Triton X-100, 0.01% sodium dodecyl sulfate (SDS), pH 7.6. (Add SDS freshly.)
4. WGA column buffer II: 50 mM Hepes, 150 mM NaCl, 0.1% Triton X-100, pH 7.6 (stable indefinitely at 4°C).
5. WGA column buffer III: 50 mM Hepes, 150 mM NaCl, 0.1% Triton X-100, 0.3M N-acetylglucoseamine, pH 7.6 (add N-acetylglucoseamine freshly).
6. Binding assay buffer A: 50 mM Hepes, 150 mM NaCl, 0.16% BSA, 1.6 mg/mL bacitracin, pH 7.9. Prepare freshly or store frozen at –20°C.
7. Binding assay buffer B: 50 mM Hepes, 150 mM NaCl, 0.1% BSA, pH 7.8. Prepare freshly or store frozen at –20°C.
8. Bovine γ-globulin: 0.3% (w/v) in water (store at 4°C).
9. Polyethylene glycol (PEG): 25% in 50 mM Hepes, 150 mM NaCl, 0.1% Triton X-100, pH 7.8. Store at 4°C.
10. PEG: 12.5% in 50 mM Hepes, 150 mM NaCl, 0.1% Triton X-100, pH 7.8. Store at 4°C.
11. ^{125}I-insulin and unlabeled insulin as in Section 2.2.
12. ^{125}I-IGF-I and unlabeled IGF-I as in Section 2.2.

2.4. Structural Studies

1. Disuccinimidyl suberate (DSS) 3.75 mg/mL in DMSO.
2. 100 mM Tris-HCl, 10 mM EDTA, pH 7.4.
3. Buffer A: 5 mM 2-(N-morpholino) ethanesulfonic acid and 1 mM CaCl$_2$ with 2 mM PMSF, 10 µg/mL leupeptin, 0.5 U/mL α_2-macroglobulin, 10 µg/mL pepstatin A, and 1 mM N-ethylmaleimide (NEM).
4. Buffer B: 0.1M sodium phosphate, pH 6.1, with 50 mM EDTA, 1.0% Triton X-100, 0.1% SDS, 1% 2-mercaptoethanol, 2 mM PMSF, and 4 µg/mL aprotinin.
5. Buffer C: 0.3M sodium citrate, pH 5.5.
6. Buffer D: 1% SDS, 10 mM dithiothreitol (DTT), 10 mM sodium phosphate, pH 7.0.
7. 10% TCA.

8. Ethanol/ether (1:1).
9. Neuraminidase (Type X) from *Clostridium perfringens.*
10. Endoglycosidase H (from *Streptomyces plicatus,* purchased from ICN Biomedicals).
11. Endoglycosidase F (from *Flavobacterium meningosepticum* purchased from New England Nuclear, Boston, MA).

2.5. Phosphorylation in a Cell-Free System

1. WGA purified receptors (*see* Section 3.).
2. 75 mM Hepes, pH 7.6.
3. 50 mM Hepes, pH 7.6.
4. Poly (Glu,Tyr)4:1: 10 mg/mL in 50 mM Hepes, pH 7.6.
5. Buffer A: 50 mM Hepes, 150 mM NaCl, 0.1% BSA, pH 7.8.
6. [γ^{32}-P]-ATP (3000 Ci/mmol; 10 mCi/mL).
7. ATP: 5 mM in 100 mM Hepes, pH 7.6.
8. CTP: 25 mM in 500 mM Hepes, pH 7.6.
9. Manganese acetate, 0.1M in water.
10. MgCl$_2$, 0.67M in water.
11. 10% TCA containing 10 mM sodium pyrophosphate.
12. Stop solution: 40 mM Hepes, pH 7.5, 0.4% Triton X-100, 20 mM EDTA, 200 mM NaF, 40 mM sodium pyrophosphate, 40 mM sodium phosphate, and 40 mM EDTA.
13. Insulin or IGF-I standards (0–7 × 10^{-6}M in buffer A).

2.6. Phosphorylation in Intact Cells

1. Phosphate-free, serum-free Eagle's Medium (MEM or DMEM).
2. ^{32}P-orthophosphoric acid.
3. Solubilization solution: 50 mM Hepes, pH 7.4, 1% Triton X-100, 2 mM sodium orthovanadate, 10 mM sodium pyrophosphate, 10 mM NaF, 4 mM EDTA, and 2 mM PMSF.
4. Antibodies: antiphosphotyrosine antibody (anti P-Tyr) (gift of Y. Zick, Weizmann Institute, Rehovot, Israel) and antireceptor antibodies (gifts of P. Gorden and S. I. Taylor, Diabetes Branch, NIDDK, NIH).
5. Pansorbin: 20% suspension (purchased from CalBiochem, La Jolla, CA).
6. Wash solution: 50 mM Hepes, pH 7.4, 1% Triton X-100, 0.1% SDS, 150 mM NaCl, 10 mM NaF, 4 mM EDTA, 2 mM sodium orthovanadate.
7. WGA columns: 0.5 mL, disposable.
8. WGA column wash solution: 50 mM Hepes, pH 7.4, 0.1% Triton X-100, 10 mM sodium pyrophosphate, 10 mM NaF, 4 mM EDTA, 2 mM sodium orthovanadate, 2 mM PMSF.
9. *p*-Nitrophenylphosphate (PNPP): 10 mM in 50 mM Hepes, pH 7.6, 0.1% Triton X-100.

2.7. Phosphoamino Acid Analysis

1. Bio-Rad AG1-X8 resin.
2. $6M$ HCl, $1M$ HCl, $0.1M$ HCl, $1M$ NaOH.
3. 200 mM NH$_4$OH.
4. Phosphoamino acid standards: 1 mg/mL stock of phosphotyrosine, phosphoserine, and phosphothreonine.
5. Thin-layer electrophoresis running buffer: 70 mL glacial acetic acid, 8 mL formic acid, 2.4 mL pyridine. Adjust pH to 3.11 with additional formic acid, and dilute to 500 mL final vol.
6. Thin-layer electrophoresis plate (cellulose MN300, 20 cm × 20 cm, 250 µm).
7. 0.5% Ninhydrin in acetone.

2.8. Phosphopeptide Mapping

1. Dried gel and autoradiogram.
2. Buffer A: 50 mM Hepes, 150 mM NaCl, 0.1% Triton X-100, 0.1% SDS.
3. *S. aureus* V8 protease (purchased from ICN Biomedicals).

3. Methods

3.1. Preparation of Neuronal and Astrocytic Glial Cells from the Brain in Primary Culture

In this section, we describe procedures to prepare and maintain astrocytic glial cells and neuronal cells from the brain in primary culture. The reader is referred to Fig. 1 for an outline of the methods to be described in the following sections.

3.1.1. Precoating of Culture Dishes with Poly-L-Lysine

Add 10 mL, 5 mL, and 3 mL poly-L-lysine solution in 100-, 60-, and 35-mm culture dishes before starting the cell culture procedure. Prepare enough plates for anticipated cell recovery (approx $3.5 \times 10^7/$ brain). Leave them in the sterile hood until ready for plating cells. Just before plating, remove poly-L-lysine solution, rinse dishes with 5–10 mL of isotonic salt solution, and proceed to plate cells. Do not allow the dishes to dry out before plating.

3.1.2. Dissociation of Neonatal Brain Cells

1. Sacrifice 1-d-old rats using 0.25 mL of pentobarbital, remove brains under sterile conditions, and place them in 100-mm diameter culture dish containing 30 mL isotonic salt solution.

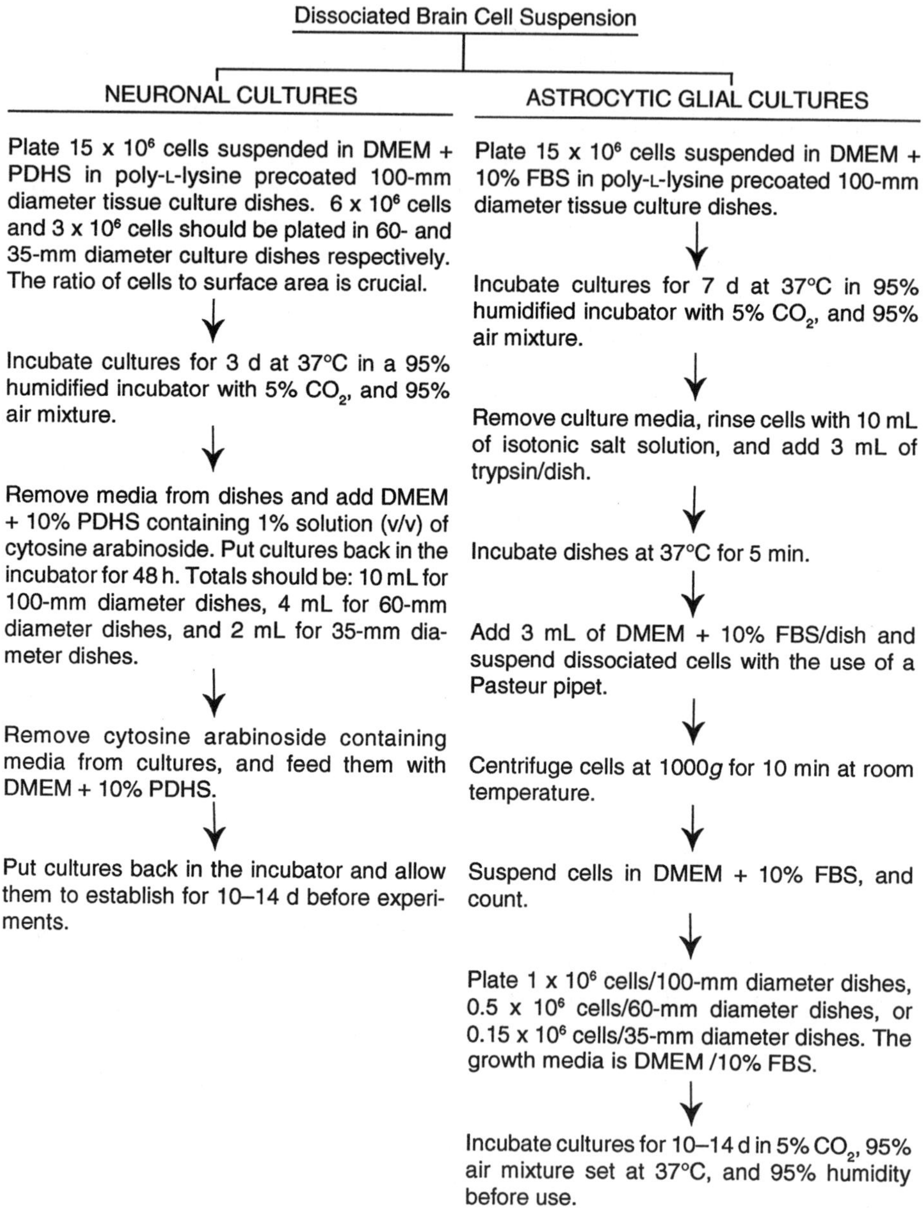

Dissociated Brain Cell Suspension

NEURONAL CULTURES

Plate 15 x 10^6 cells suspended in DMEM + PDHS in poly-L-lysine precoated 100-mm diameter tissue culture dishes. 6 x 10^6 cells and 3 x 10^6 cells should be plated in 60- and 35-mm diameter culture dishes respectively. The ratio of cells to surface area is crucial.

↓

Incubate cultures for 3 d at 37°C in a 95% humidified incubator with 5% CO_2, and 95% air mixture.

↓

Remove media from dishes and add DMEM + 10% PDHS containing 1% solution (v/v) of cytosine arabinoside. Put cultures back in the incubator for 48 h. Totals should be: 10 mL for 100-mm diameter dishes, 4 mL for 60-mm diameter dishes, and 2 mL for 35-mm diameter dishes.

↓

Remove cytosine arabinoside containing media from cultures, and feed them with DMEM + 10% PDHS.

↓

Put cultures back in the incubator and allow them to establish for 10–14 d before experiments.

ASTROCYTIC GLIAL CULTURES

Plate 15 x 10^6 cells suspended in DMEM + 10% FBS in poly-L-lysine precoated 100-mm diameter tissue culture dishes.

↓

Incubate cultures for 7 d at 37°C in 95% humidified incubator with 5% CO_2, and 95% air mixture.

↓

Remove culture media, rinse cells with 10 mL of isotonic salt solution, and add 3 mL of trypsin/dish.

↓

Incubate dishes at 37°C for 5 min.

↓

Add 3 mL of DMEM + 10% FBS/dish and suspend dissociated cells with the use of a Pasteur pipet.

↓

Centrifuge cells at 1000*g* for 10 min at room temperature.

↓

Suspend cells in DMEM + 10% FBS, and count.

↓

Plate 1 x 10^6 cells/100-mm diameter dishes, 0.5 x 10^6 cells/60-mm diameter dishes, or 0.15 x 10^6 cells/35-mm diameter dishes. The growth media is DMEM /10% FBS.

↓

Incubate cultures for 10–14 d in 5% CO_2, 95% air mixture set at 37°C, and 95% humidity before use.

Fig. 1. Culture of neuronal and astrocytic glial cells.

2. Remove blood vessels and pia matter from each brain using fine forceps under a magnifying glass.
3. Transfer all the cleaned brains into fresh isotonic salt solution.
4. Large brain pieces are chopped into 2–3 mm^3 vol chunks with the help of small sterile scissors.
5. Transfer the brain pieces into a 50-mL centrifuge tube, decant isotonic salt solution, and add trypsin solution (1.5 mL/brain).
6. Triturate 3–4 times with a Pasteur pipet and incubate at 37°C in a shaker-waterbath for 5 min, at a speed of about 200 rpm.
7. Remove the tube from the waterbath, and add 0.1 mL DNase I solution/mL trypsin under sterile conditions. Continue the incubation for an additional 5 min at 37°C.
8. Let the brain pieces settle at the bottom for 1–2 min, remove the supernatant containing dissociated cells, and add to a 50-mL centrifuge tube containing 10 mL DMEM + 10% PDHS.
9. Use a Pasteur pipet to dissociate the remaining chunks, and combine them with dissociated cells.
10. Centrifuge the cell suspension (1000g, 10 min, room temperature).
11. Discard the supernatant, and suspend cell pellet in 10 mL DMEM + 10% PDHS with the help of a fire-polished Pasteur pipet.
12. Filter the cell suspension through a sterile gauze cloth to remove undissociated tissue and bring the vol to 10 mL/brain with DMEM + 10% PDHS. Plate at 6400 cells/cm^2 in 10 mL medium/100 mm^2 dish (*see* Note 1).

3.1.3. Dissociation of Cells from 21-d-Old and/or 90-d-Old Rats

1. Brains from 21- or 90-d-old rats are removed under sterile conditions and placed in 30 mL isotonic salt solution.
2. Blood vessels and pia matter are removed, tissues minced into 2–3 mm^3 pieces, and suspended in 25 mg trypsin solution.
3. The suspension is triturated for 1 min with a 10-mL pipet (bore size 3 mm); 320 mg-DNase I is added, followed by an additional 1-min trituration.
4. Cell dissociation is continued at 37°C for 5 min. Steps 8–12 (Section 3.1.2.) are then followed to complete the dissociation process.

3.1.4. Culture of Astrocytes and Neurons

The detailed protocol is given in Fig. 1. Adult astrocytes were allowed to grow for 3–4 wk before transfer, instead of the 10 d allowed for neonatal astrocytes (*see* Notes 2–4).

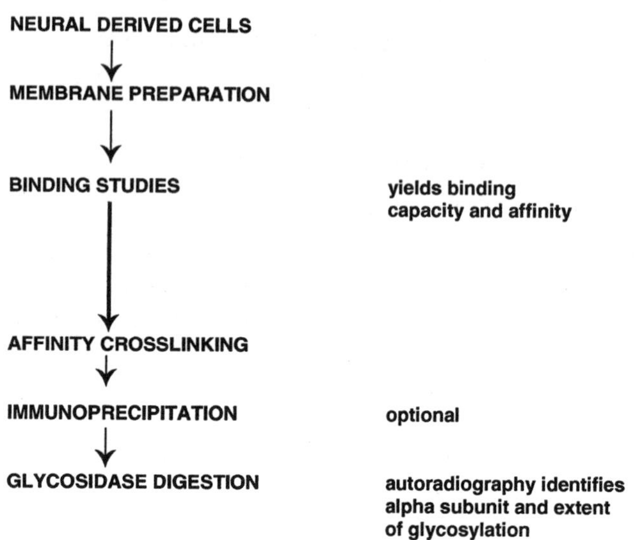

Fig. 2. Flow chart for analysis of receptor α subunit.

3.2. Membrane-Binding Studies

In this section, we describe the basic methods for obtaining preparations enriched in plasma membranes and for performing binding assays on these preparations. These methods are modifications of the method originally described by Havrankova et al. *(2)*, and have been used extensively by our groups for characterizaton of insulin and IGF-I binding to brain tissue and neurally derived cells *(3–8)*. In addition to providing information on receptor number and affinity, these methods are the "starting point" for the subsequent studies utilizing more purified receptor preparations in cell-free phosphorylation experiments as well as for studies on structural characterization (e.g., affinity crosslinking). The reader should refer to Fig. 2 for an outline of the methods to be described in the following section.

3.2.1. Membrane Preparation

1. Harvest the cells by mechanical agitation and centrifugation (600*g*, 10 min, 4°C), wash in PBS, recentrifuge, and aspirate the supernatant.
2. Homogenize the cell pellet in 40 vol of homogenization buffer at 4°C using a glass/glass homogenizer.
3. Centrifuge (600*g*, 10 min, 4°C).
4. Remove the supernatant and centrifuge (20,000*g*, 30 min, 4°C).
5. Wash the pellets once with 20–30 mL homogenization buffer.

6. Resuspend in KRP buffer, pH 7.8 to a concentration of 1.2 mg/mL protein as determined by the method of Lowry et al. *(9)*, and store until use at –70°C.

3.2.2. Binding Assay (see Notes 5–14)

Using 400-µL microcentrifuge tubes, on ice:

1. Pipet 50 µL ^{125}I-insulin (freshly diluted in buffer A, final concentration 0.2 ng/mL), followed by 50 µL unlabeled peptide at 0–10,000 ng/mL (final concentration, diluted in buffer B). The unlabeled peptide dilutions are made at 3X final dilution, i.e., 0–30,000 ng/mL and these can be stored in aliquots at –20°C.
2. Initiate the binding reaction by adding 50 µL plasma membrane in KRP, pH 7.8, at a final concentration of 0.5 mg/mL. Vortex the tubes.
3. Incubate for 18 h at 4°C.
4. Centrifuge the samples (10,000g, 5–10 min, 4°C), aspirate the supernatant, wash the pellets with 200 µL KRP, and recentrifuge. Use a razor blade to cut off the tip of the tube containing the pellet, and count the radioactivity in the pellet in a γ counter.

3.3. Receptor Solubilization and Partial Purification

For insulin and IGF-I receptors, studies of receptor autophosphorylation and tyrosine kinase activity require solubilization of receptors from membranes and partial purification on lectin affinity columns to both concentrate the receptors and remove most endogenous phosphatases. In this section, we describe the methods used for accomplishing this, as well as methods for assaying insulin receptor binding using these partially purified preparations. These methods were modified from those described in *(10)* and *(11)*, and have been used extensively in studies of neurally derived cells *(3–8)*.

3.3.1. Receptor Solubilization and WGA Purification

1. Resuspend the plasma membranes (30 mg protein) in 7 mL solubilization buffer, and incubate at 4°C for 120 min. Vortex intermittently.
2. Centrifuge the homogenate (200,000g, 50 min, 4°C).
3. Decant the supernatant containing the solubilized receptors.
4. Wash WGA column (2 mL) with 40 mL buffer I, 20 mL buffer III, and 40 mL buffer II *(see* Notes 15,16).
5. Pass solubilized receptors over a WGA affinity column and recycle twice.
6. Adsorbed material is eluted using 7 mL buffer III, and 1-mL fractions are collected.

3.3.2. Solubilized Receptor Binding Assay
(see Notes 17–22)

1. Use 12 mm × 75 mm plastic culture tubes. Incubate at 22–24°C, 125 µL of ^{125}I-ligand in assay buffer A (final concentration 0.2 ng/mL), 25 µL of solubilized WGA purified membrane proteins, and 50 µL of varying concentrations of unlabeled ligand (0–10,000 ng/mL final concentration) in assay buffer B (*see* Section 2.3.).
2. After 3 h of incubation, terminate the reaction by placing tubes on ice and adding 100 µL of 0.3% bovine γ-globulin and 300 µL of 25% polyethylene glycol. Vortex vigorously, and incubate 10 min on ice.
3. Centrifuge (2500*g*, 15 min, 4°C), and wash the pellets once with 300 µL of 12.5% polyethylene glycol.
4. Aspirate and discard the supernatants, and count the radioactivities in the tubes in a γ counter.

3.4. Structural Studies

It has been recognized for some years now that insulin and IGF-I receptors are transmembrane glycoproteins, and exist as hetero-tetramers of two ligand-binding α subunits and two β subunits that possess autophosphorylation sites and tyrosine kinase activity (Fig. 3). The structures of these disulfide bond linked subunits can be studied in a variety of ways. The more common techniques used are (a) covalent crosslinking of bound ligand to the α subunit and (b) ligand-induced autophosphorylation of the β subunit. In this section, methods are given for affinity crosslinking of ^{125}I-ligands (originally described in *12* and used in references *3–8*). In addition, techniques for enzymatic modification of carbohydrate moieties of the receptors are also given. Neuraminidase cleaves sialic acid residues, whereas endoglycosidase H cleaves high mannose linkages *(13)*, and endoglyosidase F cleaves all *N*-linked carbohydrates *(14)*. The importance of these methods lies in the fact that brain insulin and IGF-I receptor α subunits are typically smaller than those in the periphery *(15–18)*. We have used these enzymes and modifications of the methods described in *(17,19,20)* to study the glycosylation of insulin receptor α subunit in brain tissue and neurally derived cells *(6,21)*. Others have compared the glycosylation of central and peripheral insulin and IGF-I receptors using these enzymes (e.g., *18,22,23*). The reader should refer to Fig. 2 for an outline of the methods used.

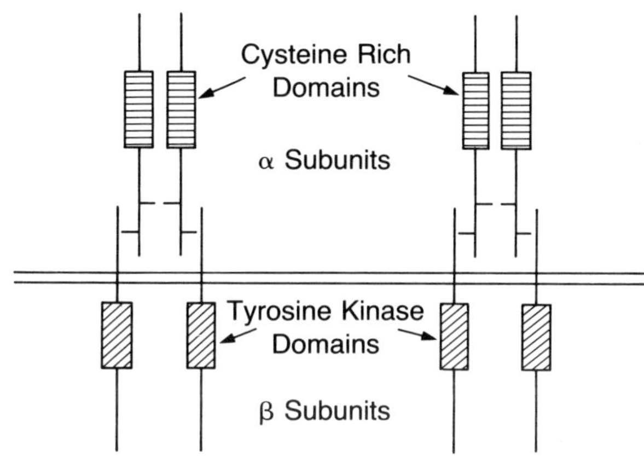

Fig. 3. Schematic of insulin and IGF-I receptor structure. Insulin and IGF-I receptors consist of two extracellular α-subunits joined by disulfide bonds to two membrane-spanning β subunits. The cysteine-rich domains (▤) of each α-subunit are believed to be important in ligand-binding, whereas the tyrosine kinase domain (▨) of each contains autophosphorylation sites and tyrosine kinase activity, believed to be important in signal transduction. α Subunits have an M_r of 130 kDa in peripheral tissue. Using the techniques described in this chapter (Section 3.4.), α subunits from brain and some neurally derived cells can be shown to have an M_r about 10 kDa less because of altered glycosylation. Phosphorylation of β subunits can be demonstrated both in cell-free systems (Section 3.5.) and in intact cells (*see* Section 3.6.). (Adapted from ref. *11a*).

3.4.1. Affinity Crosslinking (see Note 23)

1. Incubate 500 µL of ^{125}I-ligand (final concentration 10 ng/mL) with 500 µL membranes (final concentration 1.33 mg/mL) and 500 µL of unlabeled ligand (0–10,000 ng/mL final concentration) in the buffer systems described in Section 2.2.
2. Incubations are terminated by centrifugation (12,000g, 20 min, 4°C). Resuspend the pellets in 1.5 mL of BSA-free KRP, pH 7.8, and wash twice.

3. Crosslink the ligand and receptor in 1.5 mL KRP, pH 7.8, by the addition of 15 μL DSS (0.1 m*M*) for 16 min on ice.
4. Terminate the crosslinking reaction by adding 300 μL of 100 m*M* Tris with 10 m*M* EDTA, pH 7.4.
5. Wash the pellet three times in KRP buffer, pH 7.8 (1.5 mL each wash), by centrifugation (10,000*g*, 5 min, 4°C). Resuspend the washed pellet in 150 μL sodium dodecyl sulfate-polyacrylamide gel electrophoresis (SDS-PAGE) sample buffer, vortex vigorously, and heat at 95°C for 10 min.
6. The samples are then run on SDS-PAGE (5% stacking gel, 7.5% resolving gel) and autoradiography of the dried slab gel is carried out with 2 d exposure (*see* Note 24).

3.4.2. Enzymatic Digestion of Glycoprotein Receptors

3.4.2.1. NEURAMINIDASE

1. Resuspend ^{125}I-ligand crosslinked membranes (*see* Section 3.4.1.) in 1 mL of buffer A in the presence or absence of 2.5 U neuraminidase, and incubate for 30 min at 37°C.
2. Wash the pellets once in Ca^{2+} free KRP, pH 7.8.
3. Solubilize and apply to SDS-PAGE as described in Section 3.4.1., step 6.

3.4.2.2. ENDOGLYCOSIDASE F

1. Crosslink the ligand to receptors, and solubilize as described in the previous subsection. Immunoprecipitate if appropriate (*see* Notes 25,27).
2. Dilute the receptors in buffer B without or with Endo F at 60 U/mL. Incubate for 18 h at 37°C.
3. Stop the reaction by adding 10% TCA. Centrifuge and wash the precipitates with ethanol/ether.
4. Dissolve in sample buffer, denature, and apply to SDS-PAGE (*see* Fig. 4).

3.4.2.3. ENDOGLYCOSIDASE H

1. Crosslink, immunoprecipitate, and elute the receptors as described earlier for endoglycosidase F.
2. Dilute 50 μL of receptors with 200 μL buffer C, containing 1 m*M* PMSF and 10 μ*M* pepstatin A without or with 0.6 U/mL Endo H.
3. Incubate at 37°C for 6 h. Stop the reactions with 10% TCA.
4. Wash precipitates with ethanol/ether.
5. Dissolve in sample buffer, denature, and apply to SDS-PAGE.

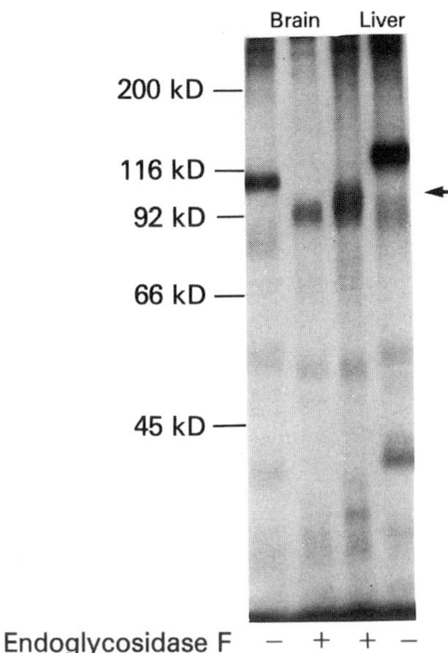

Fig. 4. Effect of endoglycosidase F on the insulin receptor from guinea pig liver and brain. Membranes prepared as in Section 3.2. from guinea pig brain and liver was crosslinked to ^{125}I-insulin, immunoprecipitated, and digested with endoglycosidase F (all as described in Section 3.4.) prior to SDS/PAGE. Autoradiography revealed that brain α subunits were ~120 kDa, whereas liver α subunits were ~135 kDa. Endo F, which should cleave all N-linked glycosylation, resulted in reduced (and similar) mobilities of brain and liver α subunits (Taken from ref. *21*.)

3.5. Phosphorylation in a Cell-Free System

Insulin and IGF-I receptors (Fig. 3) are tyrosine kinases, exhibiting increased autophosphorylation and tyrosine kianse activity toward endogenous and exogenous substrates when stimulated by ligand binding (reviewed in *24*). The methods below permit both a functional and structural analysis of this property, which appears to be important in signal transduction. Using a cell-free system, one can determine both the size of the autophosphorylated β subunit, and the regulation of its autophosphorylation and tyrosine kinase activity. The reader is referred to Fig. 5 for an outline of the methods and to references *3–8*, *21*, and *25–28* for applications of their use in neurally derived cells and brain tissue.

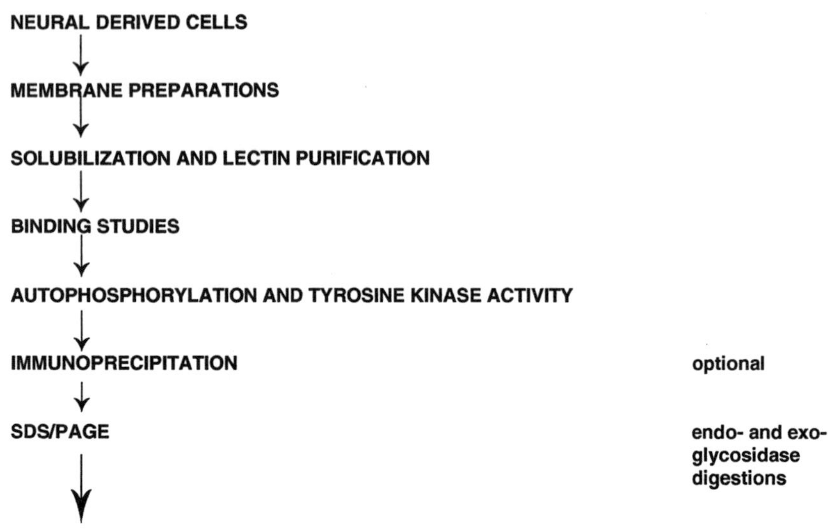

Fig. 5. Analysis of receptor β subunits and tyrosine kinase activity in a cell-free system.

3.5.1. Receptor Autophosphorylation (see Notes 29–42)

1. Add the following to a 2-mL microcentrifuge tube:
   20 μL 75 mM Hepes;
   20 μL 50 mM Hepes;
   10 μL buffer A with insulin or IGF-I at $0–7 \times 10^{-6} M$; and
   20 μL WGA purified receptor.
   Incubate-the tubes for 30–60 min at room temperature to allow ligand binding.
2. While the incubation is proceeding, prepare the reaction mixtures as follows:

Component	μL per Assay Tube
5 mM ATP	0.9
25 mM CTP	3.6
0.1M manganese acetate	2.7
H_2O	11.3
γ-^{32}P-ATP	1.5

3. At the end of the binding, initiate phosphorylation by adding 20 μL reaction mix to each tube. Incubate for the desired time (usually 1–30 min).

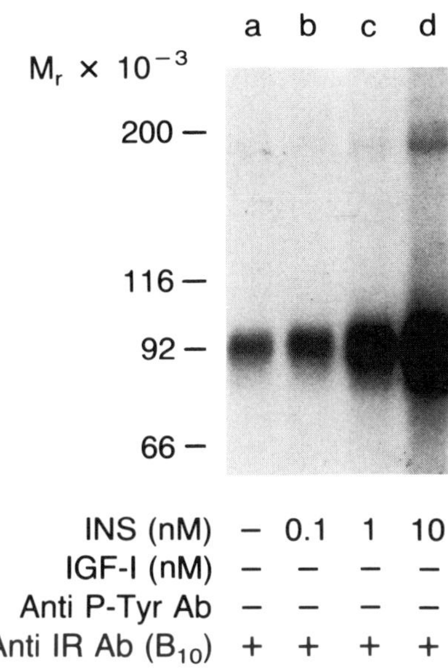

Fig. 6. Phosphorylation of insulin receptor in a cell-free system. Insulin receptors from N-18 neuroblastoma cells were partially purified on WGA columns as described in Section 3.3., and aliquots were assayed for insulin-induced autophosphorylation as described in Section 3.5. Autoradiography revealed the phosphorylated β subunit (95 kDa). Two lines of evidence, shown in this figure, indicate that this is the insulin receptor β subunit and not that of the IGF-I receptor. First, autophosphorylation was clearly stimulated by insulin at a concentration as low as 0.1 nM. Secondly, the insulin-induced phosphorylated protein was immunoprecipitated by the specific insulin receptor antibody B10 (Adamo et al., unpublished data).

4. Stop the phosphorylation by adding 60 µL of a 1:1 solution of 5X sample buffer, and stop solution. Place the tubes on ice.
5. Denature the samples at 95–100°C for 5–10 min, and apply to SDS-PAGE. *See* Fig. 6 for typical results.

3.5.2. Exogenous Substrate Phosphorylation

1. Add the following to a 15-mL conical centrifuge tube:
   20 µL 75 mM Hepes, pH 7.6;
   20 µL 50 mM Hepes, pH 7.6, without (endogenous) or with poly(Gly,Tyr) 4:1;
   10 µL buffer A with insulin or IGF-I at 0–$7 \times 10^{-6}M$; and
   20 µL WGA purified receptor.

Incubate the tubes 30–60 min at room temperature to allow insulin binding.

2. During the incubation, prepare the reaction mix as follows:

Component	µL per Assay Tube
5 mM ATP	0.9
25 mM CTP	3.6
0.67M MgCl$_2$	2.7
H$_2$O	11.3
γ-^{32}P-ATP	1.5

3. Initiate the phosphorylation by adding 20 µL of reaction mix. Incubate for the desired time (usually 15–30 min).
4. Stop the reaction by spotting 75 µL onto 3 × 3 cm squares of Whatman 3MM filter paper. Immerse in TCA/pyrophosphate solution, and wash extensively. After a final rinse in 95% ethanol, dry the filters and count in a liquid scintillation counter.

3.6. Phosphorylation in Intact Cells

The cell-free phosphorylation techniques described in the previous section are obviously useful, but they have the drawback, common to all in vitro systems, of creating an artificial system that may not faithfully mimic the in vivo situation. Metabolic labeling of intact cells, followed by stimulation with hormones and extraction and analysis of resultant phosphorylated proteins, has the advantage of allowing the study of signal transduction in a more in vivo system, and furthermore, allows one to detect potential endogenous substrates of the insulin and IGF-I kinases even if they do not copurify with the receptors. An excellent example of this is pp185, a 185-kDa nonglycoprotein phosphoprotein that was identified initially as a result of intact cell phosphorylation studies (29–31). The reader is referred to Fig. 7 for a detailed outline of the methods we have used in neurally derived cells (31,32).

3.6.1. Method for Whole Cell Extraction

1. Cells are grown to confluency in 175-mm^2 dishes. Twelve hours prior to the phosphorylation experiment, change the culture medium to serum-free medium with 2 mM glutamine.
2. Label the cells with 8 mL phosphate-free, serum-free minimum Eagle's medium (MEM or DMEM) containing carrier-free [^{32}P] orthophosphate (0.5 mCi/mL) for 2 h at 37°C and 5–10% CO$_2$ (*see* Note 43).
3. Ligand (diluted in serum-free, phosphate-free medium) is added to each plate at concentrations varying from 0–10^{-7}M (final concentration).

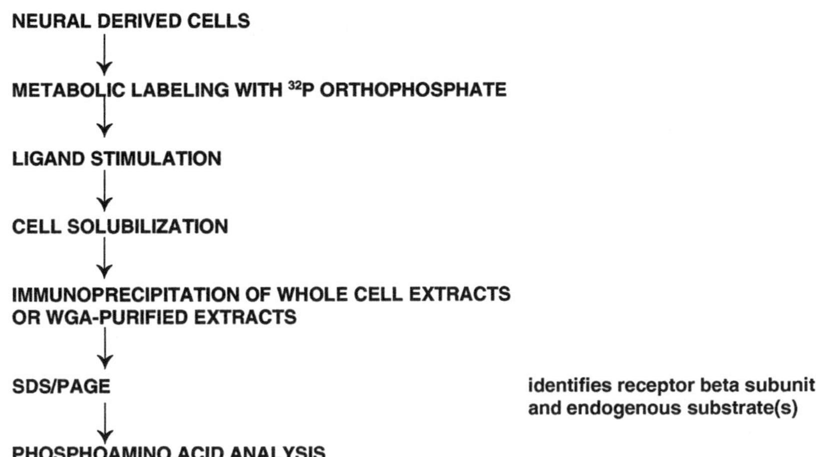

Fig. 7. Analysis of receptor and endogenous substrate phosphorylation in intact cells.

4. Stop the reaction at varying times (0–30 min) by rapidly removing the media and freezing the cells using liquid nitrogen (*see* Note 44).
5. Solubilize the cells by adding 1.4 mL of solubilization solution, allowing the cells to thaw at 4°C for 60 min.
6. Centrifuge the suspension (12,000g, 60 min, 4°C).
7. Add anti-P-tyr antibodies to the resultant supernatant at a serum dilution of 1:50, and incubate for 2 h at 4°C (*see* Note 45).
8. Immobilize the immune complexes by adding pansorbin (e.g., 200 µL) and incubating for 1 h at 4°C. Centrifuge (12,000g, 10 min, 4°C) and rinse five times (750 µL each wash) with wash solution (*see* Note 46).
9. Elute the phosphoproteins from the anti-P-TyrPansorbin complex using 60 µL PNPP. Incubate for 1 h at 4°C, vortexing every 15 min (*see* Note 47).
10. Recover P-Tyr containing phosphoproteins by adding 60 µL of 50 mM HEPES, 0.1% Triton X-100 and centrifugation. Run on 7.5% SDS-PAGE under reducing conditions. *See* Fig. 8 for results using N18 neuroblastoma cells *(31)*.
11. Reimmunoprecipitate the other half-sample with antireceptor antibodies.

3.6.2. Method for WGA-Purified Preparations

1. Carry out steps 1–5 as in Section 3.6.1.
2. After removing the solubilized cells from the incubation plates, centrifuge (40,000g, 50 min, 4°C).

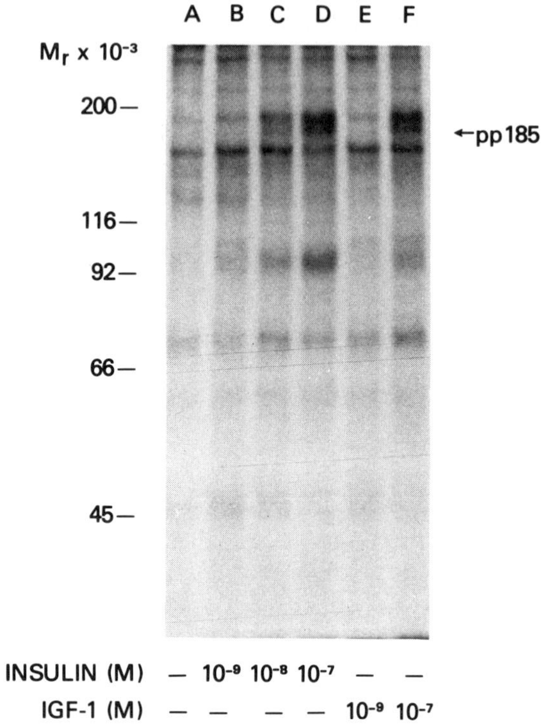

Fig. 8. Insulin and IGF-I stimulated phosphorylations in intact neuroblastoma cells. N-18 mouse neuroblastoma cells were metabolically labeled with ^{32}P-orthophosphate, and stimulated with insulin or IGF-I. After 1 min, reactions were stopped, and whole cell extracts were prepared as described in Section 3.6.1. Phosphoproteins were immunoprecipitated with antiphosphotyrosine antibody. Autoradiography revealed insulin receptor β subunits of 95 kDa, IGF-I β subunit subtypes of 95 and 105 kDa, and an endogenous substrate of 185 kDa (pp185) whose phosphoryation was stimulated by insulin and IGF-I. (Taken from ref. *31*.)

3. Apply the supernatant to 0.5 mL WGA-agarose columns. Recycle the flow through twice.
4. Wash the columns with 90 mL of washing solution.
5. Elute the adsorbed glycoproteins using 1-mL fractions of washing solution with 0.3*M* GlcNAc.
6. Immunoprecipitate those fractions with the highest radioactivity (usually fractions 1 and 2) using anti-P-Tyr or antireceptor antibodies as described in Section 3.6.1., steps 7 and 8. *See* Fig. 9 for results using intact neuronal cells in primary culture *(32)*.

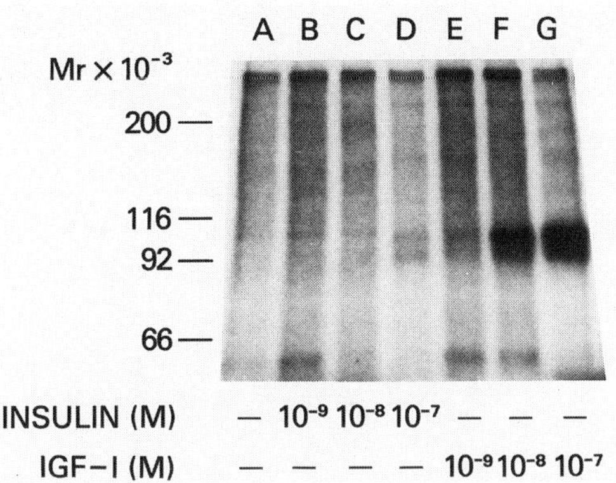

Fig. 9. Insulin and IGF-I stimulated phosphorylation in intact neuronal cells. Neuronal cells (prepared as described in Section 3.) were metabolically labeled with ^{32}P-orthophosphate, and stimulated with insulin or IGF-I, as described in Section 3.6. WGA purified extracts were prepared as described in Section 3.6.2. and immunoprecipitated with antiphosphotyrosine antibody as described (*see also* Section 3.6.1.). Autoradiography revealed the insulin and IGF-I receptor β subunits. (Taken from ref. *32*.)

3.7. Phosphoamino Acid Analysis

The identification of insulin and IGF-I receptors as tyrosine kinases obviously was aided by use of tyrosine-containing artificial substrates *(24)*; however, determining which amino acids become phosphorylated may be important in exploring the steps in the signal transduction pathway in vivo (*see* Note 48).

For example, following insulin and IGF-I stimulation of intact cells, significant phosphorylation of serine and threonine residues may occur, in addition to tyrosine phosphorylation *(31)*. This suggests that in vivo, insulin and IGF-I may also activate serine/threonine kinases. The methods described below were modified from those described in *(33)*, and have been used for analysis of insulin receptor and IGF-I receptor β subunits, and endogenous substrate in neurally derived cells *(4,31)*.

3.7.1. Preparation of Dowex

Suspend 300 mL of packed resin in 300 mL of water. Filter this in a Buchner funnel using a vacuum. Wash the resin with 2 L of 1.0M NaOH and 2 L of H_2O. Then wash the resin with 2 L of 1M HCl followed by 2 L of H_2O. Collect the resin from the Buchner funnel and resuspend it in 300 mL of H_2O. Store the activated, hydrated resin at room temperature.

3.7.2. Preparation of Sample

1. Locate the bands of interest on dried SDS-PAGE gels, using the autoradiogram as a template. Excise the band and place in a small Petri plate. Pipet onto the gel slice ~50 µL of H_2O. Allow the gel slice to hydrate (i.e., swell), and then remove the backing paper (*see* Notes 49,50).
2. Place gel pieces in the bottom of a 2.0-mL screw-cap tube.
3. Add 1.5–2 mL 6M HCl, and gently crush the gel slice. Screw on the top of tube, making it finger tight.
4. Hydrolyze proteins at 110°C for 1.5 h.
5. Remove the tubes and chill on ice. Pour the contents of the tubes into iced 15-mL conical centrifuge tubes. Ensure that the solution and residual gel pieces are completely transferred by rinsing the hydrolysis tube twice with 1.5 mL (each rinse) H_2O. Add the rinses to the 15-mL conical tubes.
6. Centrifuge the acrylamide residue in a desk-top centrifuge (2000 rpm, 5 min, 4°C).
7. Remove the supernatant, and place in a suitably sized tube on ice (e.g., plastic 12×75 mm tubes).
8. Add 4 mL H_2O to the acrylamide residue in 15-mL conical tubes, mix, and centrifuge (2000 rpm, 5 min, 4°C). Transfer this wash supernatant to the original supernatant.
9. Lyophilize extracts to dryness using a freeze-drier.
10. The dried residue will appear yellowish-brown. Resuspend this residue in 7.5 mL (total) H_2O containing 1 mL of each of the three phosphoamino acid standard solutions.
11. Thoroughly resuspend the activated and washed Dowex, and add 0.55 mL of the slurry to each tube.
12. Neutralize the hydrolysate by adding aliquots of 200 mM NH_4OH until the pH reaches 7–8 as determined with pH paper.
13. Make certain that the tubes are tightly capped, and mix on tilter or rotary shaker at room temperature for a least 6 h. This ensures quantitative adsorption of phosphoamino acids to ion-exchange resin.
14. After the thorough mixing, centrifuge the tubes (2000 rpm, 5 min, 4°C), to pellet the ion-exchange resin. Aspirate the supernatant, and resuspend the resin in 1 mL H_2O.

15. Pour the slurry into a small (e.g., 10 mL) disposable column.
16. Wash the resin twice with H_2O, using 1 mL for each wash.
17. Elute the phosphoamino acids from resin with three applications of 0.5 mL of 0.1 *M* HCl.
18. Centrifuge the eluate briefly to pellet insoluble material. Remove the supernatant.
19. Lyophilize the supernatant.
20. Resuspend the dried pellets in 10 µL of H_2O. Samples can be frozen at this point.

3.7.3. Separation of Phosphoaminoacids on Thin-Layer Electrophoresis

1. Apply samples to thin-layer electrophoresis (TLE) plates in 1-µL aliquots, allowing the spot to dry completely before applying next aliquot (*see* Notes 51,52).
2. After samples and standards have been applied, spray the TLE plate, lightly but evenly, with electrophoresis buffer. Do not spray around application spot. Rather, pipet buffer around these spots and allow it to diffuse (*see* Note 53).
3. Run TLE plates as follows:
    a. Orient edges of TLE plate against grid lines on electrophoresis plate. Using a pencil, mark off the positive (+) and negative (–) poles on the top and bottom left corners of the TLE plate.
    b. Fill the two wells in the electrophoresis unit with the electrophoresis running buffer. Then using wicks (obtained from Bio-Rad, Richmond, CA), place one wick in each well and allow it to wet thoroughly. Refill the wells with buffer, if necessary. Leave one end of the wick in the well, and place the other end of the wick along the edge of the TLC plate. Take care not to place the wick over the sample application spots (*see* Note 54).
    c. Run electrophoresis for 40–60 min at 950 V.
4. After electrophoresis, remove the plate and dry using a hand-held hair drier.
5. Locate the phosphoamino acid standards by spraying plate with ninhydrin (*see* Note 55).
6. After the ninhydrin has dried (leave plates overnight), locate the sample phosphoamino acids by autoradiography, and confirm their identity by comparing autoradiographic images to ninhydrin-stained known phosphoamino acid standards (*see* Note 56). *See* Fig. 10 for results of the phosphoamino acid analysis of β subunit from insulin receptor from intact N18 neuroblastoma cells.

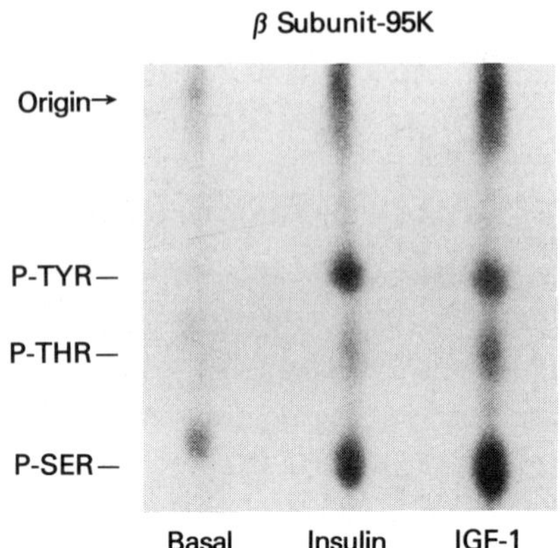

Fig. 10. Phosphoamino acid analysis of β subunit from insulin-stimulated N-18 cells. ^{32}P-labeled, insulin-stimulated intact N-18 cells were extracted and immunoprecipitated as described in Section 3.6.1. Following SDS/PAGE, bands corresponding to the 95-kDa β subunit were excised, and phosphoamino acid analysis was conducted as described in Section 3.7. As can be seen, insulin stimulated ^{32}P-incorporation into serine, threonine, and tyrosine residues (taken from ref. *31*.)

3.8. Phosphopeptide Mapping

A simple way of determining sites of phosphorylation on proteins is by phosphopeptide mapping. In this technique, one obtains a spectrum of differently sized phosphopeptides by hydrolyzing the protein using varying concentrations of a protease, such as V8 from *S. aureus*. The collection of smaller phosphopeptides is then resolved on a 15% denaturing gel *(34)*. We have used this technique (as modified from *34* and *35*) to analyze β subunits of insulin and IGF-I receptors in neuroblastoma cells *(25)*.

3.8.1. Phosphopeptide Mapping by V8 Proteolysis

1. Locate the protein bands of interest on the dried gel using the corresponding autoradiographic images.
2. Cut the gel pieces out and chop them into small pieces.
3. Soak pieces in buffer A for 16 h at 4°C.
4. Remove gel by centrifugation (12,000*g*, 10 min, 4°C). Divide supernatant into six equal aliquots, and add V8 protease (in buffer A) at 0, 0.2, 10, 20, and 200 µg/mL.
5. Incubate for 30 min at 37°C.
6. Denature in 50–100 µL sample buffer, and apply half to 15% SDS-PAGE.

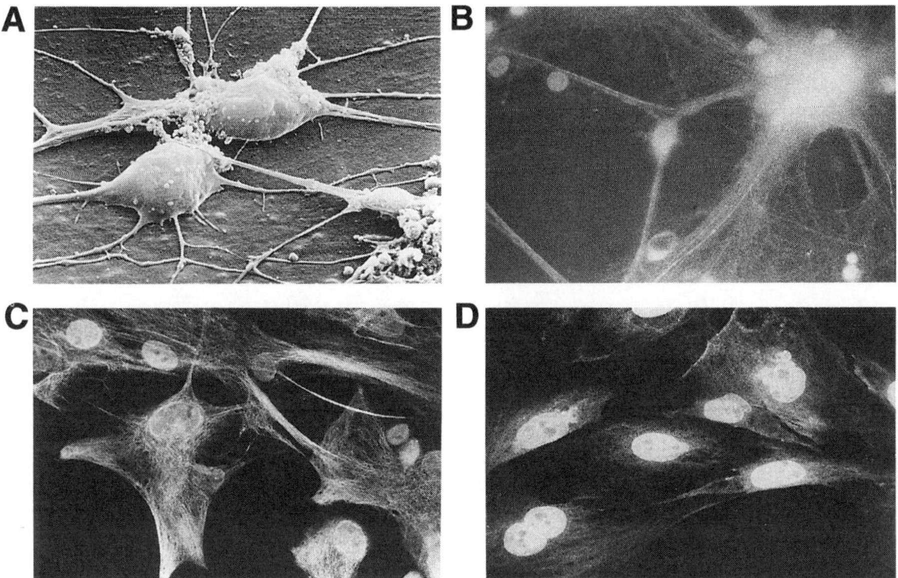

Fig. 11. Photomicrographs of neuronal and glial cells in primary culture from rat brain. **A.** Scanning electromicrograph of neuronal cells in culture. **B.** Immunofluorescent staining of neuronal cultures with antibody against neurofilament (DA_2B1). **C.** Immunofluorescent staining of glial culture from 1-d-old rat brain with antiglial fibrillary acidic protein (GFAP). **D.** Immuno-fluorescent staining of glial cultures from 21-d-old rat brain with anti-GFAP.

4. Notes

4.1. Notes on Cell Culture

1. All the procedures are conducted in a sterile hood with sterile solutions, glass-, and plasticware. All the solutions should be at 37°C before use in cell culture. The amount of time it takes from the removal of brain to placing the cells in culture is crucial. Generally, 2–3 h for 8–10 brains is appropriate. Longer times result in low recovery of total cells, and neurons in particular.
2. Figure 11 shows photomicrographs of neuronal and glial cultures. Staining with neurofilament antibodies and antiglial fibrillary acidic protein indicates the presence of 80–90% neuronal cells in neuronal and 95–98% astrocytic glial cells in glial cultures. Refer to other publications for the biochemical and pharmacological properties of these cultures *(36–40)*.

3. Identical procedures have been used to culture neurons and astrocytes from specific brain regions *(41)*, and from the brains of normotensive and genetically hypertensive rats *(42)*.

4. Studies have shown that biochemical and pharmacological alterations seen in the angiotensin-catecholamine systems of the brains of spontaneously hypertensive (SH) rats are maintained in neuronal cells in culture *(43–45)*. This indicates that the primary brain cell culture system has great potential to be used as a model system for the cellular and molecular biology of the brain.

4.2. Notes on Membrane-Binding Studies

5. Binding assays described here utilize ^{125}I-ligands. Highly purified, high specific activity monoiodinated ^{125}I-insulin and ^{125}-IGF-I are available commercially, but these can also be iodinated and purified in house *(46)*. Minimally, one needs a high specific activity (e.g., at least ~100 μCi/μg) and at least separation of free and bound ^{125}I. Investigators have to determine in their systems any additional needs (e.g., higher specific activities, monoiodinated preparations, and so on).

6. Calculate specific binding by subtracting nonspecific binding (NSB, i.e., counts remaining bound in the presence of 10,000 ng/mL unlabeled peptide) from total binding at each point. This NSB should be reasonably low, e.g., 10% of total counts. We cannot give here absolute standards on what NSB should be. This will vary with the membrane preparation, quality of tracer, degradation, and so forth. However, it will become obvious to workers that prohibitively high NSB (e.g., 40–50% of total binding) will make binding data difficult to interpret, and may indicate excessive tracer degradation and/or a poor quality of tracer. Alternatively, high NSB may be a property of the membranes, and workers may have to use lectin-purified preparations to obtain reasonable binding data. The reader is referred to ref. *46a* for further information regarding this possibility.

7. When using membrane preparations, a useful control is to determine the degradation of the ^{125}I ligand by precipitating with 10% TCA. Again, absolute allowable levels cannot be given, but in our experience, using the antiproteases and methods described, degradation is rarely >5%.

8. Investigators may want to perform preliminary experiments on time, temperature, membrane protein concentration, and pH dependence of binding. For example, we have successfully used 5-h incubations at room temperature. Overnight at 4°C is, however, a convenient time-course.

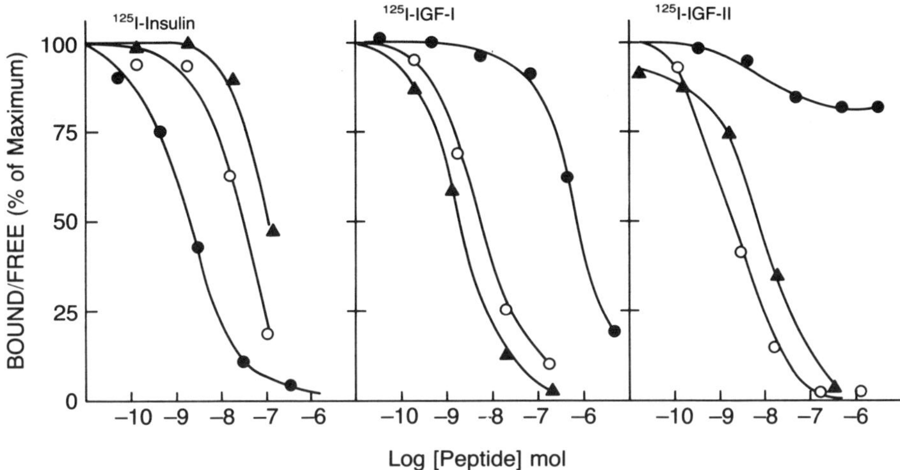

Fig. 12. Specificity of ligand binding. Competition/inhibition curves for ^{125}I-insulin, ^{125}I-IGF-I, and ^{125}I-IGF-II binding to rat brain membranes are shown. The rank order of displacement of ^{125}I-insulin by unlabeled hormones is insulin (•) >IGF-II (o) >IGF-I (▲), whereas for ^{125}I-IGF-I binding, it is IGF-I (▲) >IGF-II (o) > insulin (•), and for ^{125}I-IGF-II, it is IGF-II (o) >IGF-I (▲) >>insulin (•). Thus, such curves, whose generation is described in Sections 3.2. and 3.3., indicate the specificity of binding. Additionally, the IC50s are an index of relative affinity. For example, the IC50 for ^{125}I-insulin binding inhibition by unlabeled insulin is ~4 n*M*, whereas the IC50 for unlabeled IGF-I inhibition of ^{125}I-IGF-I binding is also ~4 n*M* (adapted from ref. *47*).

9. Binding data may be analyzed in a number of ways. A plot of percent inhibition of binding vs unlabeled ligand concentration (Fig. 12; ref. *47*) allows determination of the concentration of unlabeled ligand producing 50 inhibition of specific binding (IC_{50}), which is taken as an index of affinity. When comparing different groups of receptors, if the IC_{50}s are similar, then the percentage of maximal bound may be regarded as an index of receptor number. For more rigorous analysis, investigators may use Scatchard plots *(48)*. Scatchard plots for insulin are generally curvilinear, whereas for IGF-I, they are generally linear.

10. Occasionally, competition/inhibition plots show an apparent increase in specific tracer binding at low concentrations of competitor (so-called "hook phenomenon"; *see,* e.g., ref. *49*). Investigators should be aware of the possibility of tracer degradation (or in the case of IGF-I, binding proteins) as an explanation for such results.

11. Repeated thawing and refreezing of membranes is not recommended. Membrane preparations should be frozen in small aliquots at −70°C and each aliquot used once. Membranes are stable for at least 6 mo.

12. Use of antiproteases is straightforward, except for PMSF. This material should be prepared fresh daily as a stock solution of 0.2M in 2-propanol, and then diluted into the buffer to the desired concentration immediately before use.

13. After scraping cells and washing in PBS, the cell pellets may be stored at −70°C prior to use.

14. In addition to neuronal and glial cells, these methods can be applied to transformed neuroblastoma cell lines *(5,6)*, synaptosomes from adult rat brain *(7)*, and mechanically dissociated cells from rat brain *(8)*.

4.3. Notes on Receptor Solubilization and Partial Purification

15. The wash protocol is designed to "clean" the columns after use so they may be reused.

16. We have found the source of WGA to be critical. In our hands, ICN Biomedicals is an outstanding source.

17. The same caveats as described for membrane-binding studies also apply to solubilized receptor preparations as regards the sources and preparations of unlabeled and ^{125}I-ligands.

18. In initial experiments, the ^{125}I-ligand binding and protein concentration of the solubilized receptors prior to application, as well as the column flow-through and each of the column elution fractions, should be determined. Such "bookkeeping" also helps to ensure that columns are not being overloaded with solubilized protein. Once the method and results have become standardized, one can collect a pool of the peak fractions of receptor-binding activity. Protein concentration of solubilized and WGA purified receptors should be determined using the Coomassie blue dye binding assay of Bradford *(50)*.

19. As with membrane-binding assays, it is advisable to initially determine time, temperature, and pH dependence of binding by lectin-purified receptors. Although degradation of tracer by WGA purified receptors is expected to be less of a problem than with membranes, it may be assessed by the TCA precipitability test.

20. Calculation of specific binding and analysis of binding data are done as described earlier for membrane-binding studies. Nonspecific binding should be low (e.g., 10% of total counts), and investigators should be aware that increases in NSB may reflect tracer degradation and/or poor preparations of tracer.

21. In addition to membrane binding and solubilized lectin purified receptor binding, binding assays can be conducted using intact cells in culture *(38,39)*. This has the advantage of allowing determination of cell surface receptors.

22. WGA purified receptor preparations should be stored at –70°C for maximal stability. In our experience, thawing and refreezing does not deleteriously affect binding, but it is best to store in multiple small aliquots that will be used once. WGA preparations retain their stability at least 6 mo and probably longer.

4.4. Notes on Structural Studies

23. Affinity crosslinking can be carried out on WGA receptor preparations. Use the WGA purified receptor binding protocol, and omit the postcrosslinking solubilization step.

24. Many of the procedures described in this article require separation of proteins on SDS-PAGE. This is such a common technique that we have not included a separate section, but refer the reader to the original reference *(51)* for a description of sample buffer, gel components, and so on. For proteins in the size range of insulin receptor subunits, 7.5% resolving gels is recommended, along with 5% stacking gels. Mol-wt markers should be the high relative molecular mass proteins with a range 45–200 kDa.

25. Often, it is desirable to immunoprecipitate receptors. We have used antiinsulin receptor antibody B10 to immunoprecipitate insulin receptors from brain tissue and neurally derived cells *(3,6,21)*. If immunoprecipitation is desired, then after 3.4.1., step 4, instead of solubilizing in sample buffer, spin down the membranes, wash, and solubilize with 1 mL 50 mM Hepes, 10 mM MgSO$_4$, pH 7.6, 1% Triton X-100, and 1 mM PMSF. Mix for at least 1 h at 4°C, centrifuge (12,000g, 30 min, 4°C), and recover supernatant. Add antibody (at 1:50–1:100 dilutions), incubate overnight at 4°C, and precipitate immune complexes with 100–200 µL 20% pansorbin. Wash the immunoprecipitates two to three times with 50 mM Hepes and 0.1% Triton X-100, solubilize pellets in sample buffer (e.g., 150 µL), and apply to SDS-PAGE. Receptors may be eluted from pansorbin by boiling in buffer D, centrifuging, and recovering the supernatant, if subsequent enzymatic digestions (*see* Section 3.4.2.) are desired. Antiinsulin antibodies may also be used to immunoprecipitate crosslinked receptors. Investigators should be aware that antibodies specific for rat IGF-I receptors have not yet been reported. However, we have used α IR-3 to immunoprecipitate IGF-I receptors from human neuroblastomas *(5)*.

26. The volume of sample buffer used may be reduced in order to add crosslinked receptors to gels in greater concentrations.
27. Endo F digestion can be carried out on membranes after crosslinking, but without prior immunoprecipitation. *See (17)* for method.
28. Investigators should be aware that, when using neuronal cells, the lower M_r insulin and IGF-I receptor α subunits are seen, whereas in glial cells, the size of α subunits is similar to those seen in peripheral tissues *(3,4,52)*.

4.5. Notes on Phosphorylation in a Cell-Free System

29. When comparing different receptor preparations in autophosphorylation and tyrosine kinase assays, equal numbers of receptors are added to the reactions by diluting the preparations to similar levels of specific tracer binding/unit vol. These levels of binding are calculated as the ratio of bound to free ^{125}I-ligand (B/F ratio). This method of adding equivalent numbers of receptors requires that the affinities (i.e., IC_{50}s) be the same.
30. For insulin and IGF-I receptors, cell-free autophosphorylation and tyrosine kinase require lectin column purification. Solubilization alone is not sufficient.
31. For both autophosphorylation and tyrosine kinase assays, more receptors can be added to the reaction simply by omitting the 20 µL of 75 mM Hepes. For autophosphorylation, the 50 mM Hepes can also be omitted, allowing addition of up to 60 µL, instead of 20 µL, of receptors.
32. The ATP, CTP, and poly (Gly,Tyr) 4:1 solution should be prepared freshly.
33. Mn^{2+} and Mg^{2+} are not interchangeable in the autophosphorylation and tyrosine kinase assays, although we have added both to autophosphorylation and tyrosine kinase assays *(21,27,28)*.
34. Sodium orthovanadate (at 1 mM final concentration) may be added as an inhibitor of tyrosine phosphatases *(21,27,28)*, but it may increase basal kinase activity.
35. Time-course studies should be performed. As with all enzymatic assays, tyrosine kinase assays should be conducted over a linear time-course. Receptor autophosphorylation does not have to be analyzed during the linear time-course and, in fact, is usually analyzed when maximal, i.e., at 5–10 min.
36. Immunoprecipitation of autophosphorylated receptors prior to SDS-PAGE is often desirable. One can increase the amount of ^{32}P-ATP used and reduce the stock ATP to 0.5 mM to increase the amount of ^{32}P-incorporated into β subunits. Stop autophosphorylation with 1/3

vol of stop solution, and add antibody (antireceptor or antiphospho-tyrosine antibodies) at 1:50–1:100. Incubate overnight at 4°C, and precipitate the immune complexes with 20% pansorbin (use, e.g., 100–200 µL). After 1 h at 4°C, centrifuge (12,000g, 5 min, 4°C), wash three times with 50 mM Hepes, pH 7.6, 0.1% Triton X-100 and dissolve samples in 150 µL (or a suitable smaller vol) of sample buffer. Recover supernatant, denature, and apply to SDS-PAGE.

37. For tyrosine kinase assays, other exogenous substrates, such as casein, may be used (*see 24* for review). Analyze products on SDS-PAGE.

38. For tyrosine kinase assays using poly(Glu,Tyr)4:1, filters are washed in ~500 mL TCA/10 mM pyrophosphate with ~5 changes. Each wash should be for several hours, and one of them is usually overnight.

39. To quantitate autophosphorylation, one may perform densitometry on the autoradiograms or, alternatively, excise the gel pieces and count by Cerenkov or liquid scintillation counting.

40. To quantitate poly(Glu,Tyr) 4:1 phosphorylation, subtract counts from endogenous tubes (i.e., counts on filter paper in the absence of exogenous substrate) and convert specific counts to mole ^{32}P incorporated. Endogenous counts should be low (e.g., 10–40% of maximal). Maximal ^{32}P incorporation usually occurs at $10^{-7}M$ ligand. If desired, a background tube may be assayed that omits receptors, but contains all other reagents. This tube is basically a check on the efficiency of filter washing.

41. Glycosidase digestion of gel slices containing phosphorylated proteins (e.g., β subunit) can be carried out. *See* references (*21*) and (*53*), for detailed methods.

42. We have had one case, that of chicken liver, where a very active ATPase copurified with insulin receptors on WGA columns (*27,28*). In this case, sephacryl or pea-lectin columns were used to separate the activities and allow study of insulin receptors. The ATPase activity was detected by the lack of a stable time-course in kinase assays. Investigators should be aware of this possibility, and should also be aware that other lectins (e.g., lentil, ricin) may be used to purify insulin and IGF-I receptors with unusual carbohydrate moieties (*10*).

4.6. Notes on Phosphorylation in Intact Cells

43. High levels of ^{32}P are employed; therefore, use extreme care. Investigators may discover that their systems allow use of smaller amounts of [32]P.

44. [32]P incorporation usually occurs rapidly (e.g., by 1 min). The media listed for cell culture and metabolic labeling are based on our experience. Investigators may wish to determine if other growth media are

more appropriate, or if hormones need to be added in serum-free growth media containing 0.1% BSA. There are really no "stop points" in the procedure, since phosphoproteins are labile, and especially in whole cell extracts, subject to phosphatase action.

45. Immunoprecipitation must be performed. Direct application of extracts, even after WGA chromatography, gives uninterpretable gels. We have used both antireceptor and antiphosphotyrosine antibodies. As an alternative to PNPP elution, immunoprecipitates may be eluted with sample buffer. Regardless of the antibody used, immunoprecipitations should not exceed 2 h, except in the case of WGA purified extracts, where immunoprecipitation can go overnight. Another exception to this is when antireceptor antibodies are used to reimmunoprecipitate PNPP eluates. In this case, the second immunoprecipitation may occur overnight.

46. Thorough washing of immunoprecipitates cannot be overemphasized. Each wash should thoroughly disrupt and "homogenize" the pansorbin pellet.

47. We generally prepare all solutions freshly, although the individual components (excepting PMSF) may be prepared in advance as stock solutions. Prepare PNPP freshly.

4.7. Notes on Phosphoamino Acid Analysis

48. Phosphoamino acid analysis has been performed on in vivo and in vitro labeled phosphoproteins *(4,31)*. For in vitro phosphorylations, phosphoamino acid analysis has been performed on directly applied as well as immunoprecipitated preparations *(4)*.

49. It is sometimes necessary to pool gel slices from replicate, identical lanes in order to have enough starting material so that a signal is obtained at the end of the procedure.

50. Gel slices from wet gels may be used *(4)*. In this case, simply delete the gel piece hydration described in step 1.

51. When applying samples to TLE plates, it is best if spots are separated by 3 cm from the top and side edges of the plate and from each other. A very light pencil mark may be used to identify these separations.

52. When applying samples or standards in multiple aliquots, use a hair drier to dry the spot completely prior to addition of the next aliquot.

53. Wetting the plate with buffer prior to the TLE is critical. To ensure even and light wetting, use a spray pump bottle to apply buffer.

54. When applying wicks to the edges of the TLE plate, cover as little of the edge as evenly as possible, and try not to have excess fluid.

55. As with wetting the plate, use spray pump to stain gels with the ninhydrin solution, and dry using the hair drier.

56. In addition to adding phosphoamino acid standards to samples, it is useful to run lanes that contain only the standards. There should be one lane for the mixture of all three, as well as three separate lanes for each standard alone. Mobility of the phosphoamino acids should be: origin, P-Tyr, P-Thr, P-Ser.

Acknowledgments

The technical assistance of Elizabeth Albert in preparation of neuronal and glial cells in primary culture is gratefully acknowledged. The research is supported by grants from the Washington, D.C. Affiliate of the American Diabetes Association (D. LR.), Diabetes Research and Education Foundation (D. LR.), and the National Institutes of Health (M. K. R.) and the American Heart Association-Florida Affiliate (M. K. R.). Mohan K. Raizada is an established investigator of the American Heart Association. Martin L. Adamo is the recipient of a Juvenile Diabetes Foundation Postdoctoral Fellowship.

References

1. Adamo, M., Raizada, M. K., and LeRoith, D. (1989) Insulin and insulin-like growth factor receptors in the nervous system. *Mol. Neurobiol.* **3,** 71–100.
2. Havrankova, J., Roth, J., and Brownstein, M. J. (1978) Insulin receptors are widely distributed in the central nervous system of the rat. *Nature* **272,** 827–829.
3. Lowe, W. L. Jr., Boyd, F. T., Clarke, D. W., Raizada, M. K., Hart, C., and LeRoith, D. (1986) Development of brain -insulin receptors: structural and functional studies of insulin receptors from whole brain and primary cell cultures. *Endocrinology* **119,** 25–35.
4. Shemer, J., Raizada, M. K., Masters, B. A., Ota, A., and LeRoith, D. (1987) Insulin-like growth factor receptors in neuronal and glial cells. Characterization and biological effects in primary culture. *J. Biol. Chem.* **262,** 7693–7699.
5. Ota, A., Wilson, G. L., Spilberg, O., Pruss, R., and LeRoith, D. (1988) Functional insulin-like growth factor I receptors are expressed by neural-derived continuous cell lines. *Endocrinology* **122,** 145–152.
6. Ota, A., Shemer, J., Pruss, R. M., Lowe, W. L., Jr., and LeRoith, D. (1988) Characterization of the altered oligosaccharide composition of the insulin receptors on neural derived cells. *Brain Res.* **43,** 1–11.
7. Raizada, M. K., Shemer, J., Judkins, J. H., Clarke, D. W., Masters, B. A., and LeRoith, D. (1988) Insulin receptors in the brain: structural and physiological characterization. *Neurochem. Res.* **13,** 297–303.
8. Masters, B. A., Shemer, J., Judkins, J. H., Clarke, D. W., LeRoith, D., and Raizada, M. K. (1987) Insulin receptors and insulin action in dissociated brain cells. *Brain Res.* **417,** 247–256.
9. Lowry, O. H., Rosebrough, N. J., Farr, A. L., and Randall, R. J. (1951) Protein measurement with Folin-Phenol reagent. *J. Biol. Chem.* **93,** 265–275.

10. Hedo, J. A., Harrison, L. C., and Roth, J. (1981) Binding of insulin receptors to lectins: evidence for a common carbohydrate determinants on several membrane receptors. *Biochemistry* **20**, 3385–3390.

11. Harrison, L. C. and Itin, A. (1980) Purification of the insulin receptors from human placenta by chromatography on immobilized wheat germ lectin and receptor antibody. *J. Biol. Chem.* **255**, 12,066–12,072.

11a. Morgan, D. O., Edman, J. C., Standring, D. N., Fried, V. A., Smith, M. C., Roth, R. A., and Rutter, W. J. (1987) Insulin-like growth factor II receptor as a multifunctional binding protein. *Nature* **329**, 301–307.

12. Pilch, P. F. and Czech, M. P. (1979) Interaction of cross-linking reagents with the insulin effector system of isolated fat cells. Covalent linkage of ^{125}I-insulin to a plasma membrane receptor protein of 140,000 daltons. *J. Biol. Chem.* **254**, 3375–3381.

13. Tarentino, A. L., Plummer, T. H., Jr., and Maley, F. J. (1974) The release of intact oligosaccharides from specific glycoproteins by endo-B-*N*-acetyl glucoseaminidase H. *J. Biol. Chem.* **249**, 818–824.

14. Elder, J. H. and Alexander, S. (1982) Endo-B-*N*-acetylglucoseaminidase F: endoglycosidase from *Falvobacterium meningosepticum* that cleaves both high mannose and complex glycoproteins. *Proc. Natl. Acad. Sci. USA* **79**, 4540–4544.

15. Yip, C. C., Moule, M. L., and Yeung, C. W. T. (1980) Characterization of insulin receptor subunits in brain and other tissues by photoaffinity labeling. *Biochem. Biophys. Res. Commun.* **96**, 1671–1678.

16. Heidenreich, K. A., Zahniser, N. R., Berhanu, P., Brandenburg, D., and Olefsky, J. M. (1983) Structural differences between insulin receptors in the brain and peripheral target tissue. *J. Biol. Chem.* **258**, 8527–8530.

17. Hendricks, S. A., Agardh, C.-D., Taylor, S. I., and Roth, J. (1984) Unique features of the insulin receptors in rat brain. *J. Neurochem.* **43**, 1302–1309.

18. Heidenreich, K. A., Freidenberg, G. R., Figlewicz, D. P., and Gilmore, P. R. (1986) Evidence for a subtype of insulin-like growth factor I receptor in brain. *Regul. Peptides* **15**, 301–310.

19. Hedo, J. A., Kahn, C. R., Hayashi, M., Yamada, K. M., and Kasuga, M. (1983) Biosynthesis and glycosylation of the insulin receptor. *J. Biol. Chem.* **258**, 10,020–10,026.

20. Rosenzweig, S. A., Madison, L. D., and Jamieson, J. D. (1984) Analysis of cholecystokinin-binding proteins using endo-B-*N*-acetylglucoseaminidase F. *J. Cell. Biol.* **99**, 1110–1116.

21. Lowe, W. L., Jr. and LeRoith, D. (1986) Insulin receptors from guinea pig liver and brain: structural and functional studies. *Endocrinology* **118**, 1669–1677.

22. Heidenreich, K. A. and Brandenberg, D. (1986) Oligosaccharide heterogeniety of insulin receptors. Comparison of N-linked glycosylation of insulin receptors in adipocytes and brain. *Endocrinology* **118**, 1835–1842.

23. McElduff, A., Poronnik, P., Baxter, R. C., and Williams, P. (1988) A comparison of the insulin and insulin-like growth factor I receptors from rat brain and liver. *Endocrinology* **122**, 1933–1939.

24. Zick, Y. (1989) The insulin receptor: structure and function. *Crit. Rev. Biochem. Molec. Biol.* **24,** 217–269.
25. Ota, A., Wilson, G. L., and LeRoith, D. (1989b) Insulin-like growth factor I receptors on mouse neuroblastoma cells. Two beta-subunits are derived from differences in glycosylation. *Eur. J. Biochem.* **174,** 521–530.
26. Rees-Jones, R. W., Hendricks, S. A., Quarum, M., and Roth, J. (1984) The insulin receptors of rat brain are coupled to tyrosine kinase activity. *J. Biol. Chem.* **174,** 521–530.
27. Simon, J. and LeRoith, D. (1986) Insulin receptors of chicken liver and brain. Characterization of alpha and beta subunit properties. *Eur. J. Biochem.* **158,** 125–132.
28. Simon, J., Rosebrough, R. W., McMurtry, J. P., Steele, N. C., Roth, J., Adamo, M., and LeRoith, D. (1986). Fasting and refeeding alter the insulin receptor tyrosine kinase in chicken liver but fail to affect brain insulin receptors. *J. Biol. Chem.* **261,** 17,081–17,088.
29. White, M. F., Maron, R., and Kahn, C. R. (1985) Insulin rapidly stimulates tyrosine phosphorylation of a Mr185,000 protein in intact cells. *Nature* **318,** 183–185.
30. Kadowaki, T., Koyasu, S., Nishida, E., Tobe, K., Izumi, T., Takuku, F., Sakai, H., Yahara, I., and Kasuga, M. (1987) Tyrosine phosphorylation of common and specific sets of cellular proteins rapidly induced by insulin, insulin-like growth factor I and epidermal growth factor in an intact cell. *J. Biol. Chem.* **262,** 7342–7350.
31. Shemer, J., Adamo, M., Wilson, G. L., Heffez, D., Zick, Y., and LeRoith, D., (1987) Insulin and insulin-like growth factor-I stimulate a common endogenous phosphoprotein substrate (pp 185) in intact neuroblastoma cells. *J. Biol. Chem.* **262,** 15,476–15,482.
32. Shemer, J., Adamo, M., Raizada, M. K., Heffez, D., Zick, Y., and LeRoith, D. (1989) Insulin and IGF-I stimulate phosphorylation of their respective receptors in intact neuronal and glial cells in primary culture. *J. Mol. Neurosci.* **1,** 3–8.
33. Cooper, J. A., Sefton, B. M., and Hunter, T. (1983) Detection and quantification of phosphotyrosine in proteins. *Meth. Enzymol.* **99,** 387–402.
34. Cleveland, D. W., Fischer, S. G., Kirschner, M. W., and Laemmli, U. K. (1977) Peptide mapping by limited proteolysis in sodium dodecyl sulfate and analysis by gel electrophoresis. *J. Biol. Chem.* **252,** 1102–1106.
35. Lasky, S. R., Jacobs, B. L., and Samuel, C. E. (1982) Mechanism of interferon action. Characterization of sites of phosphorylation on the interferon-induced phosphoprotein P1. *J. Biol. Chem.* **257,** 11,087-11,093.
36. Raizada, M. K. (1983) Insulin immunoreactivity in neurons of primary cultures from rat brain. *Exp. Cell. Res.* **143,** 351–357.
37. Sumners, C., Phillips, M. I., and Raizada, M. K. (1983) Rat brain cells in primary culture: visualization and measurement of catecholamines. *Brain Res.* **264,** 267–275.
38. Clarke, D. W., Boyd, F. T., Kappy, M. S., and Raizada, M. K. (1984) Insulin binds and stimulates 2-deoxy-D-glycose uptake in cultured glial cells from rat brain. *J. Biol. Chem.* **259,** 11,672–11,675.

39. Boyd, F. T., Clarke, D. W., Muther, T. F., and Raizada, M. K. (1985) Insulin receptors and insulin modulation of norepinephrine uptake in neuronal cultures from rat brain. *J. Biol. Chem.* **260,** 15,880–15,884.

40. Raizada, M. K., Phillips, M. I., Crews, F. T., and Sumners, C. (1987) Distinct angiotensin II receptors in primary culture of glial cells from rat brain. *Proc. Natl. Acad. Sci. USA* **84,** 4655–4659.

41. Sumners, C. and Raizada, M. K. (1986) Angiotensin II stimulates norepinephrine uptake in hypothalamic/brain stem neuronal cultures. *Am. J. Physiol.* **250,** C236–C244.

42. Raizada, M. K., Muther, T. F., and Sumners, C. (1984) Increased angiotensin II specific receptors in neuronal cultures of spontaneously hypertensive rat brain. *Am. J. Physiol.* **247,** C364–C372.

43. Sumners, C., Muther, T. F., and Raizada, M. K. (1985) Altered norepinephrine uptake in neuronal cultures from spontaneously hypertensive rat brain. *Am. J. Physiol.* **248,** C488–C497.

44. Feldstein, J. B., Gonzales, R. A., Baker, M. S. P., Sumners, C., Crews, F. T., and Raizada, M. K. (1986) Decreased alpha-adrenergic receptor mediated inositide hydrolysis in neurons from hypertensive rat brain. *Am. J. Physiol.* **251,** C230–C237.

45. Raizada, M. K. and Sumners, C. (1989) Lack of alpha-adrenergic receptors mediated down-regulation of Angiotensin II receptors in neuronal cultures of spontaneously hypertensive rat brain. *Mol. Cell. Biochem.* **91,** 111–115.

46. Roth, J. (1975) Methods for assessing immunological and biologic properties of iodinated peptide hormones. *Methods Enzymol.* **37,** 223–233.

46a.Adamo, M., Simon, J., Rosebrough, R. W., McMurtry, J. P., Steele, N. C., and LeRoith, O. (1987) Characterization of the chicken muscle insulin receptor. *Gen. Comp. Endocrinol.* **68,** 456–465.

47. Gammeltoft, S., Haselbacher, G. K., Humbel, R. E., Fehlmann, M., and Van Obberghen, E. (1985) Two types of receptor for insulin-like growth factors in mammalian brain. *EMBO J.* **4,** 3407–3412.

48. Scatchard, G. (1949) The attractions of protein for small molecules and ions. *Ann. NY Acad. Sci.* **51,** 660–672.

49. Lowe, W. L., Jr., Adamo, M., Werner, H., Roberts, C. T., Jr., and LeRoith, D. (1989) Regulation by fasting of rat insulin-like growth factor I and its receptor. Effects on gene expression and binding. *J. Clin. Invest.* **84,** 619–626.

50. Bradford, M. (1976) A rapid and sensitive method for the quantification of microgram quantities of protein utilizing the principle of protein-dye binding. *Anal. Biochem.* **72,** 248–254.

51. Laemmli, U. K. (1970) Cleavage of structural proteins during assembly of the head of bacteriphage T4. *Nature* **227,** 680–685.

52. Burgess, S. K., Jacobs, S., Cuatrecasas, P., and Sahyoun, N. (1987) Characterization of a neuronal subtype of insulin-like growth factor I receptor. *J. Biol. Chem.* **262,** 1618–1622.

53. McElduff, A., Watkinson, A., Hedo, J. A., and Gorden, P. (1986) Characterization of the N-linked high-mannose oligosaccharides of the insulin proreceptor and mature insulin receptor subunits. *Biochem. J.* **239,** 679–683.

CHAPTER 15

Use of Affinity Chromatography in Purification of A_1 Adenosine Receptors from Rat Brain Membranes

Hiroyasu Nakata

1. Introduction

Purification of receptor proteins has always been an important and challenging task in advancing the structural characterization and also in providing sequence data for the molecular cloning of the receptors. Some of the difficulties in the purification of receptors are:

1. Low concentrations of receptors in the tissue;
2. Solubilization of intact receptors from cell membranes; and
3. Development of an efficient affinity chromatography system.

Because there are many differences in the properties of receptors and the tissues that contain the receptors, there are no general methods for the purification, and we have to develop a specific purification method for each receptor. However, the basic sequence of the purification method for A_1 adenosine receptor from rat brain membranes described in this chapter, i.e., solubilization with mild detergents, affinity chromatography using antagonists as an immobilizing ligand, hydroxylapatite chromatography, and finally, reaffinity chromatography may be applicable to the purification of other receptors.

Adenosine, which is known to modulate varous physiological activities of many tissues and cell types, is considered to be an important endogenous modulator in central and peripheral nervous systems.

From: *Methods in Molecular Biology, Vol. 13: Protocols in Molecular Neurobiology*
Edited by: A. Longstaff and P. Revest Copyright © 1992 The Humana Press, Totowa, NJ

Most of these actions are believed to be mediated via adenosine receptors, which are present on external cell membranes. The adenosine receptors are one of the receptors that are coupled with adenylate cyclase via G protein(s). They are usually classified as A_1 and A_2. A_1 adenosine receptors are coupled to the inhibition of adenylate cyclase, and A_2 adenosine receptors are coupled to the stimulation of adenylate cyclase. The purification of A_1 adenosine receptors has been hampered by its low concentration and the lack of an efficient affinity chromatography method. Recently, two affinity adsorbates were developed for the purification of A_1 adenosine receptors. One employs xanthine amine congener (XAC), an A_1 adenosine receptor antagonist, and the other uses N^6-aminobenzyladenosine, an A_1 adenosine receptor agonist, as immobilized ligands *(1,2)*. It is of interest to note that purification of A_1 adenosine receptors by the agonist-coupled affinity gel gives a highly purified receptor preparation that is still coupled with G proteins, whereas the purification of A_1 adenosine receptor with affinity chromatography using the antagonist-coupled agarose gel yields a receptor no longer coupled with G proteins. This suggests a different mode of binding of receptors with agonists and antagonists.

The purification procedures described here were used successfully for obtaining more than 300 pmol of completely purified rat brain A_1 adenosine receptor from one series of experiments, and have also proved to be useful for the complete purification of the rat testicular and human brain A_1 adenosine receptors *(3,4)* (*see* Notes 1 and 2).

2. Materials

1. Rat brain: Whole rat brains (Sprague-Dawley, 3–6 mo, male and female) are quickly frozen after the decapitation and stored at −50°C until use (1 yr maximum).
2. Homogenizing buffer: 50 mM Tris-acetate, pH 7.2, 1 mM EDTA, 1 mM phenylmethanesulfonyl fluoride (PMSF). PMSF should be added just before use.
3. Solubilization buffer: 1% digitonin, 0.1% sodium cholate, 50 mM Tris-acetate, pH 7.2, 100 mM NaCl, 1 mM EDTA, 5 mM MgCl$_2$, 1 mM dithiothreitol, 0.1 mM PMSF, and 1 μg/mL each of pepstatin A, leupeptin, chymostatin, and antipain. The stock cocktail of protease inhibitors should be added freshly just before use (*see* Notes 3–5).

4. Buffer A: 50 mM Tris-acetate, pH 7.2, 100 mM NaCl, 1 mM EDTA, and 0.1% digitonin; filter through a 0.20-µm filter; may be stored at 4°C for 1 d.

5. Stock protease inhibitors: 200 mM PMSF is dissolved in ethanol. A cocktail of pepstatin A, leupeptin, chymostatin, and antipain is made in water at 5 mg/mL each. Stored at –20°C for 1 mo.

6. Phosphate buffers:
   a. 10 mM phosphate buffer: 10 mM potassium phosphate, pH 7.0, 100 mM NaCl, and 0.1% digitonin.
   b. 110 mM phosphate buffer: 110 mM potassium phosphate, pH 7.0, 100 mM NaCl, and 0.1% digitonin.
   c. 500 mM phosphate buffer: 500 mM potassium phosphate, pH 7.0, 100 mM NaCl, and 0.1% digitonin.

   These buffers are filtered through 0.20-µm filter and may be stored at 4°C for 1 d.

7. Elution buffer: 100 µM 8-cylopentyltheophylline (CPT), 50 mM Tris-acetate, pH 7.2, 1 mM EDTA, and 0.1% digitonin; filter through a 0.20-µm filter and use immediately. Stock CPT solution (10 mM) dissolved in dimethylsulfoxide (DMSO) is added to make 100 µM just before use.

8. XAC-agarose: 100 mL of Affi-gel 10 (Bio-Rad) is washed extensively with DMSO, and the moist gel cake is immediately resuspended in 200 mL of DMSO containing 100 mg of XAC. The gel suspension is incubated at room temperature overnight with continuous rotation. The reaction is stopped by washing the gel with DMSO extensively. The gel is further washed sequentially with water, 1M NaCl, and water. The washed gel is incubated with 200 mM Tris-acetate, pH 8.0, for 24 h. Finally, after washing with water, the XAC-agarose is stored at 4°C in 0.02% NaN₃ (*see* Note 6).

9. Polyethyleneimine (PEI) solution: 0.3% PEI solution is made fresh from stock 10% PEI.

10. Stock [³H]8-cyclopentyl-1,3-dipropylxanthine (DPCPX) solution: Dilute approx 2 µL of original [³H]DPCPX with 2 or 3 mL of 0.05% CHAPS aqueous solution to make 10 nM [³H]DPCPX, and store at 4°C.

11. Binding buffer: 50 mM Tris-acetate, pH 7.2, 1 mM EDTA.

12. Stock 0.1 mM XAC solution: Make 10 nM XAC in DMSO. Dilute 10 mM XAC solution 100-fold with water, and store at –85°C for 1 mo.

13. Amido-black solution 0.1% Amido-black (Amido-schwarz 10B) dissolved in methanol:acetic acid:water (45:10:45, vol%).

14. Destaining solution: Methanol:acetic acid:water (90:2:8, vol%).

15. Extraction solution: 25 mM NaOH, 50 µM EDTA, and 50% ethanol.

3. Methods

3.1. Purification of A$_1$ Adenosine Receptors

1. All the purification procedures were performed at 4–8°C unless otherwise indicated.
2. Thaw approx 100 g of frozen brains, and homogenize with a Waring blender for 15 s in 3–5 vol of homogenizing buffer. Centrifuge the homogenate (39,000g, 20 min, 4°C), and collect the pellet. Resuspend the pellet into 10 vol of homogenizing buffer, and centrifuge as described above. Wash the resulting pellet three times by repeating the resuspension and centrifugation (*see* Note 1).
3. Resuspend the final pellet into 3 vol of homogenizing buffer containing 2 U/mL adenosine deaminase, and incubate it for 20 min at 30°C with stirring in order to remove endogenous adenosine enzymatically. Centrifuge the suspension (39,000g, 20 min, 4°C), and resuspend the pellet in 3 vol of homogenizing buffer. The final membrane suspension can be stored at –85°C for several months without a significant loss of binding activity.
4. Centrifuge the membrane preparation, and collect the pellet. Add approximately 10 vol of solubilization buffer to the pellet, and homogenize with a Polytron homogenizer for 1 min. Stir the suspension for 1 h at 4°C, and then centrifuge (46,000g, 1 h, 4°C). Collect the supernatant as the solubilized preparation, and filter through a 0.20-μm filter (Notes 2 and 3).
5. Equilibrate a XAC-agarose column (2.5 × 14 cm) with 3 column vol of buffer A.
6. Apply the solubilized preparation (1000 mL) on the XAC-agarose column at a flow rate of 50 mL/h. Save the pass-through fractions to be used as an additive to the binding assay mixture, as described in Section 3.2.2. After the application, the column is washed with 5 column vol of buffer A at the same flow rate. The absorbance at 280 nm of the eluate should be close to the baseline after the washing.
7. Apply 3 column vol of elution buffer at a flow rate of 15 mL/h. Perform the ligand binding assay of the eluates from the column, and pool the active fractions (40–50 mL).
8. Equilibrate a small hydroxylapatite column (0.5-mL vol) with 10 mL of buffer A.
9. Apply the pooled active fractions to the hydroxylapatite column at a flow rate of 20 mL/h.
10. Wash the column sequentially with 10 mL of 10 and 110 mM phosphate buffer.

11. Elute the receptor activity with 3 mL of 500 mM phosphate buffer (*see* Notes 7 and 8).
12. Dilute the eluate (3 mL) fourfold by the addition of buffer A.
13. Equilibriate a XAC-agarose column (1 × 5 cm) with 20 mL of 110 mM phosphate buffer.
14. Apply the diluted eluate from the hydroxylapatite column to the XAC-agarose column at a flow rate of 10 mL/h. Wash the column with 5–6 column vol of buffer A.
15. Elute the receptor with 2 column vol of the elution buffer at a flow rate of 10 mL/h.
16. Divide the final receptor preparation into aliquots, and store at –85°C for several weeks.
17. The results obtained from a typical purification are summarized in Table 1. The receptor is purified approx 50,000-fold with an overall yield of 4% of the initial binding in the membrane fractions (*see* Note 9).

3.2. Receptor-Binding Assay

[³H]phenylisopropyladenosine ([³H]PIA), [³H]N^6-cyclohexyladenosine ([³H]CHA), [³H]DPCPX, and [³H]XAC, which have reasonably high affinity for the A₁ adenosine receptor, are often used for the measurement of A₁ adenosine receptor binding activity. [³H]PIA and [³H]CHA are agonists, and are available with specific activities in the range of 20–40 Ci/mmol. [³H]DPCPX and [³H]XAC are antagonists, and are available with specific activities in the range of 100–170 Ci/mmol. Because antagonist binding is usually higher and more consistent than agonist binding, [³H]DPCPX may be the first choice as an A₁ adenosine receptor ligand. For the routine measurement of the binding activity, 2 nM [³H]DPCPX is used in the presence (nonspecific binding) or absence (total binding) of 2 μM XAC. For the saturation analysis, 0.2, 0.4, 0.6, 1.0, 2.5, 5.0, and 10 nM [³H]DPCPX are required.

3.2.1. Assay for Membrane Preparations

1. Place glass tubes into ice-saturated water, and add 50 μL of stock 10 nM [³H]DPCPX solution into each glass tube. Then add 150 μL of the binding buffer (total binding) or 145 μL of the binding buffer with 5 μL stock 0.1 mM XAC solution (nonspecific binding).
2. Add 50 μL of the membrane preparation (50–200 μg of protein in the binding buffer), and mix with a vortex mixer.
3. Incubate for 1 h at 25°C or overnight at 0°C, and then filter under vacuum through Whatman GF/B filters presoaked in 0.3% PEI solution for 2–10 h at room temperature.

Table 1

Purification of A_1 Adenosine Receptor from Rat Brain Membranes

Step	Total activity, pmol	Total protein, mg	Specific activity, pmol/mg	Yield, %	Purification, -fold
Membranes	3830	8700	0.44	100	1
Solubilized	1150	2800	0.41	30	0.9
XAC-agarose	460	0.418	1100	12	2500
Hydroxylapatite	300	0.030	10,000	7.8	22,700
Re-XAC-agarose	160	0.0073	21,900	4.2	49,800

4. Wash the filters three times with 3 mL of the cold binding buffer quickly. All the filtration will be finished within 15 s using a cell harvester.
5. Measure the radioactivity in each filter with a liquid scintillation counter after incubating each filter in 10 mL of a scintillation fluid suitable for aqueous samples for 10 h. Specific binding is the difference between total and nonspecific binding.

3.2.2. Assay for Solubilized Preparations

1. Place glass tubes into ice-saturated water, and add 50 µL of stock 10 nM [³H]DPCPX to each tube. Add 5 µL of 0.1 mM XAC for the nonspecific binding. Add buffer A to make the final vol 250 µL after the receptor preparation is added. Add the soluble receptor preparation (up to 195 µL) and mix.
2. If the adenosine ligand is present in the soluble receptor preparations, such as the eluate from the affinity column, it should be removed by desalting using a small Sephadex G-50 column before the binding assay. The desalting method is described below.
3. For the binding assay of the highly purified receptor preparations, such as the eluates from the first affinity chromatography or hydroxylapatite chromatography, 10 µL (approx 5 µg of protein) of pass-through fractions (Section 3.1., step 6) of the first affinity chromatography that had been heated at 80°C for 3 min, centrifuged (10,000g, 10 min, 4°C) to remove precipitates, and desalted on Sephadex G-50 columns (*see below*) are added to the incubation mixtures in order to restore the full activity of the receptor. Make the final vol 250 µL by adjusting the volume of buffer A to be added to the incubation mixture (*see* Note 10).
4. Incubate for about 10 h at 0°C (*see* Note 11).
5. Terminate the reaction by filtration under vacuum through Whatman GF/B filters pretreated with 0.3% PEI for 2–10 h at room temperature. Wash the filters three times with 3 mL of the cold binding buffer. The filtration is performed by a cell harvester.
6. Measure the radioactivity of the filters as described in Section 3.2.1., step 5.

3.3. Desalting by Sephadex G-50 Columns

1. Make columns that have dimensions of approx 0.6 × 18 cm using disposable glass 5-mL pipets. Plug the bottom with a small amount of glass wool.
2. Swell Sephadex G-50 (fine) completely in the water, and make 50% aqueous suspension.
3. Pour the gel suspension into the columns to make 3.5 mL settled gel vol (0.6 × 13.5 cm). The use of Sephadex G-50 (fine) keeps the column

from drying or cracking. The columns can be stored at least 2 mo in a cold room (4–6°C) without covers.

4. Equilibrate each column with 10 mL of buffer A.
5. Apply the samples to be desalted in a final vol of 0.5 mL onto the column. The flow-through is discarded.
6. Put 0.6 mL of buffer A into the column and discard the flow-through again.
7. Add 1 mL of buffer A into the column and collect the flow-through. This eluate has the receptor binding activity diluted twofold and should be used immediately as a receptor sample for the binding assay.
8. The column is regenerated by washing with 15 mL of binding buffer and can be reused extensively.

3.4. Protein Determination

Protein concentration of the membrane fractions or the solubilized preparations can be determined by the Bradford method *(5)* using bovine serum albumin as a standard. The total protein of the highly purified receptor preparations is determined by the Amidoschwarz method *(6)* (*see* Note 12).

1. Prepare bovine serum albumin (BSA) aqueous solutions as assay standards in the range of 0–3 µg in a final vol of 270 µL.
2. Prepare samples to be determined by dilution with water. Maximum sample vol are 270 µL. Make a proper blank that contains the same amount of buffer, but not the protein.
3. Add 30 µL of 1 *M* Tris-HCl, pH 7.5, containing 1% SDS to the samples and standards, and mix with a vortex mixer.
4. Add 60 µL of 90% trichloroacetic acid (TCA) and mix.
5. Incubate at room temperature for more than 2 min.
6. Wet an 8 × 11 cm nitrocellulose (NC) membrane filter (0.45 µm) with water, and set it into "Bio Dot" microfiltration apparatus (Bio-Rad).
7. Transfer the sample and standard solutions into the sample wells of the apparatus, and filter under vacuum.
8. Rinse the sample tubes with 300 µL of 6% TCA, pipet it into the wells, and filter.
9. Pipet 300 µL of 6% TCA into the wells and filter.
10. Remove the NC filter from the apparatus, and incubate in 0.1% Amido-black solution for 2 min with mild shaking.
11. Rinse the NC membrane filter in distilled water (about 1 L) for 30 s and then wash it under agitation for 1 min each time in three portions (about 500 mL each) of destaining solution, and finally wash it in distilled water (about 1 L) for 2 min. The background of the NC membrane filter should be colorless.

12. Remove the water on the NC filter by blotting onto a Whatman filter paper, and cut out the stained spots. Put each spot into a plastic tube.
13. Add 600 µL of extraction solution, and incubate for 10 min at room temperature.
14. Read the absorbance of the extract at 630 nm with a spectrophotometer.
15. Determine the protein concentration of the samples using the linear standard curve obtained from the BSA standards.

4. Notes

1. The whole purification is usually finished in 3 d. Moreover, the purification can be scaled up to 1.5-fold by increasing the amount of starting brains and the size of the first affinity column by 1.5-fold. Approximately 250–300 pmol of purified receptor proteins can be obtained in the scaled-up condition.
2. The ratio of digitonin to membrane proteins is important to obtain a maximum and consistent yield of the receptor at the solubilization stage.
3. Other detergents, such as CHAPS (2) and sodium cholate (7), may be used for the solubilization of A_1 adenosine receptor from rat brain membranes. However, longer ultracentrifugation times are necessary to separate the soluble fractions from the insoluble materials when CHAPS or sodium cholate is used as a detergent, and the resulting supernatant is often still turbid. A clear supernatant can be easily obtained after the centrifugation of the solubilizing mixture (46,000g, 1 h, 4°C) when digitonin is employed as a detergent.
4. NaCl at the concentration of 100 mM should be included in all the buffers to keep the receptor solubilized.
5. Digitonin in buffers precipitates after storing for several days at 4°C. Therefore, make the minimum volume of these buffers as needed, and use them as soon as possible.
6. The efficiency of the XAC-agarose gradually decreases after several uses. It is important to regenerate the gel as soon as the chromatography is finished. Regenerate the XAC-agarose by washing successively with 10 gel vol of the following buffers:
   a. 50 mM Tris-acetate, pH 7.2, containing 1M NaCl and 0.1% sodium cholate;
   b. 100 mM Tris, pH 10.5;
   c. Water;
   d. 10 mM phosphoric acid, pH 2.5;
   e. Water; and
   f. 50 mM Tris-acetate, pH 7.2, containing 100 mM NaCl and 1 mM EDTA.
7. Using hydroxylapatite chromatography, the A_1 adenosine receptor preparation is not only approx 10-fold purified, but also highly

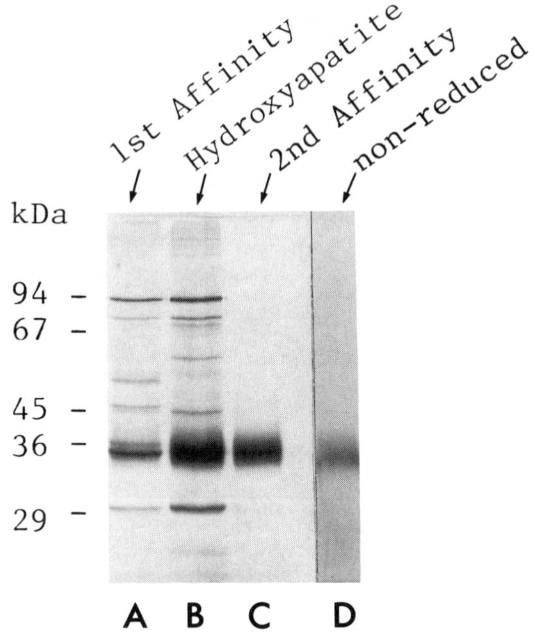

Fig. 1. SDS-Polyacrylamide gel electrophoresis of A_1 adenosine receptor preparations at various stages of the purification stained by silver. Lane A, the first XAC-agarose chromatography eluate; lane B, hydroxylapatite eluate; lane C, the second XAC-agarose eluate, lane D, the same as lane C except that the electrophoresis was performed under nonreducing conditions. Approximately 100 ng of protein were applied to each lane.

concentrated. It is also noted that the antagonist present in the receptor preparation eluted from the XAC-agarose column is removed during the hydroxylapatite chromatography, and the receptor preparation eluted from the hydroxylapatite column can be directly applied onto the second affinity column.

8. After the washing step during the first affinity chromatography, the hydroxylapatite column can be connected in series to the outflow tube of the affinity column. Then, start the elution of the XAC-agarose column by addition of 200 mL of elution buffer. After the elution, the hydroxylapatite column is separated from the XAC-agarose column, and eluted successively with 10, 110, and 500 mM phosphate buffer.

9. The final receptor preparation shows one single band on SDS-polyacrylamide gel electrophoresis as shown in Fig. 1. The protein band is broad, suggesting that the receptor protein is glycosylated.

10. The addition of pass-through fractions to the assay tubes increases the ligand binding of the highly purified receptors about fourfold and is essential for the assay of the highly purified preparations.
11. Receptor binding assays for the solublized and highly purified receptor preparations should be performed at low temperature. The binding with adenosine ligands is unstable at higher temperatures, such as 25°C *(8)*.
12. The advantage of the Amido-schwarz protein assay method is that a diluted protein solution, such as 1 μg/mL, can be analyzed. The sample volume can be increased to 1 mL, but the same volume of the standards and blanks should be prepared.

Acknowledgment

I would like to take this opportunity to thank Dr. David M. Jacobowitz for his support and encouragement during this study.

References

1. Nakata, H. (1989) Purification of A_1 adenosine receptor from rat brain membranes. *J. Biol. Chem.* **264,** 16,545–16,551.
2. Munshi, R. and Linden, J. (1989) Co-purification of A_1 adenosine receptors and guanine nucleotide-binding proteins from bovine brain. *J. Biol. Chem.* **264,** 14,853–14,859.
3. Nakata, H. (1990) A_1 adenosine receptor of rat testis membranes. Purification and partial characterization. *J. Biol. Chem.* **265,** 671–677.
4. Nakata, H. (1992) Biochemical and immunological characterization of A_1 adenosine receptors purified from human brain membranes. *Eur. J. Biochem.* in press.
5. Bradford, M. M. (1976) A rapid and sensitive method for the quantitation of microgram quantities of protein utilizing the principle of protein-dye binding. *Anal. Biochem.* **72,** 248–254.
6. Schaffner, W. and Weissman, C. (1973) A rapid, sensitive, and specific method for the determination of protein in dilute solution. *Anal. Biochem.* **56,** 502–514.
7. Nakata, H. and Fujisawa, H. (1983) Solubilization and partial characterization of adenosine binding sites from rat brainstem. *FEBS Lett.* **158,** 93–97.
8. Nakata, H. (1989) Affinity chromatography of A_1 adenosine receptors of rat brain membranes. *Mol. Pharmacol.* **35,** 780–786.

CHAPTER 16

Purification and Structure of L-Type Calcium Channels

Toni Schneider, Stefan Regulla, Wolfgang Nastainczyk, and Franz Hofmann

1. Introduction

Calcium channels are an essential part of the cellular signal transduction system, since they produce changes in cytosolic calcium. Three types of voltage-dependent calcium channels (T, L, and N channels) have been identified by electrophysiological and pharmacological techniques. L- and N-type channels have attracted intensive interest since their opening and closing is modulated by hormones, G proteins, protein kinases, toxins, and drugs. The permeation of calcium through L-type channels is blocked by several organic drugs, the calcium channel blockers (CaCB), including the dihydropyridines (DHP), phenylalkylamines, and benzothiazepines. Each class of CaCB binds to a specific site that is present on the α_1-subunit of the L-type calcium channel. The binding to these CaCB-specific sites is modulated allosterically by the membrane potential, calcium, and the occupancy of the other binding sites *(1–3)*.

A high density of CaCB-binding sites has been identified in the transverse tubules of skeletal muscle *(4)*. The physiological function of these binding sites has been clarified in so far as they may function in vivo as voltage sensors *(5–8)* and as calcium channels *(8–10)*. The purified skeletal muscle CaCB receptor contains five proteins with mol wt of 165,000 (α_1), 55,000 (β), 32,000 (γ), 135,000 (the disulfide-linked dimer of α_2), and 28,000 (δ). In vitro phosphorylation of the skeletal muscle CaCB receptor by cAMP kinase, presumably at Ser-687 of the α_1-subunit *(11,12)*, increases dramatically the open probability of the reconstituted channel *(9,13,14)*. The α_1 subunit of the CaCB receptor has been cloned from skeletal *(15)*, cardiac *(16)*, and smooth muscle *(17)*. Expression of these α_1 subunit clones in dysgenic myotubes *(7,8)*, L-cells *(10)*, and *Xenopus* oocytes *(16,17)* induces L-type calcium currents. In addition to the α_1 subunits, the primary structure of the skeletal muscle β *(18)*, γ *(19,20)*, and α_2/δ *(21,22)* subunits has been deduced. The functional role of these other subunits is unclear.

The skeletal muscle α_1 subunit contains the known binding sites for all major types of organic CaCB *(23–28)*. The DHP-binding site has been identified in the skeletal muscle CaCB receptor *(20)*. An identical site is present in skeletal, cardiac, and smooth muscle *(29)*. This chapter describes a large-scale purification *(27,29)* of the skeletal muscle CaCB receptor and methods to study the binding of two different classes of CaCBs to the pure receptor *(23–29)*. The cardiac CaCB receptor has been purified by a similar purification protocol *(30)*.

2. Materials

2.1. Purification

2.1.1. General Points

1. All buffers contain aliquots of the following protease inhibitor stock solutions:
   a. 300 m*M* iodoacetamide in water;
   b. 1*M* 1,10-phenanthroline in 100% ethanol;
   c. 100 m*M* phenyl methyl sulfonyl fluoride (PMSF) in 100% ethanol;
   d. 1 mg/mL antipain in 50% ethanol;
   e. 1 mg/mL leupeptin in 50% ethanol;
   f. 1 m*M* pepstatin in 50% ethanol; and
   g. 1*M* benzamidine in 50% ethanol.
   The stock solutions are kept at –20°C, and aliquots are added to all buffers immediately before use. The stock solutions 1 and 2–7 are diluted 1:300- and 1:1000-fold, respectively.
2. Digitonin is prepared as a 3% stock solution in water. Batches of digitonin are different in quality, and are tested for purity and solubility. The solubility is checked with a 3% solution after stirring the suspension for 24 h at 4°C. Only batches with a solubility of 90% or better should be used (*see* Note 2). All buffers that contain digitonin are filtered through cellulose acetate filter (pore size 0.45 μm).

2.1.2. Buffers

1. Buffer A (preparation of microsomes): 20 mM 3-(N-morpholino)-propanesulfonic acid (Mops)/KOH, pH 7.4 (at 4°C), 10 mM EDTA, and 300 mM sucrose.
2. Buffer B (labeling and solubilization of the CaCB receptor): 10 mM N-2-hydroxyethyl-piperazine-N-2-ethanesulfonic acid (Hepes)/KOH, pH 7.4 (at 4°C), 185 mM KCl, and 1.5 mM CaCl$_2$.
3. Buffer C (WGA-lectin chromatography, regeneration): 0.1M Tris/HCl, pH 8.5, 0.5M NaCl, and 0.05% (w/v) merthiolate.
4. Buffer D (WGA-lectin chromatography, regeneration): 0.1M sodium acetate, pH 5.0, and 0.5M NaCl.
5. Buffer E (WGA-lectin chromatography): 10 mM Hepes/KOH, pH 7.4 (at 4°C), 100 mM KCl, 1.5 mM CaCl$_2$, and 0.2% digitonin.
6. Buffer F (WGA-lectin chromatography): 10 mM Hepes/NaOH, pH 7.4, 40 mM NaCl, 1.5 mM CaCl$_2$, and 0.1% digitonin.
7. Buffer G (DEAE Ion-exchange chromatography): 10 mM Hepes/NaOH, pH 7.2 (at 4°C), 50 mM NaCl, 1.5 mM CaCl$_2$, and 0.1% digitonin.
8. Buffer H: as buffer G but containing in addition 250 mM NaCl (final concentration).
9. Buffer I (sucrose density gradient, DHP binding): 10 mM Hepes/NaOH, pH 7.4 (at 4°C), 1.5 mM CaCl$_2$, and 0.1% digitonin.
10. Buffer K (sucrose density gradient, phenylalkylamine binding): 10 mM Hepes/NaOH, pH 8.0 (at 4°C), and 0.1% digitonin.

2.1.3. Columns

1. WGA-Lectin Chromatography
   Precolumn: 10 mL (settled vol) Sepharose 4B; and
   Column: 30 mL wheat germ agglutinin (WGA) Sepharose 6 MB.
2. DEAE Ion-Exchange Chromatography
   Precolumn: TSK DEAE-5PW, 6 × 10 mm; and
   Column: TSK DEAE-5PW, 21.5 × 150 mm.

2.2. Buffers for the Identification of CaCB-Binding Sites

2.2.1. Reversible Binding of DHP

1. Buffer L: 2 mM Mops, pH 7.4 (at 4°C), 0.6 mM CaCl$_2$, 8% sucrose, and 0.05% digitonin.
2. Buffer M: 100 mM Hepes/NaOH, pH 7.4 (at 4°C), 1.2 mM CaCl$_2$, 16% sucrose, and 0.05% digitonin.
3. Solution N: 50 nM [^{3}H] (+) isradipine in 5% ethanol (70–80 cpm/fmol).
4. Buffer P: 100 mM Hepes/NaOH, pH 7.4, 30% (m/v) polyethylene-glycol (PEG) 6000.

5. Solution Q: 5 mg/mL IgG, 5 mg/mL bovine serum albumin.
6. Buffer R: 100 mM Hepes /NaOH, pH 7.4, 8.5% PEG 6000.

2.2.2. Reversible Binding of Phenylalkylamines

1. Buffer S: 30 mM Hepes/NaOH, pH 8.0 (at 4°C), 0.075% digitonin, 5 mM CaEGTA, and 5 mM EGTA.
2. Buffer T: 20 mM Hepes/NaOH, pH 7.2 (at 4°C), and 0.05% digitonin.
3. Solution U: 240 nM [^{3}H] (–)desmethoxyverapamil (20–23 cpm/fmol) in 5% ethanol.
4. Buffer V: 22.5 mM acetate/NaOH, pH 5.3, 280 mM sucrose, 5.63 mM CaCl$_2$, 5.63 mM EDTA, and 11.25% polyethyleneglycol 6000.

2.2.3. Irreversible Binding of DHP

1. UV light: 200-W lamp (260–320 nM).
2. Buffer W: 20 mM Hepes/NaOH, pH 7.4 (at 4°C), 1.5 mM CaCl$_2$, 5% sucrose, and 0. 05% digitonin.
3. Buffer X: 10 mM Hepes/NaOH, pH 7.4 (at 4°C), and 0.05% digitonin.
4. Radioligand: 400 nM [^{3}H]azidopine in 10% ethanol (48 cpm/fmol).
5. Unlabeled ligand: 200 µM (±)isradipine.

2.2.4. Irreversible Binding of Phenylalkylamines

1. Radioligand: 200 nM [^{3}H]Lu 49888 in 5% ethanol.
2. Unlabeled ligand: 200 µM (±)devapamil.

3. Methods

3.1. Purification of Skeletal Muscle CaCB Receptor

The following scheme outlines the purification steps that allow the purification of 3.0 mg pure CaCB receptor *(24,27,29)*.

	protein, mg
The back and leg muscle of two rabbits	700,000
↓	
Crude microsomes	500
↓	
Solubilized CaCB receptor	220
↓	
WGA-column eluate	4
(6× WGA eluates)	24
↓	
DEAE ion-exchange column eluate	8
↓	
Sucrose gradients	3

3.1.1. Preparation of Microsomes

Microsomal membranes are prepared from white rabbit skeletal muscle.

1. Two rabbits are sacrificed and bled. The back and leg muscles are immediately excised and kept on ice. All further steps are carried out at 4°C.
2. The connective tissue is removed, and the muscle minced with scissors. The minced muscle (700 g muscle) is homogenized in a Waring blender for 30 s at low speed and 30 s at high speed with 2100 mL buffer A. The homogenate is centrifuged (8600*g*, 10 min, 4°C). The supernatant is filtered through cheese cloth.
3. The pellet is rehomogenized in a Waring blender for 15 s at low and 15 s at high speed with 1050 mL buffer A. The homogenate is centrifuged as in step 2.
4. The supernatants (1700 mL) are combined, and solid KCl is added to a final concentration of 0.6*M*. The solution is stirred until the KCl is completely dissolved. The solution is then centrifuged (125,000*g*, 60 min, 4°C).
5. The microsomal pellet is suspended in 800 mL buffer A with a glass/ Teflon™ homogenizer by two to three up and down strokes. The microsomes are resedimented (15,000*g*, 40 min, 4°C). The pellet is resuspended in 50 mL buffer A (final protein concentration between 15– 20 mg/mL), and stored in 10-mL aliquots at –70°C.

This procedure yields 70–100 mg microsomal protein/100 g wet wt muscle containing 10.5 ± 1.9 pmol ($n = 13$) isradipine-binding sites/mg protein determined as in *(24)*. (*See* Section 3.2.1.1.)

3.1.2. Labeling and Solubilization of the CaCB Receptor

About 5% of the binding sites of the CaCB receptor are labeled by the DHP [^{3}H] (+) isradipine ([+]PN200-110) to follow the CaCB receptor during the purification.

1. Radioactive labeling is carried out immediately before solubilization of the CaCB receptor. The microsomal membranes of step 1 (500 mg) are thawed and diluted with cold buffer B containing 1 m*M* EDTA to a vol of 200 mL.
2. 200 μL of 1 μ*M* [^{3}H] (+) isradipine in 10% ethanol are added (final concentration 1 n*M*). The suspension is mixed and incubated for 90 min at 4°C.
3. The suspension is centrifuged (114,000*g*, 30 min, 4°C).
4. The resulting pellet is resuspended in 30 mL of buffer B containing 1% digitonin and disrupted in a glass/Teflon™ homogenizer by six up-and-down strokes.

5. The suspension is diluted further with buffer B containing 1% digitonin to a final vol of 200 mL.
6. After 40 min, the suspension is centrifuged (114,000*g*, 30 min, 4°C). The supernatant is retained.
7. The protein concentration and amount of bound isradipine is determined in aliquots.

3.1.3. WGA-Lectin Chromatography

3.1.3.1. COLUMN HANDLING

1. The precolumn material is renewed for each run. The precolumn and column are connected, and are equilibrated overnight with 200 mL of buffer B containing 0.3% digitonin.
2. The WGA column can be used about 20 times. Twenty runs correspond to the solubilized protein from 10 g of microsomes. The column can be used thereafter if the amount of solubilized protein loaded onto the column is decreased.
3. After each run, the WGA column is washed at room temperature with 50 mL of buffer C containing 1% digitonin. The buffer is circulated overnight through the column. Thereafter, 200 mL each of buffers C, D, and B are passed through the column at 4°C.

3.1.3.2. COLUMN CHROMATOGRAPHY

1. The supernatant from the solubilized microsomes (*see* Section 3.1.2.) is passed through a cellulose acetate filter (pore size 0.45 μm) and diluted 1.5-fold with buffer B containing 0.1% digitonin.
2. The solution is then pumped through the tandem column system at a flow rate of 3 mL/min.
3. The columns are washed with 150 mL of buffer E. The precolumn is disconnected.
4. The WGA column is washed further with 50 mL of buffer F.
5. The CaCB receptor is eluted with 100 mL of buffer F containing 0.3*M* *N*-acetyl-D-glucosamine. Fractions of 4 mL are collected, and their radioactivity and protein concentration are determined.
6. Peak fractions are pooled and stored at –70°C in 10-mL aliquots. A total of 500 mg of microsomal protein yields 4.4 mg ± 0.9 (*n* = 8) mg of partially purified CaCB receptor.

3.1.4. DEAE Ion-Exchange Chromatography

3.1.4.1. COLUMN HANDLING

1. The columns are stored in water containing 0.05% sodium azide. Before use, 60 mL of buffer G are pumped over the column at a flow rate of 2 mL/min.

2. At the end of each run, 60 mL of buffer G containing $1\,M$ NaCl (final concentration), followed by 500 mL of water containing 0.05% sodium azide are passed through the columns. The content of the precolumn is renewed when the pressure increases.
3. The columns are extensively regenerated after seven runs as recommended by the manufacturer. A total of 500 mL of the following gradients are used at a flow rate of 2 mL/min (a gradient of 1%/min), each gradient being separated by a wash with 500 mL water: $0.1–1\,M$ NaCl, water, 0–20% methanol, water, $0–6\,M$ guanidine/HCl, water, $0–0.1\,M$ acetic acid and water.

3.1.4.2. COLUMN CHROMATOGRAPHY

The DEAE-5PW column chromatography is carried out with a two-pump HPLC system.

1. The eluate of six WGA columns (24 mg protein in 200 mL) are thawed and pumped through the HPLC-DEAE columns at a flow rate of 2 mL/min.
2. The columns are washed and eluted by a salt gradient generated by the pump controller according to the scheme shown in Table 1. Fractions of 3 mL (every 1.5 min) are collected. The CaCB receptor elutes at 78 min and at a NaCl concentration of 175 mM.
3. The radioactive peak fractions, which contain 8.0 ± 1.3 ($n = 5$) mg protein, are pooled and concentrated 30–100-fold by ultrafiltration using a Centricon-30 microconcentrator.
4. The concentrated sample is diluted 10-fold with buffer I (*see* step 5) and concentrated again in order to lower the NaCl concentration below 20 mM.
5. The concentrated sample (0.3–1 mL) is either stored at –70°C or layered immediately on to the top of one to four sucrose gradients.

3.1.5. Sucrose Density Gradient

Two or four 5–20% continuous sucrose gradients are prepared by mixing 2×19.8 mL of buffer I containing 20 and 5% sucrose. The gradient is pumped through a needle into 40-mL centrifugation tubes that must be sealed tightly as recommended by the manufacturer. The concentrated and desalted CaCB receptor from step 4 (about 8 mg) is thawed, and 0.3–0.5 mL aliquots containing at most 2.5 mg of protein are layered onto individual sucrose gradients. The tubes are sealed by melting and centrifuged in a vertical rotor (242,000g, 90 min, 4°C). The gradient is pumped off from the bottom through a needle and fractionated by a fraction collector into 2-mL aliquots. The peak frac-

Table 1
Protocol for DEAE-5PW Column Chromatography

Time, min:	0	6	12	60	72	92	100	110	120	
% Buffer G	100	100	62	56	56	28	28	0	100	pump A
% Buffer H	0	0	38	44	44	72	72	100	0	pump B
Step		←————— wash —————→				←elution→		←—wash—→		

tions of protein-bound radioactivity (peak at fraction 10; tube bottom is fraction 1) are determined. The peak fractions (fractions 8–11) are pooled and stored at –70°C. These fractions contain 3 mg of purified CaCB receptor (*see* Fig. 1 and Note 1).

The purified CaCB receptor can be used directly for DHP-binding studies. However, this preparation is not optimal for the study of the phenylalkylamine-binding site, since the high calcium concentration inhibits binding to this site. For the latter purpose, the sucrose gradient buffer I is replaced by buffer K, which does not contain added calcium (*see* Note 3).

3.2. Identification of the CaCB-Binding Sites

3.2.1. Reversible Binding

3.2.1.1. THE DHP-BINDING SITE

1. The purified CaCB receptor is diluted 50-fold with buffer L to a digitonin concentration of 0.05% and a protein concentration of 1–10 μg/mL. The solubilized receptor (*see* Section 3.1.2.) and the WGA-column eluate (*see* Section 3.1.3.1.) are diluted 20-fold.
2. For a binding experiment, the following solutions are mixed in a plastic (polycarbonate or polypropylene) tube:
   - 75 μL buffer M
   - 30 μL solution N, containing variable concentrations (5–50 nM) of the radioligand
   - 15 μL 10% ethanol or 15 μL 50 μM (±) isradipine in ethanol
   - 30 μL CaCB receptor diluted in buffer L
3. The binding is started by the addition of the receptor. Nonspecific binding is determined in the presence of 5 μM unlabeled (±) isradipine in alternate tubes.
4. The incubation is terminated after 2 h at 4°C or 1 h at 20°C by adding 3 × 40 μL of the binding assay mixture to 200 μL ice-cold buffer P followed by the addition of 60 μL solution Q. The mixture is kept for 10 min on ice and is diluted further with 3.5 mL ice-cold buffer R.

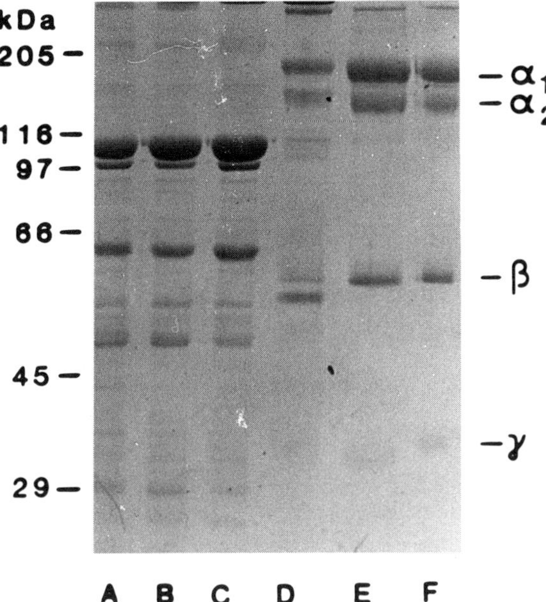

Fig. 1. SDS gel electrophoresis of each step of the skeletal muscle CaCB receptor purification. Each lane contains 15 µg protein: **A,** microsomes (Section 3.1.1.); **B,** solubilized receptor (Section 3.1.2.); **C,** flow-through fraction of WGA column (Section 3.1.3.); **D,** WGA-eluate (Section 3.1.3.); **E,** DEAE column (Section 3.1.4.); **F,** sucrose gradient (Section 3.1.5.). The gel was stained with Coomassie Blue.

5. The precipitated receptor–ligand complex is collected on a Whatman GF/C filter. The filter is washed twice with 3.5 mL ice-cold buffer R.
6. The filter is placed into minivials (5-mL) and a Triton-X-100/toluene-based scintillation cocktail is added. The retained radioactivity is determined in a scintillation counter.

3.2.1.2. The Phenylalkylamine-Binding Site

1. The purified CaCB receptor (*see* Note 3) is diluted to a digitonin concentration of 0.05% and a protein concentration of 2 µg/mL with cold buffer T.
2. For a binding experiment, the following solutions are mixed in a plastic tube:
   120 µL buffer S
   30 µL solution U containing variable concentrations (8–240 nM) of the radioligand

30 µL 5% ethanol or 30 µL 500 µM desmethoxverapamil in 5% ethanol

60 µL CaCB receptor diluted in buffer T

3. The binding is started by the addition of the receptor. Nonspecific binding is determined in the presence of 20–70 µM unlabeled desmethoxyverapamil in alternate tubes.

4. The incubation is terminated after 2 h at 4°C by adding 2 mL ice-cold buffer V followed by the addition of 50 µL solution Q. The mixture is kept for 10 min on ice.

5. The precipitated receptor–ligand complex is collected on a Whatman GF/C filter. The filter is washed twice with 4 mL ice-cold buffer V.

6. The filters are transferred to minivials, to which 4.5 mL of a Triton-X-100/toluene-based scintillation cocktail is added. The radioactivity is determined in a scintillation counter.

3.2.2. Irreversible Binding

3.2.2.1. THE DHP-BINDING SITE

1. The purified CaCB receptor is diluted with cold buffer X to a protein concentration of 80 µg/mL.

2. The receptor is incubated with the azido analog in the dark at 4°C. Use a red light when handling the azido compounds!

3. The following buffers are mixed together in a well of a titertek plate:

70 µL buffer W

35 µL radioligand (azidopine)

35 µL CaCB receptor diluted in buffer X

4. The binding is started by the addition of 35 µL CaCB receptor (80 µg/mL). Nonspecific incorporation is determined in the presence of 20 µM (±) isradipine.

5. After 90 min, the plate is placed on ice under the UV-lamp for 5 min. The plate is rotated during photolysis.

6. The photolysed samples are denatured for 15 min at 37°C in the presence of 1.2% SDS, and in the absence or presence of 250 mM 2-mercaptoethanol.

7. The peptides are separated on a 7.5% sodium dodecylsulfate polyacrylamide gel that is crosslinked by 0.3% diallyltartardiamide. The incorporated radioligand is determined either by autoradiography of a stained gel or by cutting its lanes into 2-mm slices. Individual slices are cleaved by oxidation with 0.2 mL 2% sodium metaperiodate in 4% acetic acid at pH 3.0 for 3 h at room temperature.

8. Thereafter, 4.5 mL of a Triton-X-100/toluene-based scintillation cocktail are added, and the radioactivity is determined in a scintillation counter.

3.2.2.2. THE PHENYLALKYLAMINE-BINDING SITE

The following buffers are mixed together in a well of a titertek plate:
    70 µL buffer S
    35 µL radioligand (the azido analog Lu 49888)
    35 µL CaCB receptor diluted in buffer X (*see* Note 3)
Binding and photoincorporation are carried out as described for the DHP site using the azido analog of devapamil and buffer S. Nonspecific incorporation is determined in the presence of 20 µ*M* (±)devapamil (*see* Note 4).

4. Notes

1. Figure 1 shows the purification achieved at each step. The major contaminating proteins have M_r of over 300 (ryanodine receptor?), 110 (CaATPase?), and 60 (calsequestrin?), and are removed during the WGA-lectin chromatography. The purified receptor contains five peptides. The α_1 subunit (apparent M_r 165 kDa) contains the high-affinity binding sites for dihydropyridines (isradipine), phenylalkylamines (devapamil), and benzothiazepines (diltiazem), and is thought to constitute, together with the β and the γ subunits, the ion channel. The α_1, β, and γ subunits are present in a 1:1:1 stoichiometry *(24)*, whereas the stoichiometry of the α_2/δ subunit is unclear at present *(24)*.
2. The solubility of digitonin is determined graphimetrically. An aliquot of the initial water suspension of digitonin is dried in a speed vac concentrator, and its dry wt is determined. An equal aliquot of the digitonin solution which has been stirred for 24 h, is centrifuged, dried in a speed vac concentrator, and its dry wt determined. Comparison of the two values yields the apparent solubility of digitonin.
3. The optimal binding of DHP requires 0.1–1.0 m*M* free calcium, whereas phenylalkylamine binding occurs only in the presence of micromolar calcium and is inhibited by more than 50% at 1 m*M* free calcium. The K_D values for isradipine and devapamil are somewhat lower with the purified CaCB receptor than with the membrane-bound receptor *(1,23,24,27)*. However, the stoichiometry of each binding site is between 0.7 and 0.9 sites/mol α_1 subunit. These values are obtained if the binding experiments are carried out at 4°C. At higher temperature, the binding depends on the calcium concentration, is often instable, and requires determination of the equilibrium time *(1,2,27)*.
4. The CaCB receptor and ligand concentrations given for the photoincorporation experiments result in a 1:1 stoichiometry of receptor and ligand. The described condition leads to the covalent

modification of 5–9% *(25,29)* and 3% *(29)* of the DHP- and phenyl-alkylamine-binding sites, respectively. Optimal incorporation rates are only observed if the thickness of the receptor ligand solution is kept between 1 and 2 mm during photoincorporation.

Acknowledgments

Work carried out in the laboratory of the authors has been supported by grants from Deutsche Forschungsgemeinschaft and Fonds der Chemischen Industrie.

References

1. Hofmann, F., Flockerzi, V., Nastainczyk, W., Ruth, P., and Schneider, T. (1990) The molecular structure and regulation of muscular calcium channels. *Curr. Top. Cell. Regul.* **31**, 225–239.
2. Glossmann, H. and Striessnig, J. (1988) Calcium channels. *Vitamins and Hormones* **44**, 155—328.
3. Catterall, W. A., Seagar, M. J., and Takahashi, M. (1988) Molecular properties of dihydropyridine-sensitive calcium channels in skeletal muscle. *J. Biol. Chem.* **263**, 3535–3538.
4. Fosset, M., Jaimovich, E., Delpont, E., and Lazdunski, M. (1983) [^{3}H]Nitrendipine receptors in skeletal muscle. *J. Biol. Chem.* **258**, 6086–6092.
5. Ríos, E. and Pizarró, G. (1988) Voltage sensors and calcium channels of excitation-contraction coupling. *News Physiol. Sci.* **3**, 223–227.
6. Leung, A.T., Imagawa, T., Block, B., Franzini-Armstrong, C., and Campbell, K. P. (1988) Biochemical and ultrastructural characterization of the 1,4-dihydropyridine receptor from rabbit skeletal muscle. *J. Biol. Chem.* **263**, 994–1001.
7. Adams, B. A., Tanabe, T., Mikami, A., Numa, S., and Beam, K. G. (1990) Intramembrane charge movement restored in dysgenic skeletal muscle by injection of dihydropyridine receptor cDNAs. *Nature* **346**, 569–572.
8. Tanabe, T., Beam, K. G., Adams, B. A., Niidome, T., and Numa, S. (1990) Regions of the skeletal muscle dihydropyridine receptor critical for excitation contraction coupling. *Nature* **346**, 567–569.
9. Flockerzi, V., Oeken, H.-J., Hofmann, F., Pelzer, D., Cavaliè, A., and Trautwein, W. (1986) The purified dihydropyridine binding site from skeletal muscle T-tubules is a functional calcium channel. *Nature* **323**, 66–68.
10. Perez-Reyes, E., Kim, H. S., Lacerda, A. E., Horne, W., Wei, X., Rampe, D., Campbell, K. P., Brown, A. M., and Birnbaumer, L. (1989) Induction of calcium currents by the expression of the α_1 subunit of the dihydropyridine receptor from skeletal muscle. *Nature* **340**, 233–236.
11. Röhrkasten, A., Meyer, H. E., Nastainczyk, W., Sieber, M., and Hofmann, F. (1988) cAMP-dependent protein kinase rapidly phosphorylates Ser 687 of the rabbit skeletal muscle receptor for calcium channel blockers. *J. Biol. Chem.* **263**, 15325–15329.

12. Jahn, H., Nastainczyk, W., Röhrkasten, A., Schneider, T., and Hofmann, F. (1988) Site-specific phosphorylation of the purified receptor for calcium-channel blockers by cAMP- and cGMP-dependent protein kinases, protein kinase C, calmodulin-dependent protein kinase II and casein kinase II. *Eur. J. Biochem.* **178,** 535–542.
13. Hymel, L., Striessnig, J., Glossmann H., and Schindler, H. (1988) Purified skeletal muscle 1,4-dihydropyridine receptor forms phosphorylation-dependent oligomeric calcium channels in planar bilayers. *Proc. Natl. Acad. Sci. USA* **85,** 4290–4294.
14. Nunoki, K., Florio, V., and Catterall, W. (1989) Activation of purified calcium channels by stoichiometric protein phosphorylation. *Proc. Natl. Acad. Sci. USA* **86,** 6816–6820.
15. Tanabe, T., Takeshima, H., Mikami, A., Flockerzi, V., Takahashi, H., Kangawa, K., Kojima, M., Matsuo, H., Hirose, T., and Numa, S. (1987) Primary structure of the receptor for calcium channel blockers from skeletal muscle. *Nature* **328,** 313–318.
16. Mikami, A., Imoto, K., Tanabe, T., Niidome, T., Mori, Y., Takeshima, H., Narumiya, S., and Numa, S. (1989) Primary structure and functional expression of the cardiac dihydropyridine-sensitive calcium channel. *Nature* **340,** 230–233.
17. Biel, M., Ruth, P., Bosse, E., Hullin, R., Stühmer, W., Flockerzi, V., and Hofmann, F. (1990) Primary structure and functional expression of a high voltage activated calcium channel from rabbit lung. *FEBS Lett.* **269,** 409–412.
18. Ruth, P., Rohrkasten, A., Biel, M., Bosse, E., Regulla, S., Meyer, H. E., Flockerzi, V., and Hofmann, F. (1989) Primary structure of the subunit of the DHP-sensitive calcium channel of skeletal muscle. *Science* **245,** 1115–1118.
19. Bosse, E., Regulla, S., Biel, M., Ruth, P., Meyer, H. E., Flockerzi, V., and Hofmann, F. (1990) The cDNA and deduced amino acid sequence of the subunit of the L-type calcium channel from rabbit skeletal muscle. *FEBS Lett.* **267,** 153–156.
20. Jay, S. D., Ellis, S. B., McCue, A. F., Williams, M. E., Vedvick, T. S., Harpold, M. M., and Campbell, K. P. (1990) Primary structure of the γ subunit of the DHP-sensitive calcium channel from skeletal muscle. *Science* **248,** 490–492.
21. Ellis, S. B., Williams, M. E., Ways, N. R., Brenner, R., Sharp, A. H., Leung, A. T., Campbell, K. P. McKenna, E., Koch, W. J., Hui, A., Schwartz, A., and Harpold, M. M. (1988) Sequence and expression of mRNAs encoding the α_1 and α_2 subunits of a DHP-sensitive calcium channel. *Science* **241,** 1661–1664.
22. DeJongh, K. S., Warner, C., and Catterall, W. A. (1990) Subunits of purified calcium channels. α_2 and δ are encoded by the same gene. *J. Biol. Chem.* **265,** 14738–14741.
23. Flockerzi, V., Oeken, H.-J., and Hofmann, F. (1986) Purification of a functional receptor for calcium channel blockers from rabbit skeletal muscle microsomes. *Eur. J. Biochem.* **161,** 217–224.
24. Sieber, M., Nastainczyk, W., Zubor, V., Wernet, W., and Hofmann, F. (1987) The 165-kDa peptide of the purified skeletal muscle dihydropyridine receptor contains the known regulatory sites of the calcium channel. *Eur. J. Biochem.* **167,** 117–122.

25. Glossmann, H., Ferry, D. R., Striessnig, J., Goll, A., and Moosburger, K. (1987) Resolving the structure of the Ca^2+ channel by photoaffinity labeling. *Trends Pharmacol. Sci.* **8**, 95–100.
26. Takahashi, M., Seagar, M. J., Jones, J. F., Reber, B. F. X., and Catterall, W. A. (1987) Subunit structure of dihydropyridine-sensitive calcium channel from skeletal muscle. *Proc. Natl. Acad. Sci. USA* **84**, 5478–5482.
27. Schneider, T., Regulla, S., and Hofmann, F. (1990) The devapamil binding site of the purified skeletal muscle CaCB receptor is modulated by micromolar and milimolar calcium. *Eur. J. Biochem.* **200**, 245–253.
28. Kim, H. S., Wei, X., Ruth, P., Perez-Reyes, E., Flockerzi, V., Hofmann, F., and Birnbaumer, L. (1990) Studies on the structural requirements for the actiity of the skeletal muscle dihydropyridine receptor/slow Ca^{2+} channel. *J. Biol. Chem.* **265**, 11858–11863.
29. Regulla, S., Schneider T., Nastainczyk W., Meyer H. E., and Hofmann, F. (1991) Identification of the site of interaction of the dihydropyridine channel blockers nitrendipine and azidopine with the calcium-channel α_1 subunit. *EMBO J.* **10**, 45–49.
30. Schneider, T. and Hofmann, F. (1988) The bovine cardiac receptor for calcium channel blockers is a 195 kDa protein. *Eur. J. Biochem.* **174**, 369–375.

CHAPTER 17

Purification and Reconstitution of the Ryanodine-Sensitive Ca²⁺ Release Channel Complex from Muscle Sarcoplasmic Reticulum

F. Anthony Lai and Gerhard Meissner

1. Introduction

Muscle contraction is affected by the rapid release of Ca^{2+} ions through Ca^{2+}-conducting channels localized in an intracellular membrane compartment, the sarcoplasmic reticulum (SR). The rabbit skeletal muscle SR Ca^{2+} release channel has been identified as a high-affinity receptor for the plant alkaloid ryanodine, and purified to apparent homogeneity as a 30S protein complex comprising four identical high-mol-wt polypeptide subunits of relative mol mass (M_r) ~560,000 *(1–4)*. Morphological and immunolocalization studies of the ryanodine receptor complex have indicated that the subunits of the tetramer are arranged in the form of a four-leaf clover *(1,5,6)*, and are identical to the previously described bridging structures ("feet") *(7)* that span the transverse tubular-SR junctional gap *(8)*.

Although the mechanism of physiological SR Ca^{2+} release in skeletal and cardiac muscle has not yet been fully defined *(9,10)*, rapid mixing vesicle–ion flux *(11–14)* and planar lipid bilayer–single channel *(15,16)* measurements with isolated SR vesicles have shown that SR Ca^{2+} release is mediated by a high-conductance divalent cation-conducting channel that is activated by Ca^{2+} and modulated by the endogenous ligands ATP, Mg^{2+}, and calmodulin.

From: *Methods in Molecular Biology, Vol. 13: Protocols in Molecular Neurobiology*
Edited by: A. Longstaff and P. Revest Copyright © 1992 The Humana Press, Totowa, NJ

Using the high-affinity ligand [³H]ryanodine as a channel-specific marker, the detergent-solubilized Ca²⁺ release channels from skeletal and cardiac muscle have been isolated by immunoaffinity chromatography *(17,18)*, sequential column chromatography *(19,20)*, and density gradient centrifugation *(1,21,22)*, as a polypeptide with an apparent M_r of 350,000–450,000 on sodium dodecyl sulfate (SDS) polyacrylamide gels. Cloning and sequencing of the complementary DNA of the rabbit and human skeletal muscle ryanodine receptors have revealed an open reading frame encompassing 15 kb and encoding a 5032-amino acid residue polypeptide of M_r 563,584 *(3,4)*.

We describe here the procedures routinely employed in our laboratory for isolation of a fully functional, detergent-solubilized rabbit skeletal muscle SR Ca²⁺ release channel by density gradient centrifugation, and reconstitution of the purified 30S channel complex into planar lipid bilayers (Fig. 1). The sucrose gradient centrifugation procedure is relatively simple and straightforward, and has been employed in this laboratory also to isolate the 30S ryanodine receptor from canine cardiac muscle SR *(22)* and frog skeletal muscle SR *(23)*, as well as to identify a 30S channel complex from porcine aorta *(24)* and rat and bovine brain *(25)*. Other laboratories that have employed the sucrose gradient centrifugation procedure have described the purification of the rabbit skeletal *(26)*, porcine skeletal *(27)*, and canine cardiac *(28,29)* muscle ryanodine receptors, as well as the inositol trisphosphate receptor from bovine aorta *(30)*.

2. Materials

2.1. Preparation of Heavy SR Vesicles

1. Homogenization buffer: $0.1 M$ NaCl, 0.5 mM EDTA, 10 mM sodium Hepes, pH 7.5, and protease inhibitors (100 nM aprotinin, 1 μM leupeptin, 1 μM pepstatin, 1 mM benzamidine, 0.2 mM phenylmethylsulfonylfluoride [PMSF]).
2. Extraction buffer: $0.6 M$ KCl, 100 μM EGTA, 90 μM CaCl₂, 10 mM potassium Pipes, pH 7, 0.2 mM PMSF, and 1 μM leupeptin.
3. Gradient solutions: 20% and 40% (w/w) sucrose in extraction buffer. Storage medium: $0.3 M$ sucrose and 5 potassium Pipes, pH 7.

2.2. Isolation of 30S Ca²⁺ Release Channel Complex

1. NaCl stock solution (2X): $2.0 M$ NaCl, 200 μM EGTA, 300 μM CaCl₂, 10 mM AMP, 2 mM dithiothreitol (DTT), 2 mM diisopropylfluorophosphate (DIFP), and 40 mM sodium Pipes, pH 7.1.

Rabbit skeletal muscle tissue

↓ Waring blender 2 x 30 seconds

Crude homogenate

↓ Centrifuge 25 min at 2,600 x g
Discard pellet

Supernatant

↓ Centrifuge 30 min at 33,000 x g

Microsomal membranes

↓ Extraction buffer
Centrifuge 45 min at 105,000 x g

KCl-extracted membranes

↓ 20-45% sucrose gradient
Centrifuge 12-16 hr at 110,000 x g

"Heavy" sarcoplasmic reticulum membranes

↓ Chaps/NaCl buffer
Centrifuge 25 min at 57,000 x g

Solubilized heavy SR proteins

↓ 5-20% sucrose gradients in SW28
Centrifuge 16 hr at 120,000 x g

> 95% Pure ryanodine receptor

↓

1. SDS-gel analysis
2. [^{3}H]Ryanodine binding studies
3. Single channel recordings

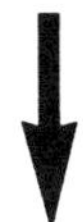

Fig. 1. Summary of the experimental steps involved in rabbit skeletal muscle ryanodine receptor purification.

2. Chaps stock solution: 150 mM 3-[(3-cholamidopropyl) dimethyl-ammonio]-1-propanesulfonate in water.
3. Phospholipid stock solution: 100 mg/mL phospholipid (95% soybean phosphatidylcholine, solubilized with 150 mM Chaps in 0.1M NaCl, 100 μM EGTA, 100 μM CaCl$_2$, 1 mM DTT, 1 mM DIFP, and 20 mM sodium Pipes, pH 7.1.
4. Gradient solutions: 5% (w/w) sucrose in NaCl stock solution (1X), containing 16 mM Chaps and 5 mg/mL phospholipid; 20% (w/w) sucrose in NaCl stock solution (1X), containing 8 mM Chaps and 2.5 mg/mL phospholipid.
5. Ryanodine stock solution: [9,21-^{3}H]ryanodine (30–60 Ci/mmol) in ethanol.

2.3. Characterization of Purified Ca^{2+} Release Channel Complex

2.3.1. SDS Polyacrylamide Gel Analysis

1. Acrylamide stock solution (20%): 195 mg/mL acrylamide and 5 mg/mL N,N'-methylenebisacrylamide.
2. Separating gel stock solution (4X): 1.5M Tris-HCl, pH 8.8, 0.4% SDS.
3. Stacking gel stock solution (4X): 0.5M Tris-HCl, pH 6.8, 0.4% SDS. (All above chemicals are purchased as "Ultrapure" grade, and solutions are filtered through Whatman Grade 2 filters [8 μm], and stored in the dark at 4°C for up to 2 mo).
4. Gel sample buffer (2X): 0.125M Tris-HCl, pH 6.8, 4% SDS, 4% β-mercaptoethanol, 20% glycerol, 50 μg/mL bromophenol blue.
5. Gel running buffer (5X): 0.95M glycine, 0.125M Tris, 0.5% SDS, pH 8.3.

2.3.2. [^{3}H]Ryanodine Binding

1. Binding medium (2X): 2.0M NaCl, 200 μM EGTA, 300 μM CaCl$_2$, 10 mM AMP, 2 mM DIFP, 40 mM sodium Pipes, pH 7.1.
2. Ryanodine stock solutions (5X): 10 nM–100 μM [^{3}H]ryanodine and 10 mM unlabeled ryanodine in 0.1M NaCl, and 5 mM sodium Pipes, pH 7.1.
3. Polyethyleneimine: 1% and 5% (w/w) in H$_2$O.

3. Methods

3.1. Preparation of Heavy SR Vesicles

Heavy- and light-density SR membranes *(31,32)* derived from rabbit leg and back muscle are prepared in the presence of protease inhibitors as follows:

1. Forty grams of minced muscle are homogenized in 300 mL homogenization buffer at 4°C for two 30-s periods (high-speed setting) in a Waring blender. The homogenate is centrifuged ($2600g$, 25 min, 4°C), and the supernatant is poured through two layers of cheesecloth to remove floating fat particles and then recentrifuged ($33,000g$, 30 min, 4°C).
2. The resulting pellets are resuspended in 100 mL extraction medium with a Dounce homogenizer (10 passes), kept for 1 h on ice to solubilize contaminating myofibrillar proteins, centrifuged ($105,000g$, 45 min, 4°C), and resuspended as described earlier in 18 mL extraction medium.
3. These KCl-extracted membranes are then subfractionated according to their buoyant density by layering 6-mL aliquots at the top of three 20–45% linear sucrose gradients in extraction medium. The gradients are centrifuged ($110,000g$, 12–16h, 2°C) and the resulting gradients fractionated into 3.0-mL fractions. The position of each fraction in the gradient is estimated by measurement of "% sucrose" using a refractometer (*see* Notes 1 and 2).
4. Gradient fractions are diluted with 2 vol of $0.4M$ KCl, centrifuged ($105,000g$, 45 min, 4°C), and resuspended as described earlier to give a protein concentration of 15–20 mg/mL in storage medium. They are then rapidly frozen and stored at −80°C.

3.2. Isolation of 30S Ca²⁺ Release Channel Complex

The membrane-bound Ca²⁺ release channel is first solubilized using the zwitterionic detergent Chaps and high ionic strength, and then purified by centrifugation through a linear sucrose gradient (Fig. 2). Use of a Beckman SW28 rotor to accommodate six 33-mL gradients of 5–20% sucrose allows the purification of ~0.4 mg of channel protein (Table 1) (*see also* Notes 3–6).

1. SR membranes (1–1.5 mg protein/mL) are solubilized in 18 mL NaCl stock solution (1X) containing 27 mM Chaps and 5 mg/mL phospholipid. The extent of ryanodine receptor solubilization, as well as the migration distance of the solubilized [³H]ryanodine receptor on the sucrose gradients, is monitored by adding [³H]ryanodine (final concentration 2 nM) to an aliquot (3 mL) of the solubilized sample.
2. Incubate for 2 h at 23°C, and take 20-μL aliquots to determine (a) protein concentration by the method of Kaplan and Pederson *(33)* using Amido Black and 0.45-μm Millipore filters, and (b) total and bound [³H]ryanodine (Section 3.3.2.).
3. Small amounts of Chaps-insoluble material (<10% of protein) are removed by aliquoting the samples into six centrifuge tubes (five tubes

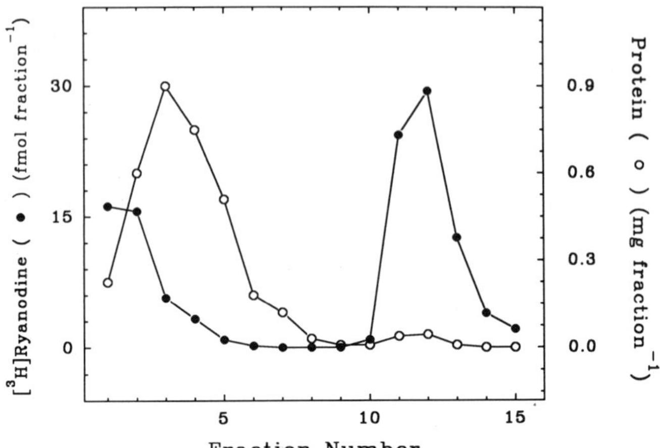

Fig. 2. Sedimentation profile of Chaps-solubilized heavy SR proteins
and ryanodine receptor. Heavy SR membranes (Section 3.1.) were solubilized in Chaps and centrifuged through linear sucrose gradients (Section
3.2.). Fractions (~2 mL) were then analyzed for protein and radioactivity
content. The majority of the solubilized proteins sediment between fractions 1–6, whereas the [³H]ryanodine receptor peak is present in fractions
11–13 comigrating with a small protein peak. The radioactivity remaining
at the top of the gradient (fractions 1–4) represents unbound [³H]ryanodine.

Table 1
Purification of Rabbit Skeletal Muscle Ryanodine Receptor[*]

		[³H]Ryanodine binding	
Fraction	Protein, mg	pmol/mg Protein	Purification
Crude homogenate	15,000	0.4	1
Heavy SR membranes	22	15	37.5
Purified ryanodine receptor	0.4	450	1125

[*]Heavy SR membranes isolated from rabbit skeletal muscle homogenates
(Section 3.1.) were solubilized in Chaps at 1.3 mg protein/mL (Section 3.2.).
The Chaps supernatant (18 mL) was then loaded (3 mL each) onto six 33-mL 5–
20% linear sucrose gradients and centrifuged for 16 h at 120,000g, 2°C. The
fractions, which sedimented at ~30S corresponding to the ryanodine receptor (*see*
Figs. 1,2), were pooled and assayed for protein content and [³H]ryanodine-binding
activity (Section 3.3.2.).

for sample without [³H]ryanodine, and one tube for sample with [³H]ryanodine) and centrifuged (57,000g, 25 min, 4°C).

4. The supernatants containing the solubilized ryanodine receptor complex are collected, and after taking 20-μL aliquots for determination of protein concentration and [³H]ryanodine binding (Section 3.3.2.), are layered (3 mL each) at the top of six 33-mL, 5–20% linear sucrose gradients, centrifuged (120,000g 16 h, 2°C), and the gradients fractionated into 2.0-mL fractions.

5. An aliquot (50 μL) of each fraction from the gradient containing the [³H]ryanodine-labeled sample is used to determine the position of the [³H]ryanodine-labeled receptor on the gradients by scintillation counting. Receptor peak fractions, which sediment with an apparent sedimentation coefficient of ~30S are found in the lower third of the gradient (Fig. 2).

3.3. Characterization of Purified Ca²⁺ Release Channel Complex

3.3.1. SDS Polyacrylamide Gel Analysis

The polypeptide composition of native and solubilized SR membrane vesicles (Section 3.1.) and of the various solubilized fractions obtained from sucrose density gradients (Section 3.) is analyzed by size separation of constituent proteins on 1.5-mm polyacrylamide slab gels *(34)* following reduction and denaturation in the presence of β–mercaptoethanol and SDS (Fig. 3) (*see* Notes 7–9).

1. A separating gel (27 mL) with a linear gradient of 3–12% acrylamide in 1X separating gel solution containing 1 μL/mL N,N,N',N'-tetramethylethylenediamine (TEMED) and 0.15 mg/mL ammonium persulfate is poured in a vertical gel casting unit forming a gel with dimensions of 144 × 120 mm.

2. After polymerization of the separating gel (~1 h), a stacking gel (12 mL) of 3% acrylamide in 1X stacking gel solution containing 1 μL/mL TEMED and 0.5 mg/mL ammonium persulfate is added, forming a gel with 15 wells.

3. Samples to be analyzed are denatured in 1X gel sample buffer at 95°C for 3 min, then loaded onto the gel in a final vol of 60 μL and electrophoresed in 1X gel running buffer at 30 mA constant current, 15°C.

4. Gels are then stained in 0.1% Coomassie brilliant blue R250 (in 50% methanol, 10% acetic acid) for >1 h and destained in 10% methanol, and 15% acetic acid. If the gel samples have a low protein concentration, a more sensitive silver stain should be used *(35)*.

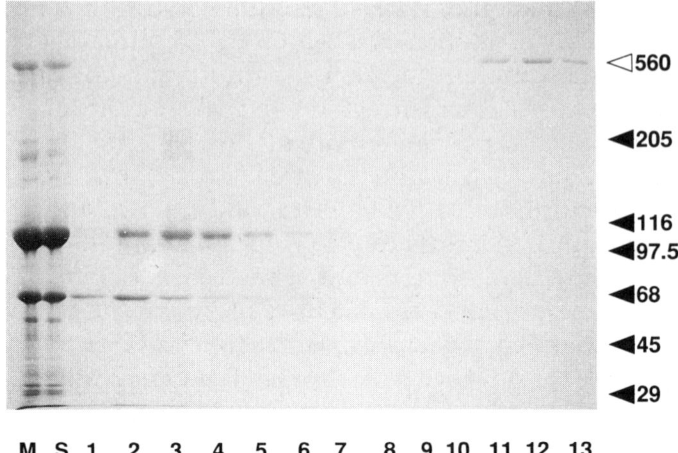

Fig. 3. SDS polyacrylamide gel analysis. Heavy SR membranes (lane 1, 30 mg protein) were solubilized in Chaps (lane 2, 28 mg) and centrifuged through linear sucrose gradients as described in Fig. 2. An aliquot (30 µL) of each gradient fraction was then analyzed by SDS gel electrophoresis on a 3–12% linear polyacrylamide gradient gel and stained with Coomassie blue (Section 3.3.1.). The top gradient fractions contain the major heavy SR proteins, Ca-ATPase and calsequestrin. Fractions 11–13 specifically contain a single polypeptide that comigrates with the ryanodine receptor peak (Fig. 2). Molecular-weight standards ($\times 10^{-3}$) are shown to the right (closed arrowheads). Open arrowhead represents the predicted mol wt of the ryanodine receptor polypeptide from cDNA sequence analysis *(3,4)*.

3.3.2. [^{3}H] Ryanodine Binding

We describe here a binding assay that is useful for quantitating the extent and affinity of Ca^{2+}-dependent ryanodine binding to membrane-bound, and Chaps-solubilized receptors, by Scatchard plot analysis (*see* Note 10).

1. Total [^{3}H]ryanodine binding to SR membranes (0.3–0.5 mg of protein/mL) is determined in 200 µL of 1X binding medium containing 2 nM–20 µM [^{3}H]ryanodine. Nonspecific binding is estimated by incubating an identical sample with an excess of unlabeled ryanodine (2 mM).
2. The samples are incubated for 4 h at 37°C.
3. Each total [^{3}H]ryanodine binding assay sample is now divided and treated as follows:
   a. A 10-µL aliquot is placed into a scintillation vial and counted to determine total radioactivity.

    b. Three 50-μL aliquots, after 25-fold dilution with ice-cold water, are passed through a Whatman GF/B filter soaked in 1% polyethyleneimine. After rinsing with three 5-mL vol of ice-cold water under vacuum, the filters are soaked overnight in a liquid scintillation cocktail and counted.

    c. The remainder is centrifuged (90,000g, 30 min, 4°C) to enable determination of the free [³H]ryanodine concentration in the supernatant fraction by counting a 10-μL aliquot.

4. [³H]Ryanodine binding to the Chaps-solubilized, purified receptor is determined as described earlier for SR membranes, but with the following modifications. First, the gradient solutions contain, in addition, detergent (8–16 mM Chaps) and phospholipid (2.5–5 mg/mL) (cf. Section 3.2.). Second, binding is determined at 22°C for 10–15 h and not at 37°C, since the latter temperature results in rapid inactivation of the solubilized receptor. Partial inactivation of the receptor during the binding reaction can be estimated by determination of receptor binding after longer times of incubation and extrapolation to zero time (Fig. 4). Third, the solubilized receptor is not readily sedimented by centrifugation, and it is therefore more practical to estimate free [³H]ryanodine as the difference between total and bound ryanodine. Fourth, Whatman glass fiber filters are soaked in 5% instead of 1% polyethyleneimine to optimize retention of the receptor by the filters *(37)*.

3.4. Reconstitution of the Purified 30S Ca²⁺ Release Channel Complex into a Planar Lipid Bilayer

The Chaps-solubilized 30S Ca²⁺ release channel complex purified in the absence of [³H]ryanodine (Section 3.2.) readily incorporates into planar lipid bilayers of the Mueller-Rudin type. In the presence of monovalent cations, large single-channel current fluctuations are observed upon incorporation of the channel complex into the bilayer (Fig. 5). The equipment and methods for recording and analyzing single-channel current fluctuations have been described *(36)*.

1. Mueller-Rudin lipid bilayers (phosphatidylethanolamine, phosphatidylserine, phosphatidylcholine, in a ratio of 5:3:2, at 50 mg/mL in decane solution) are formed by transferring a small aliquot (1–2 μL) of dissolved lipid on to a circular aperture (250–300 μm diameter) through a polyvinylidene difluoride wall separating two chambers (3-mL vol each). The two chambers contain a buffered solution (e.g., 20 mM Na or KPipes, pH 7), a monovalent cation at 100–500 mM concentration (K⁺ or Na⁺) as the current carrier, and micromolar free concentrations of Ca²⁺ (10–100 μM) to activate the incorporated channels partially (*see* Note 11).

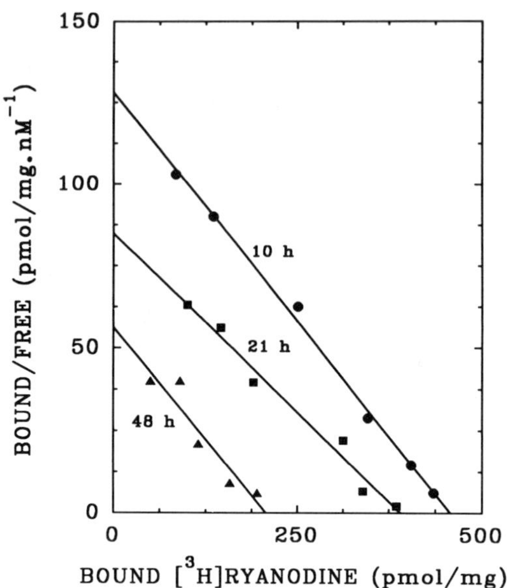

Fig. 4. Scatchard analysis of ryanodine binding to the purified ryanodine receptor. Purified 30S ryanodine receptor (15 µg protein/mL) was incubated for 10, 21, and 48 h at 22°C with [^{3}H]ryanodine as described in Section 3.3.2. After 10 h of incubation, the B_{max} is ~450 pmol/mg protein with a K_d of 3–5 nM. Increasing time of incubation reduces the B_{max} value, but does not significantly change the K_d, indicating a progressive inactivation of receptor binding with time.

2. Upon addition of a small aliquot of the purified channel protein (0.2–3 µL) close to one side of the bilayer (designated the *cis* side and corresponding to the cytoplasmic side of the SR membrane), channels incorporate spontaneously into the lipid bilayer and are detected as stepwise increases in bilayer conductance *(1,18)*.

3. Observation of a Ca^{2+}-conducting channel can be made by perfusion of the *trans* (corresponding to the SR lumenal side) chamber with a Ca^{2+} solution and of the *cis* chamber (SR cytoplasmic side) with a slowly permeating cation, such as Tris or choline (Fig. 5) (*see* Note 12).

4. Notes

1. Skeletal SR isolated by the above procedure includes vesicles either containing, or lacking, the ryanodine-sensitive Ca^{2+} release channel protein. A majority (75–90%) of the heavy vesicles, which are recovered from the 35–40% sucrose region of gradients, contains the Ca^{2+}

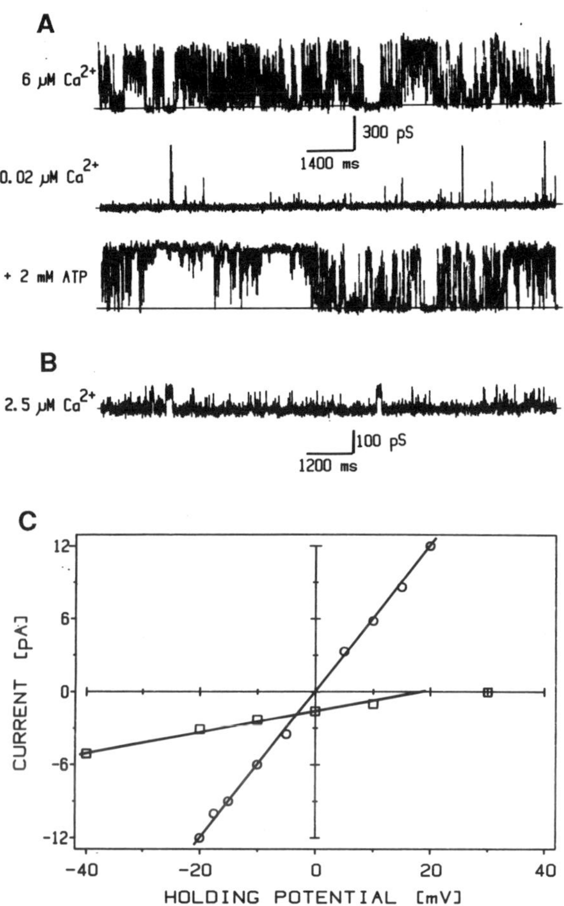

Fig. 5. Reconstitution of the 30S ryanodine receptor complex into planar lipid bilayers. (A) Single channel currents, shown as upward deflections, were recorded in symmetric 0.5M NaCl, 10 mM NaPlPES, pH 7, with 6 μM free Ca²⁺ *cis* (100 μM EGTA, 100 μM CaCl₂) (top trace), 0.02 μM free Ca²⁺ *cis* (2.1 μM EGTA, 100 μM CaCl₂) (second trace), or 0.02 μM free Ca²⁺ plus 2 mM ATP in the *cis* chamber (bottom trace); holding potential (HP), –15 mV. (B) Single-channel current recorded after perfusion with 50 mM Ca(OH)₂/250 mM HEPES, pH 7.4, 10% glycerol *trans,* and 125 mM Tris/ 250 mM HEPES, pH 7.4, 10% glycerol, plus 2.5 μM free Ca²⁺ (100 mM EGTA, 100 mM CaCl₂) *cis.* HP, O mV. (C) Current-voltage relationship for recordings A (top trace) and B. Values of unit conductance: γ, 595 pS with 0.5M Na⁺ (O); and γ, 91 pS with 50 mM Ca²⁺ (□) as the conducting ion. Recordings were filtered at 300 Hz and sampled at 2 kHz. (Taken with permission from ref. *1.*)

release channel (1–10 channels/vesicle, ref. *9*), whereas only a small proportion (5–20%) of the light vesicles recovered from the 25–30% sucrose region contain the channel *(32)*.

2. Measurement of dihydropyridine receptor content by [³H]PN200-110 binding and Ca²⁺-dependent, high-affinity [³H]ryanodine binding to determine Ca²⁺ release channel content (Section 3.3.2.), has indicated that our heavy SR vesicle preparations contain a relatively low content of transverse tubular membranes.

3. Early important observations by Pessah and coworkers indicated that the high-affinity ryanodine-binding activity of skeletal SR membranes could be solubilized in the presence of Chaps and high salt concentrations *(37)*. The Chaps-solubilized ryanodine receptor was found to behave as a large protein upon gel permeation chromatography. A study of ryanodine "binding" to various detergents also indicated Chaps to have the lowest nonspecific binding *(19)*. In our studies, density gradient centrifugation of solubilized SR membranes through linear gradients of sucrose results in optimal separation of the 30S ryanodine receptor complex from other solubilized SR proteins, because of its much faster sedimentation rate *(1,38)*. Higher concentrations of solubilized SR protein applied to the gradient will enable a higher yield of ryanodine receptor to be obtained in the peak fractions, although a concomitant contamination of the receptor protein with Ca²⁺ ATPase (M_r 110,000) and myosin (M_r 205,000) is observed *(38)*. Using lower protein concentrations of solubilized heavy SR (<1.5 mg/mL), we find that the ryanodine receptor peak fractions contain a more than 95% pure M_r 560,000 polypeptide (Fig. 3). If further removal of the minor contamination, mostly oligomerized M_r 110,000 Ca-ATPase, is required, the concentrated peak fractions can be recentrifuged on a second sucrose gradient in exactly the same manner as the first gradient. This protocol appears to result in deoligomerization of the Ca-ATPase, which remains at the top of the second gradient, whereas the 30S ryanodine receptor again sediments to the bottom one-third of the gradient.

4. The presence of exogenous phospholipid in the sucrose gradients is required for maintaining optimal ryanodine-binding activity and for preservation of single-channel activity in planar lipid bilayers that exhibit sensitivity to known modulators of the native channel. This lipid requirement is most notably observed in experiments with canine cardiac SR membranes, which have been used to isolate the cardiac ryanodine receptor for ryanodine-binding measurements and single channel recordings *(22)*. For structural analysis of the purified receptor by electron microscopy, however, exogenous lipid is

omitted from sucrose gradients, since it interferes with visualization by negative staining *(1)*. Previously, highy purified, but expensive, phospholipid mixtures of phosphatidylcholine, phosphatidylethanolamine, and phosphatidylserine were used in the solubilization and gradient media *(1)*. For routine studies, we now use the less expensive, 95% pure soybean phosphatidylcholine (lecithin), with the added precaution of including millimolar dithiothreitol in all solutions to counteract the possible presence of low levels of oxidized lipids in the 95% pure extract. Sedimentation of the 30S receptor complex has previously been shown to be unaffected by dithiothreitol *(38)*.

5. The use of detergents other than Chaps in the sucrose gradient medium, which would allow the exchange of the Chaps-solubilized receptor into an alternative detergent, so that it is possible to study, for example, the effect of detergent type on ryanodine binding and receptor reconstitution, also results in the sedimentation of the ryanodine receptor as a 30S complex. An exception to note is Zwittergent 3-14, a hydrophobic zwitterionic detergent that has been shown to dissociate the 30S receptor into its four subunits and simultaneously destroy high-affinity ryanodine binding *(2)*.

6. Reduction of the salt concentration in the gradients below $0.5\,M$ NaCl results in a less pure receptor preparation, caused by an increase in the degree of SR Ca ATPase oligomerization to large species that are found smeared throughout the gradient, as well as a lower yield of 30S ryanodine receptor protein, owing to aggregation of receptors resulting in formation of a small pellet at the bottom of the gradient tube.

7. Gel electrophoresis is an integral part of the procedure for SR membrane and ryanodine receptor isolation, since it allows the monitoring of changes in various membrane protein components thoughout the heavy SR membrane isolation and extraction steps, and also enables an estimation of the purity of the M_r 560,000-containing ryanodine receptor fractions collected from the sucrose gradients, as well as assessment of the degree of proteolytic degradation of the receptor polypeptide (Fig. 3). Previous studies have shown that the M_r 560,000 ryanodine receptor is highly susceptible to attack by endogenous and exogenous proteases *(39,40)*. In particular, Ca^{2+}-activated neutral protease (calpain) has been shown to act selectively on the ryanodine receptor polypeptide *(39)*.

8. A gradient of polyacrylamide in the separating gel from low (3%) to intermediate (12%) concentration is useful for separating proteins with a broadly disparate relative mol masses in that it allows a sufficient migration of the high M_r receptor polypeptide into the gel ($R_f\sim0.15$),

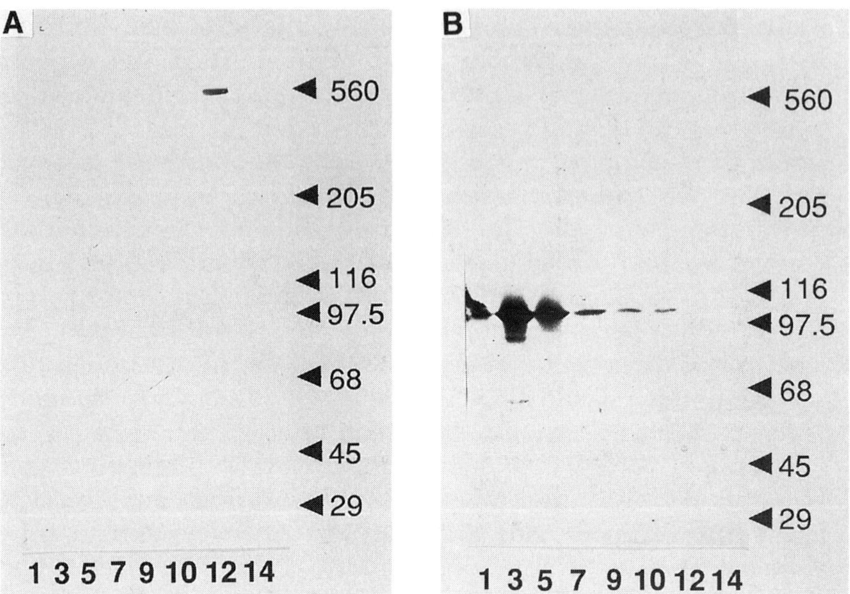

Fig. 6. Immunoblot analysis. Aliquots (30 µL) of gradient fractions were electrophoresed through SDS gels as described in Fig. 3, and the separated proteins were transferred to Immobilon membrane *(39)*. The blots were blocked with 5% nonfat dried-milk proteins, and then incubated with either rabbit antirat ryanodine receptor antisera (1:2000, Fig. 4A), or rabbit antisera to a synthetic rabbit skeletal SR Ca-ATPase oligopeptide (amino acids 192–205, a gift of D. G. Ferguson) (1:1000, Fig. 4B), for 1 h at room temperature. Following incubation with peroxidase-conjugated goat antirabbit IgG antisera (1:2000) for 1 h at room temperature, the blots were developed using the substrate 3,3'-diaminobenzidine and H_2O_2. All incubations were in phosphate-buffered saline containing 0.02% Tween 20. Panel A shows a single band at $M_r \sim 560,000$, specifically in fraction 12, which comigrates with the ryanodine receptor peak (Fig. 2). Panel B shows staining of the M_r 110,000 Ca-ATPase, which is most highly enriched in fractions 3–5 and displays decreasing intensity toward the bottom of the gradient.

and yet still resolves proteins with M_r down to ~25,000. Further, in gradient gels, the leading edge of migrating zones is retarded more than the trailing edge, resulting in a marked sharpening of protein bands as compared to separations using nongradient gels. In addition, the electrophoretic transfer of proteins of large and small M_r is more uniform in gradient gels (Fig. 6).

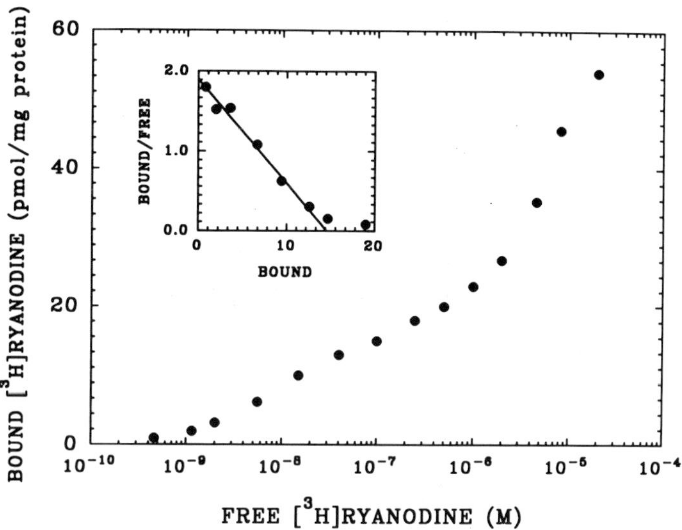

Fig. 7. High- and low-affinity ryanodine binding to heavy SR membranes. Binding of heavy SR membranes to [³H]ryanodine was determined as described in Section 3.3.2. Scatchard analysis (inset) reveals a curvilinear slope indicating the presence of both high-affinity (K_d = 7 nM) and low-affinity sites *(2)*.

9. The immunoblot analysis of Fig. 6 indicates that the major protein component of the heavy SR membrane, the Ca-ATPase, remains predominantly in the top half of the sucrose gradient. The M_r 560,000 polypeptide, in contrast, is found only in the bottom one-third of the gradient comigrating with the [³H]ryanodine receptor peak (*see* Fig. 2).

10. Both high- and low-affinity [³H]ryanodine-binding sites in rabbit skeletal muscle SR have been discerned (Fig. 7; ref. *2*). The rate of binding of [³H]ryanodine to the skeletal and cardiac receptors is "slow," and has been found to be affected by Ca^{2+}, ionic strength, adenine nucleotides, and Mg^{2+}, as well as other ligands that have been shown to modulate Ca^{2+} release channel activity (for review, *see* ref. *9*).

11. Since the Ca^{2+} release channel is highly selective for cations over anions, these experiments can be carried out using the chloride salts of K^+, Na^+, Ca^{2+}, or $Tris^+$ *(41)*.

12. The single-channel recordings of reconstituted ryanodine receptor displayed in Fig. 8 show the characteristic effects of low and high

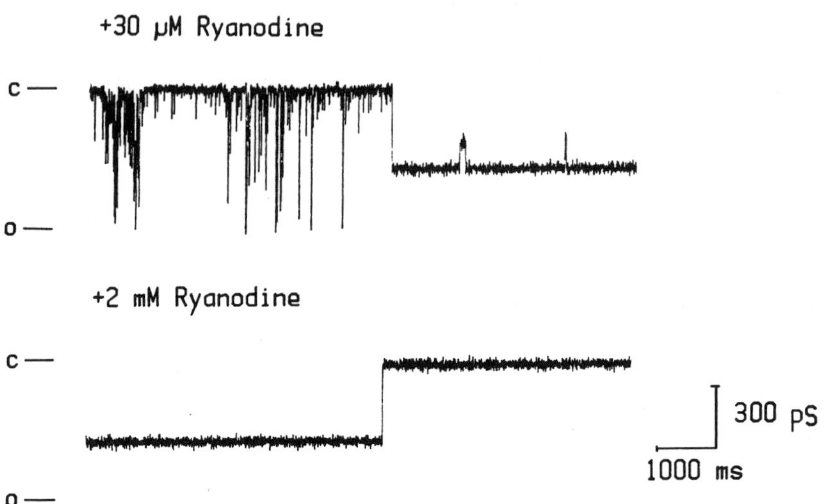

Fig. 8. Effect of ryanodine on a single reconstituted purified ryanodine receptor channel. Single-channel recordings of K$^+$ current of purified ryanodine receptor protein incorporated into a planar lipid bilayer in symmetric 250 mM KCl buffer (20 mM K-Pipes, pH 7.0, 150 µM CaCl$_2$, 100 µM EGTA,, 250 mM KCl) with 50 µM free Ca^{2+}. Unitary conductance, 700 pS; HP, 20 mV. The upper trace shows the appearance of the subconducting state with an open probability of ~1, several minutes after *cis* addition of 30 µM ryanodine. An additional, infrequently observed substate can also be noticed. The lower trace illustrates the sudden transition from the subconductance state to a fully closed state within 1 min after *cis* addition of 2 mM ryanodine. In these single-channel measurements, relatively high ryanodine concentrations of 30 µM and 2 mM are used to reduce the time required to observe the otherwise very slow interaction of ryanodine with the channel. Bars on the left represent the open (o) and closed (c) channel. (Taken with permission from ref. *2.)*

concentrations of ryanodine that induce a permanently open, reduced conductance state (upper trace) and the completely closed state (lower trace), respectively. This unique effect of ryanodine on the SR Ca^{2+} release channel corroborates similar observations previously described for vesicle ^{45}Ca^{2+}-flux studies *(42,43)*, and provides an unambiguous identification of this ryanodine-sensitive channel in planar lipid bilayer recordings.

Acknowledgments

This work was supported by NIH grants AR18687, HL27430, and HL38835, and the Carl M. Pearson Fellowship from the Muscular Dystrophy Association (F. A. L.).

References

1. Lai, F. A., Erickson, H. P., Rousseau, E., Liu, Q.-Y., and Meissner, G. (1988) Purification and reconstitution of the calcium release channel from skeletal muscle. *Nature* **331**, 315–319.

2. Lai, F. A., Misra, M., Xu, L., Smith, H. A., and Meissner, G. (1989) The ryanodine receptor-Ca^{2+} release channel complex of skeletal muscle sarcoplasmic reticulum. Evidence for a cooperatively coupled, negatively charged homotetramer. *J. Biol. Chem.* **264**, 16,776–16,785.

3. Takeshima, H., Nishimura, S., Matsumoto, T., Ishida, H., Kangawa, K., Minamino, N., Matsuo, H., Ueda, M., Hanaoka, M., Hirose, T., and Numa, S. (1989) Primary structure and expression from complementary DNA of skeletal muscle ryanodine receptor. *Nature* **339**, 439–445.

4. Zorzato, F., Fujii, J., Otsu, K., Phillips, M., Green, N. M., Lai, F. A., Meissner, G., and MacLennan, D. H. (1990) Molecular cloning of cDNA encoding human and rabbit forms of the Ca^{2+} release channel (ryanodine receptor) of skeletal muscle sarcoplasmic reticulum. *J. Biol. Chem.* **265**, 2244–2256.

5. Saito, A., Inui, M., Radermacher, M., Frank, J., and Fleischer, S. (1988) Ultrastructure of the calcium release channel of sarcoplasmic reticulum. *J. Cell. Biol.* **107**, 211–219.

6. Wagenknecht, T., Grassucci, R., Frank, J., Saito, A., Inui, M., and Fleischer, S. (1989) Three-dimensional architecture of the calcium channel/foot structure of sarcoplasmic reticulum. *Nature* **338**, 167–170.

7. Franzini-Armstrong, C. (1980) Structure of sarcoplasmic reticulum. *Fed. Proc.* **39**, 2403–2409.

8. Kawamoto, R. M., Brunschwig, J.-P., Kim, K.-C., and Caswell, A. H. (1986) Isolation, characterization, and localization of the spanning protein from skeletal muscle triads. *J. Cell. Biol.* **103**, 1405–1414.

9. Lai, F. A. and Meissner, G. (1989) The muscle ryanodine receptor and its intrinsic Ca^{2+} channel activity. *J. Biomemb. Bioenerg.* **21**, 227–246.

10. Fleischer, S. and Inui, M. (1989) Biochemistry and biophysics of excitation–contraction coupling. *Annu. Rev. Biophys. Chem.* **18**, 333–364.

11. Nagasaki, K. and Kasai, M. (1983) Rapid Ca^{2+} release rate from sarcoplasmic reticulum vesicles by stopped flow method of chlortetracycline fluorescence. *J. Biochem. Tokyo* **94**, 1101–1109.

12. Ikemoto, N., Antoniu, B., and Meszaros, L. Y. (1985) Rapid flow chemical quench studies of calcium release from isolated sarcoplasmic reticulum. *J. Biol. Chem.* **260**, 14,096–14,100.

13. Meissner, G., Darling, E., and Eveleth, J. (1986) Kinetics of rapid Ca^{2+} release by sarcoplasmic reticulum. Effects of Ca^{2+}, Mg^{2+} and adenine nucleotides. *Biochemistry* **25**, 236–244.

14. Meissner, G. (1986) Evidence of a role for calmodulin in the regulation of calcium release from skeletal muscle sarcoplasmic reticulum. *Biochemistry* **25**, 244–251.

15. Smith, J. S., Coronado, R., and Meissner, G. (1985) Sarcoplasmic reticulum contains adenine nucleotide-activated calcium channels. *Nature* **316**, 446–449.

16. Smith, J. S., Coronado, R., and Meissner, G. (1986) Single channel measurements of the calcium release channel from skeletal muscle sarcoplasmic reticulum. *J. Gen. Physiol.* **88**, 573–588.

17. Imagawa, T., Smith, J. S., Coronado, R., and Campbell, K. P. (1987) Purified ryanodine receptor from skeletal muscle sarcoplasmic reticulum is the Ca^{2+} permeable pore of the calcium release channel. *J. Biol. Chem.* **262**, 16,636–16,643.

18. Smith, J. S., Imagawa, T., Ma, J., Fill, M., Campbell, K. P., and Coronado, R. (1988) Purified ryanodine receptor from rabbit skeletal muscle is the calcium release channel of sarcoplasmic reticulum. *J. Gen. Physiol.* **92**, 1–26.

19. Inui, M., Saito, A., and Fleischer, S. (1987) Purification of the ryanodine receptor and identity with feet structures of junctional terminal cisternae of sarcoplasmic reticulum from fast skeletal muscle. *J. Biol. Chem.* **262**, 1740–1747.

20. Inui, M., Saito, A., and Fleischer, S. (1987) Isolation of the ryanodine receptor from cardiac sarcoplasmic reticulum and identify with the feet structures. *J. Biol. Chem.* **262**, 15637–15642.

21. Lai, F. A., Anderson, K., Rousseau, E., Liu, Q. Y., and Meissner, G. (1988) Evidence for a Ca^{2+} channel within the ryanodine receptor complex from cardiac sarcoplasmic reticulum. *Biochem. Biophys. Res. Commun.* **151**, 441–449.

22. Anderson, K., Lai, F. A., Liu, Q. Y., Rousseau, E., Erickson, H. P., and Meissner, G. (1989) Structural and functional characterization of the purified cardiac ryanodine receptor-Ca^{2+} release channel complex. *J. Biol. Chem.* **264**, 1329–1335.

23. Liu, Q. Y., Lai, F. A., Xu, L., Jones, R. V., LaDine, J. K., and Meissner, G. (1989) Comparison of the mammalian and amphibian skeletal muscle ryanodine receptor-Ca^{2+} release channel complexes. *Biophys. J.* **55**, 85a.

24. Herrmann-Frank, A., Darling, E., and Meissner, G. (1990) Single channel measurements of the Ca^{2+}-gated ryanodine-sensitive Ca^{2+} release channel of vascular smooth muscle. *Biophys. J.* **57**, 156a.

25. Lai, F. A., Xu, L., and Meissner, G. (1990) Identification of a ryanodine receptor in rat and bovine brain. *Biophys. J.* **57**, 529a.

26. Hawkes, M. J., Diaz-Munoz, M., and Hamilton, S. L. (1989) A procedure for purification of the ryanodine receptor from skeletal muscle. *Membr. Biochem.* **8**, 133–145.

27. Mickelson, J. R., Litterer, L. A., and Louis, C. F. (1990) Isolation and reconstitution of the ryanodine receptor from malignant hyperthermia susceptible and normal pigs. *Biophys. J.* **57**, 277a.

28. Imagawa, T., Takasago, T., and Shigekawa, M. (1989) Cardiac ryanodine receptor is absent in Type I slow skeletal muscle fibers: immunochemical and ryanodine binding studies. *J. Biochem. (Tokyo)* **106**, 342–348.

29. Rardon, D. P., Cefali, D. C., Mitchell, R. D., Seiler, S. M., and Jones, L. R. (1989) High molecular weight proteins purified from cardiac junctional sarcoplasmic reticulum vesicles are ryanodine-sensitive calcium channels. *Circ. Res.* **64**, 779–789.

30. Chadwick, C. C. and Fleischer, S. (1990) Purification of the inositol 1,4,5 trisphosphate receptor from smooth muscle microsomes. *Biophys. J.* **57**, 285a.

31. Meissner, G. (1975) Isolation and characterization of two types of sarcoplasmic reticulum vesicles. *Biochim. Biophys. Acta* **389**, 51–68.

32. Meissner, G. (1984) Adenine nucleotide stimulation of Ca^{2+}-induced Ca^{2+} release in sarcoplasmic reticulum. *J. Biol. Chem.* **159**, 1365–1374.

33. Kaplan, R. A. and Pederson, P. L. (1985) Determination of microgram quantities of protein in the presence of milligram levels of lipid with Amido Black 10B. *Anal. Biochem.* **150**, 97–104.

34. Laemmli, U. K. (1970) Cleavage of structural proteins during the assembly of the head of bacteriophate T4. *Nature* **227**, 680–685.

35. Oakley, B. R., Kirsch, D. R., and Morris, N. R. (1980) A simplified ultrasensitive silver stain for detecting proteins in polyacrylamide gels. *Anal. Biochem.* **105**, 361–363.

36. Smith, J. S., Coronado, R., and Meissner, G. (1988) Techniques for observing calcium channels from skeletal muscle sarcoplasmic reticulum in planar lipid bilayers. *Meths. Enz.* **157**, 480–489.

37. Pessah, I. N., Francini, A. O., Scales, D. J., Waterhouse, A. L., and Casida, J. E. (1986) Calcium-ryanodine receptor complex. Solubilization and partial characterization from skeletal muscle junctional sarcoplasmic reticulum vesicles. *J. Biol. Chem.* **261**, 8643–8648.

38. Lai, F. A., Erickson, H. P., Block, B. A., and Meissner, G. (1987) Evidence for a junctional feet-ryanodine receptor complex from sarcoplasmic reticulum. *Biochem. Biophys. Res. Commun.* **143**, 704–709.

39. Seiler, S., Wegener, A. D., Whang, D. D., Hathaway, D. R., and Jones, L. R. (1984) High molecular weight proteins in cardiac and skeletal muscle sarcoplasmic reticulum vesicles bind calmodulin, are phosphorylated, and are degraded by Ca^{2+}-activated protease. *J. Biol. Chem.* **259**, 8550–8557.

40. Meissner, G., Rousseau, E., and Lai, F. A. (1989) Structural and functional correlation of the trypsin-digested Ca^{2+} release channel of skeletal muscle sarcoplasmic reticulum. *J. Biol. Chem.* **264**, 1715–1722.

41. Liu, Q. Y., Lai, F. A., Rousseau, E., Jones, R. V., and Meissner, G. (1989) Conductance heterogeneity of the purified calcium release channel complex from skeletal sarcoplasmic reticulum. *Biophys. J.* **55**, 415–424.

42. Meissner, G. (1986) Ryanodine activation and inhibition of the Ca^{2+} release channel of sarcoplasmic reticulum. *J. Biol. Chem.* **261**, 6300–6306.

43. Lattanzio, F. A., Schlatterer, R. G., Nicar, M., Campbell, K. P., and Sutko, J. L. (1987) The effects of ryanodine on passive calcium fluxes across sarcoplasmic reticulum membranes. *J. Biol. Chem.* **262**, 2711–2718.

CHAPTER 18

Identification of a Ligand-Gated Ion Channel by Photoaffinity Labeling and Microsequencing

Ferdinand Hucho

1. Introduction

1.1. Classification and General Architecture

It is generally agreed that ion channels are operationally composed of two functional domains: The selectivity filter comprises the part of the protein that determines which ion may pass and which is retained; the gate determines under which conditions the selected ions may pass. Named according to the mechanisms that regulate the gate, the two groups of ion channels are today known as the ligand-gated and the voltage-gated channels.

With the advent of recombinant DNA techniques in neurobiology, we have learned that this subdivision of channel types is not just semantic, but that structural principles underlie the distinction (Fig. 1): The primary structure of voltage-gated channels contains a motif of six presumed membrane-spanning α-helices. (This motif can be present once only as found with voltage-dependent K^+ channels [1,2] or four times as with voltage-dependent Na^+ [3] and Ca^+ channels [4]). Ligand-gated ion channels are predicted to have only one motif of four transmembrane helices (5–7).

Gate and filter are parts of a channel protein, and these are not easily identified by biochemical means. The following chapter

From: *Methods in Molecular Biology, Vol. 13: Protocols in Molecular Neurobiology*
Edited by: A. Longstaff and P. Revest Copyright © 1992 The Humana Press, Totowa, NJ

307

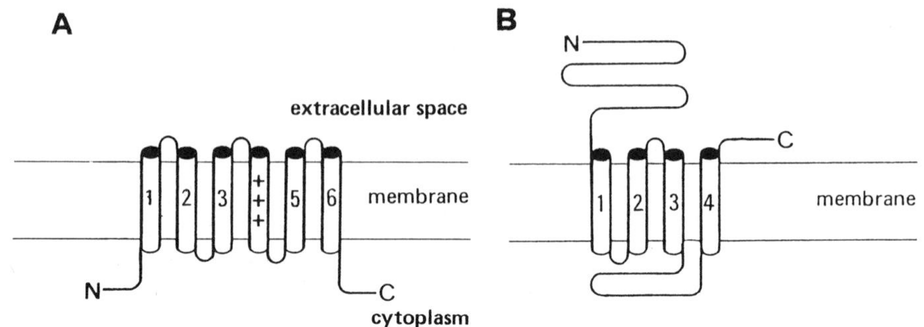

Fig. 1. Predicted pattern of membrane-spanning helices of **(A)** voltage-dependent and **(B)** ligand-gated ion channels.

therefore deals with experimental approaches toward elucidating the structure of a ligand-gated ion channel.

The prototype of the group of ligand-gated ion channels is the nicotinic acetylcholine receptor (AChR). It is a cation-selective channel that is controlled by the neurotransmitter acetylcholine. Before delving into our experiments, the biochemistry of this integral membrane protein has to be briefly summarized. The AChR occurs in the central nervous system, in peripheral ganglia, and at the neuromuscular synapse. Here I focus on the latter. More specifically, I describe the AChR from the electric tissue of the electric ray, *Torpedo*. This receptor has become a model for many receptors of its kind, mainly because it is easily available in milligram quantities. There is ample evidence that the fish receptor is very similar to, and in many basic aspects even identical with, the peripheral AChR from higher vertebrates, including humans.

1.2. Experimental Approaches

The AChR (Fig. 2) is presently the most thoroughly investigated membrane receptor protein *(8–11)*. The most direct method of observing the channel is by means of the electron microscope (Fig. 3). In connection with computer-aided image processing, electron microscopy yielded a fairly comprehensive view of the overall AChR molecule. With respect to the ion channel, this method *(12,13)* provided data on the diameter of the channel entrance and its total length, but could not resolve the most interesting part, the selectivity filter hidden in the membrane. Electrophysiology has been successful in providing data concerning the number of functional (and thereby

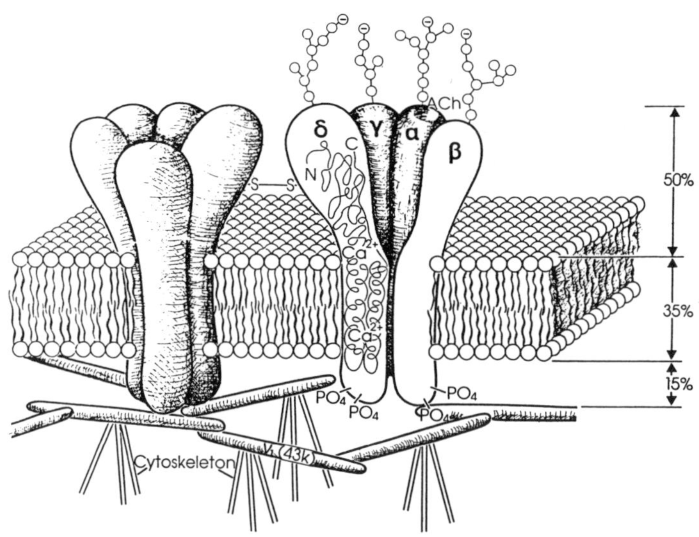

Fig. 2. Biochemical model of the nicotinic acetylcholine receptor from *Torpedo sp. (11).*

structural) states, the kinetics, and even dimensions of the channel *(14).* A more biochemical/biophysical approach was the reconstitution of purified receptor proteins in vesicular or planar artificial lipid membrane systems *(15).* These methods were favored in the 1970s and early 1980s; their most significant result was the proof that the signal receiving and transducing moieties and the effector, the ion channel, were indeed integral components of one and the same protein, which had been purified by affinity chromatography, a method directed toward the ligand-binding site only.

The AChR has a relative mol mass of approx 290,000, including carbohydrate residues. Its heteropentameric quaternary structure is $\alpha_2\beta\gamma\delta$. The subunit's primary structures have been deduced from cloned cDNAs encoding the respective subunit precursors. Based on these amino acid sequences, several alternative and controversial secondary structures have been predicted, among which those models that portray four membrane-spanning helices (called M1–M4) are the most widely accepted *(5,16).*

All five receptor subunits span the membrane, all are glycosylated on the extracellular surface, and all extend both their C- and N-terminal ends to this side of the plasma membrane. Possibly with the excep-

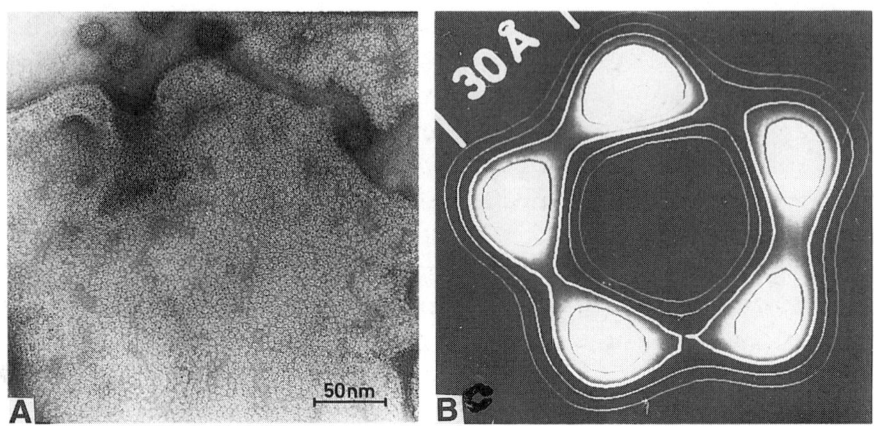

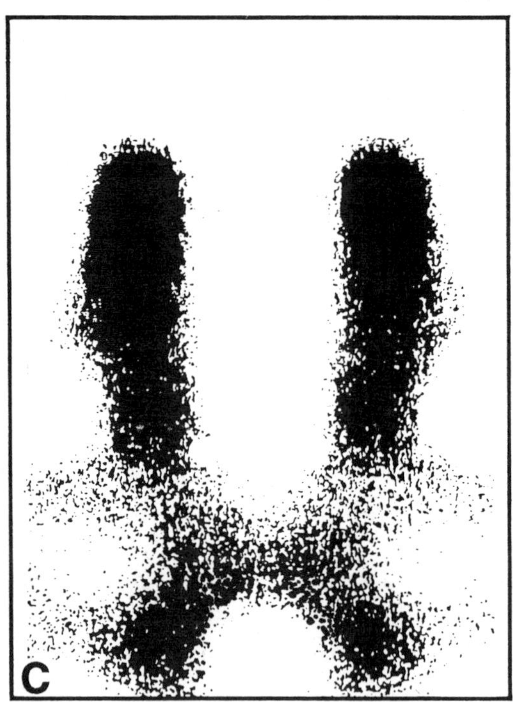

Fig. 3. Electron microscopy of nicotinic acetylcholine receptors. **(A)** A receptor-rich membrane from *Torpedo* electric tissue (negative stain). The ring-like structures represent AChR embedded in the membrane. **(B)** An individual receptor molecule as obtained by image processing of **(A)** *(13)*. The five light spots probably represent the receptor subunits. The dark area in the center is assumed to be the channel entrance. **(C)** Side view of the receptor as reconstructed from an unstained vitrified membrane preparation *(12)*.

tion of the α subunits, all subunits are multiply phosphorylated. The receptor contains large amounts of Ca^{2+}. The functional roles of these posttranslational additions to the receptor protein (including its sialic and fatty acid moieties) are at present unclear *(11)*.

The development of molecular biology has allowed many advances in receptor research. The primary structures of the receptor subunits have been deduced from cDNAs, not only for *Torpedo* AChR, but also for AChRs from many other sources and for an ever-increasing number of other receptors *(17)*. By comparison of the sequence data, common structural principles were recognized among different receptors. Site-directed mutagenesis with expression of the mutated receptor subunits in cells accessible to patch-clamp investigations has begun to reveal the relationship between structure and function in these molecules. For example, it has been shown that the δ subunit plays a special role in determining the channel conductance *(18)*. Within the δ subunit, the helix M2 has been singled out as an important protein domain *(19)*, supporting the channel model derived from our photoaffinity experiments *(20)* described below. Also, by removing or replacing individual amino acids, a very detailed picture of the channel architecture has been obtained *(21,22)*.

Seemingly old-fashioned photoaffinity labeling in connection with microsequencing can give the most direct and accurate picture of the AChR's ion channel *(20,23)*. I now describe this approach in more detail as a valuable example of a technique with wide applicability in molecular neurobiology.

1.3. Photoaffinity Labeling in AChR Research (24)

Both modified (e.g., azido or diazo) and unmodified ligands have been used in receptor research. For example, benzodiazepine-binding proteins were identified in brain membranes by photolabeling with radioactive flunitrazepam; strychnine was the tool used for identifying (and later isolating) the glycine receptor from mammalian spinal cord. With the nicotinic acetylcholine receptor, photoaffinity labeling was successfully applied in various investigations.

In 1970, Kiefer et al. *(24a)* introduced 4-azido-2-nitrobenzyltrimethylammoniumfluoroborate. With its quaternary ammonium group, it possesses some structural analogy to the cholinium groups of the cholinergic agonist acetylcholine and other agonists and antagonists. Accordingly, it was shown to inactivate (upon irradiation) AChR in frog satorius muscle. With the radioactive arylazide *(25)*, we obtained evidence that the smallest (α) polypeptide chains of AChR contained the binding sites for cholinergic antagonists and that the three large chains (β, γ, and δ) also contained binding sites for quaternary ammonium compounds.

Later this was confirmed with arylazido derviatives of α neurotoxins from snake venoms *(26–28)*. Photoaffinity labeling in this case was found mainly in the α-chains, but some crosslinking of toxin to the δ-chain could also be obtained. This was interpreted as indicative of a quaternary structure placing the δ-polypeptide chain next to a α-subunit.

Dennis and coworkers from the Institut Pasteur used [^{3}H]-*p*-(diemthylamino)-benzenediazonium fluoroborate (DDF) in an attempt to photolabel amino acids that form the acetylcholine binding site of the receptor *(29)*. More sophisticated applications of photoaffinity labeling are attempts to perform labeling with a high time resolution in order to examine short-lived receptor states.

1.4. Pharmacology of the AChR and Its Ion Channel

Much effort focused on the AChR's ion channel as one of the basic functional domains of the receptor. It seems to be the site of action of a large variety of drugs called noncompetitive antagonists *(30)*. Among this group of receptor ligands are local anesthetics, such as lidocaine, detergents, such as Triton-X 100, neuroleptics, such as chlorpromazine, natural toxins, including histrionicotoxin from the skin of certain frogs, and triphenylmethylphosphonium (TPMP$^+$), the one compound that we added to this list *(31,32)* and that was shown to be an especially useful tool for elucidating channel components *(23,31,33,34)* (Fig. 4).

2. Materials

2.1. Solutions and Tissue

1. Sodium buffer used in the preparation of receptor-rich membranes: 0.4M NaCl, 20 mM Na$_2$HPO$_4$/NaH$_2$PO$_4$, pH 7.4, 2 mM Na$_4$EDTA, 0.1 mM, phenylmethylsulfonyl-fluoride (PMSF).
2. Solution used in density gradient centrifugation: 25–50% sucrose in H$_2$O, 0.02% NaN$_3$.
3. Ringer solution: 0.160M NaCl, 5 mM KCl, 3 mM Na$_2$HPO$_4$/NaH$_2$PO$_4$, pH 7.0, 2 mM CaCl$_2$, 2 mM MgCl$_2$.
4. Sample buffer used for electrophoretic separation of labeled subunits: 0.25M Tris-HCl, pH 6.8, 10% glycerol, 5% β–mercaptoethanol, 3% sodiumdodecyl sulfate.

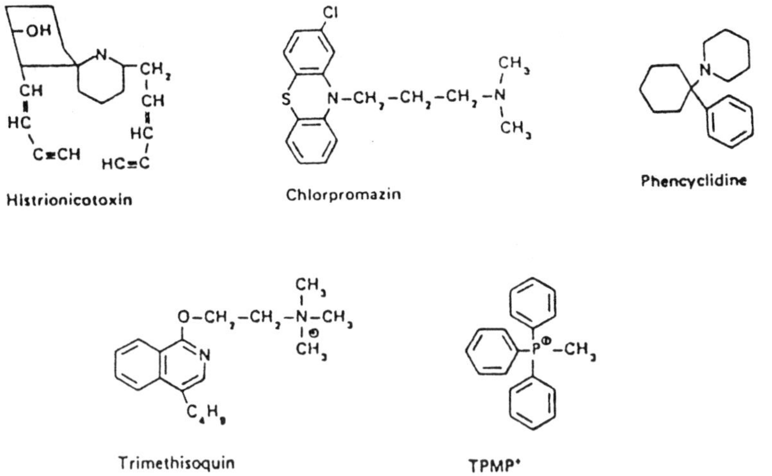

Histrionicotoxin

Chlorpromazin

Phencyclidine

Trimethisoquin

TPMP⁺

Fig. 4. Examples of channel-blocking molecules that have been used without further modification as photoaffinity labels.

5. Elution buffer used in preparative electrophoresis: 0.576M glycine in 0.075M Tris-HCl buffer, pH 8.3, 0.1% sodiumdodecyl sulfate.
6. The tissue used in these investigations is the electric organ from *Torpedo californica*. It was purchased as tissue frozen in liquid N_2. Tissue excized from live specimen of the electric fish was used in some cases; the results were essentially the same.

2.2. Apparatus

2.2.1. HPLC

The gradient liquid chromatograph was composed of two HPLC pumps, model 64.00, controlled by a 50 B microprocessor programmer and a variable-wavelength UV detector, model 87.00 (all from Knauer, Berlin), an automatic sampler Wisp 710 B from Waters Assoc., USA, a date processor Chromatopac C-R3A from Shimadzu, Japan, a recorder, model 2210, from LRB, Munich, a degasser model RC-3320 from ERC, Regensburg, and a fraction collector microcol TCD 80 from Abimed, Düsseldorf.

2.2.2. Sequencer

A high-quality sequencer is required, which should give good results in the 100-pmol protein range. In our experiments, we used a model 470 A sequencer from Applied Biosystems.

3. Methods

3.1. Photolabeling with Millisecond Time Resolution (35)

Three groups proposed methods for photoaffinity labeling nicotinic acetylcholine receptors with a resolution in the 2–100 ms time range (Fig. 5), in order to identify protein domains involved in channel structure and function. The receptor ion channel opens for a few milliseconds only, a time span too short for classical photolabeling. However, even without this high time resolution, receptor domains forming the channel can be identified for reasons discussed later.

Fahr and Hucho *(35)* combined a stopped-flow apparatus with a high-energy pulse laser (Fig. 5C). The laser is triggered by a switch that is closed by the movement of the pistons of the syringes containing the protein and ligand, respectively. The trigger can be delayed electronically to start the photoreaction at variable time intervals after mixing protein and ligand in the mixing chamber of the stopped-flow apparatus. The delay time can be varied from 0 ms to 500 s. A detailed description of the experimental setup has been published elsewhere *(35);* first results from photoaffinity labeling experiments have proven its usefulness *(36)*.

A different device for photolabeling after rapid mixture was developed by Karlin and coworkers *(37,38)* (Fig. 5A). In this case, the main aim was to accomplish labeling of large amounts of protein by means of a continuous-flow, rapid-mixing technique. This resembled the experimental setup designed by Changeux and his coworkers *(39)* (Fig. 5B) for photolabeling the nicotinic acetylcholine receptor with tritiated chlorpromazine and other drugs with similar properties. The main difference between the three methods (Table 1) is the time resolution. Photolabeling with normal W light sources requires longer irradiation, and the design of the mixing chamber and tubing system determines the delay time between mixing and photolysis. In the method used by Cox et al. *(37)*, 20 ms are the obtainable resolution. Heidmann and Changeux *(39)* apply their apparatus in the 100 ms to seconds range. The delay time of Fahr and Hucho's apparatus *(35)* is 2.4 ms. Another important parameter is the yield, which depends largely on the reactants and which has not been compared for the three methods under similar conditions.

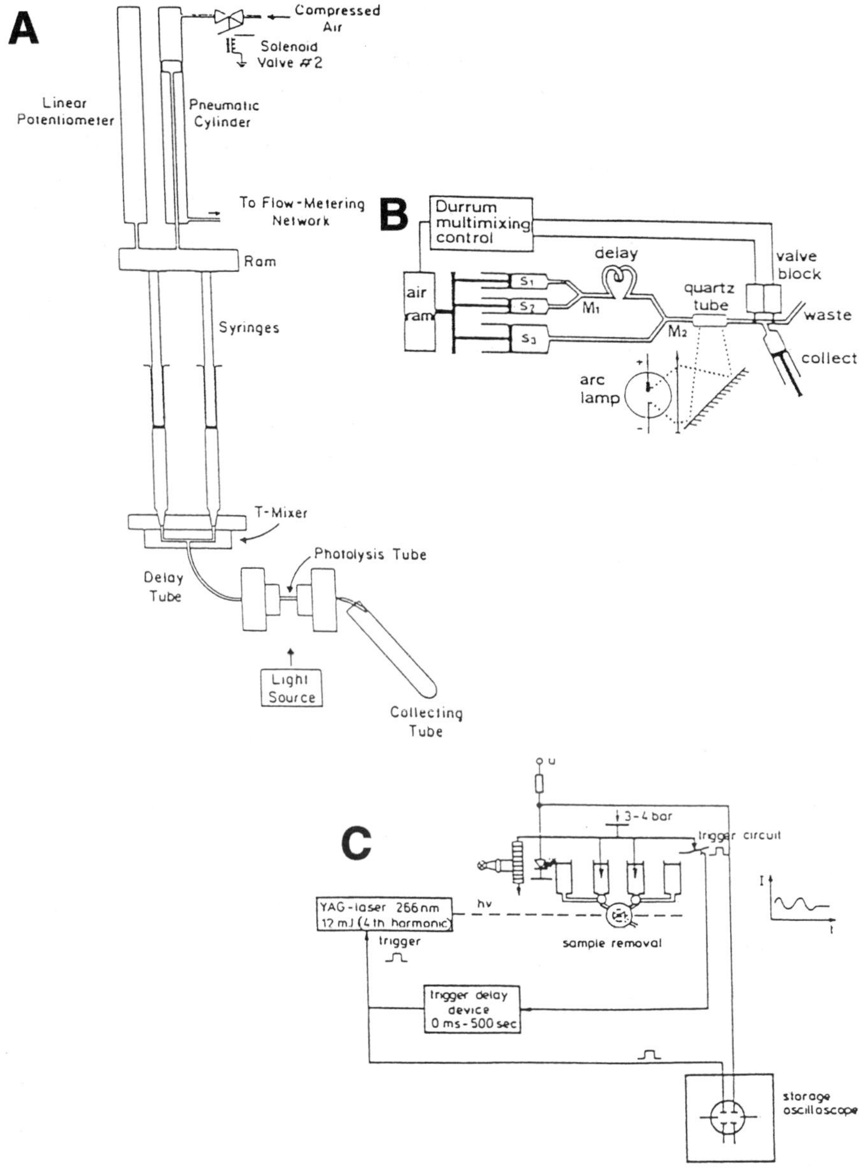

Fig. 5. Three alternative apparatus for rapid photoaffinity labeling *(34):* **(A)** continuous-flow rapid mixing apparatus according to Cox et al. *(37),* **(B)** rapid-mixing photolabeling apparatus used by Heidmann and Changeux *(39),* and **(C)** laser-flash apparatus for photoaffinity labeling with high time resolution, designed by Fahr and Hucho *(35).*

Table 1
Comparison of Three Rapid Photolabeling Methods

	Cox et al. (37)	Heidmann and Changeaux (39)	Fahr and Hucho (35)
Delay time	20 ms	50 ms	2.4 ms
Photolysis	1 ms	<20 ms	5 ns
Lower limit of time resolution	~20 ms	~100 ms	2.4 ms

3.2. Photolabeling at Equilibrium Conditions

3.2.1. Preparation of AChR-Rich Membranes

Detergent extraction and affinity purification may alter the AChR structure significantly. All experiments aimed at elucidating the native structure, especially of the AChR channel, are therefore performed with the receptor in its membrane-bound state. Membranes especially rich in AChR are routinely obtained as follows *(40)*:

1. Homogenize 1 vol frozen electric organ from *Torpedo californica* with 2 vol of ice-cold sodium buffer for 3 min at high speed in a Waring blender.
2. Centrifuge the homogenate (27,000g, 90 min, 4°C). Rehomogenize the pellet in 1 vol of sodium-free buffer for 3 min at low speed. Centrifuge this homogenate (37,000g, 90 min, 4°C), and homogenize the resulting pellet again in the sodium-free buffer (3 min, at low speed).
3. After centrifugation, scrape the soft pellet off the denser bottom part, and resuspend in the sodium-free buffer (1 mL/3 g electric tissue). Centrifuge the suspension at low speed (1000g, 10 min, 4°C), and load the supernatant onto a gradient of 25–50% sucrose made up in water containing 0.02% NaN_3. The supernatant obtained from 100 g tissue is layered into six tubes, each containing 25 mL of the sucrose gradient.
4. Centrifuge the gradient tubes (59,000g, 8 h, 4°C), and then fractionate them (1-mL fractions). Usually, fractions 5–15 (from bottom to top) contain the AChR-rich membranes, as determined by a binding assay using radioactive α-bungarotoxin.
5. Dilute the pooled fractions fivefold with water, centrifuge, and take up the pellet in Tris buffer to about 1 mg protein/mL.

3.2.2. Radioactive Labeling with the Channel Blocker [³H]TPMP⁺

Covalent incorporation of [³H]TPMP⁺ into the receptor-rich membranes by UV irradiation is achieved in the following way: Before use, the sodium Ringer solution is extensively deaerated with N_2 (30 min).

1. Mix the AChR-membrane suspension (33 μL) in sodium Ringer solution with 11 μL of [³H]TPMP⁺ (specific radioactivity should be at least 50 Ci/mmol); add either cholinergic agonist (e.g., carbamoyl choline) in sodium Ringer solution or further sodium Ringer solution to a final vol of 66 μL.
2. Leave the incubation mixtures to equilibrate on ice for 8 min.
3. Subsequently add carbamoylcholine ($10^{-4}M$) and [³H]TPMP⁺ ($10^{-5}M$), and leave the mixture for binding equilibration. For UV irradiation, place the whole sample mixture on a cover slide that is mounted on an ice-cooled steel plate. Irradiation is for 2 min with a prewarmed (30 min) UV lamp (0.35 W) without a filter, positioned 8 cm above the sample.

3.2.3. Purification of the [³H]TPMP⁺-Labeled Polypeptide Chains

Dissolve the photolabeled receptor-rich membranes in sample buffer *(41)*, and separate the polypeptide chains in two steps by preparative SDS-polyacrylamide gel electrophoresis. The dimensions of the gel are: diameter 0.11 cm, length 5.5 cm, 7.5% polyacrylamide (lower gel) plus 3 cm, 3% polyacrylamide (upper gel). In the first run, the upper gel is 3% SDS and the lower 7.5%. About 4 mg protein is applied. Electrophoresis is performed at 6 mA and 150 V. The elution buffer is as described in Section 2.1. The elution rate is 15 mL/h.

Assess the purity of the chains in the fractions by analytical SDS-polyacrylamide gel electrophoresis. Pool the fractions containing predominantly δ-chains and reelectrophorese as described earlier, after dialysis against water and lyophilization, but on a 10% gel.

3.2.4. CNBr Cleavage

After reducing the lyophilized chains collected from the preparative electrophoresis with β-mercaptoethanol, treat with approx 100-fold molar excess of CNBr over the methionine residues present, in 70% formic acid up to 24 h *(42)*. The reaction takes place under nitrogen and in the dark.

3.2.5. Tryptic Digestion

Digest the dialyzed chains for up to 20 h with about 10% (w/w of protein) L-1-tosylamide-2-phenylethylchloromethyl (TPCR)-trypsin in a large volume (5–20 mL) of water adjusted to pH 9 with ammonia. Stop the digestion by adding an excess of concentrated formic acid. Reduce the volume of the hydrolysate in a vacuum concentrator before HPLC. The protein concentration during digestion should be 15 μg/mL.

3.2.6. Reversed-Phase Chromatography (HPLC)
and Microsequencing

1. The separation was performed on a prepacked steel column purchased from Knauer, Berlin, 250 × 4.0 mm id, filled with Organogen HP-Gel-RP-7, pore size 300 Å, particle size 7 μm.
2. Elute the proteins at 60°C. with gradients from buffer A (0.1% trifluoroacetic acid [TFA] in water) and buffer B (0.03% TFA in 2-propanol:acetonitrile 70:30).
3. Perform automatic microsequencing by liquid-phase sequencing on a gas-phase sequencer. Identification of the released phenylthiohydantoin (PTH) amino acids was made on line, employing isocratic, recycling HPLC or by injecting in a separate isocratic system. Count one-third of each PTH-amino acid fraction in 5 mL Supertron in a liquid scintillation counter.

The result *(30)* of this type of experiment is the localization of the label to amino acids Ser 262 in the δ chain, Ser 254 in the β chain, and Ser 248 in the α chain (*see* Section 4.1.).

3.2.7. Labeling of Different Receptor States (34)

In principle, the AChR can exist in many conformational and functional states. The channel is open for permeating cations only transiently, for a few milliseconds at most. Photolabeling of the open channel under equilibrium conditions is therefore not possible. A rapid labeling method as described in Section 3.1. might be feasible for this purpose. At equilibrium, three receptor states can be produced and kept for slow labeling techniques: the so-called resting, desensitized, and antagonist states *(30,34,43)*, depending on whether no agonist, an excess of agonist, or an antagonist is present, respectively. In all of these states, the channel is closed, and since the same labeling sites were found in each case, it was postulated that the binding site for channel blockers, such as TPMP$^+$, must be located in a wider part of the funnel-shaped channel mouth. The selectivity filter must be located deeper down in the funnel, closer to the cytoplasmic side of the membrane.

3.2.8. Channel Modeling

The positions photolabeled by the channel blocker [^{3}H]TPMP$^+$ turned out to be located in homologous sequences of the α, β, and δ subunits of the AChR. These sequences are therefore assumed to form the channel wall. The identified sequences are predicted to form membrane-spanning helices, termed the M2 domain. The results of photoaffinity labeling led therefore to the so-called M2-helix model of

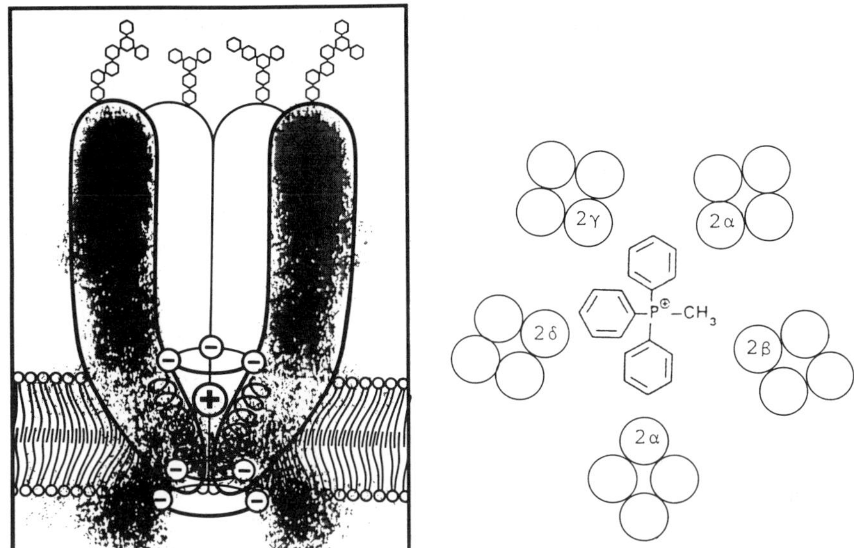

Fig. 6. The Helix-M2 model of the nicotinic acetylcholine receptor (8,20). Left: The figure represents a superimposition of results from electron microscopy *(12)*, electrophysiology *(21)*, and biochemistry *(20)*. Right: Cross-section through the model at the plane of the site labeled by TPMP⁺. Each group of four circles represents the four membrane-spanning helices of one of the subunits. The helices M2 form the inner wall of the channel.

the ligand-gated AChR-ion channel (Fig. 6). The model was nicely supported by site-directed mutagenesis in combination with patch-clamp measurements *(21)* and by electron microscopy in combination with computer-aided image processing *(12)*. Results from these three methods are superimposed in Fig. 6. Computer modeling using the information obtained from protein chemistry led to the postulation that the selectivity filter of the channel is formed by the negatively charged side chains of amino acids α-Glu 241, β-Glu 244, and δ-Glu 255, and the polar amide side chain γ-Gln 250 *(8)*.

4. Notes

In the case of [³H]TPMP⁺-labeled CNBr peptides from the δ chain of AChR, a maximum of radioactivity was released in the fifth Edman step, which could be tentatively assigned to position Ser 262 following Met 257, the site of CNBr cleavage. Since the yield of photolabeling was very low (about 1%), one could not be sure whether the radioactivity was truly associated with this position or was contained in another peptide being present as a contamination. Therefore, in a second experiment peptides obtained by tryptic digestion were sequenced. Now the radioactivity maximum was at the sixth Edman step. From this it was concluded that the site of [³H]TPMP⁺ labeling was located five steps "downstream" from a CNBr and six steps from a tryptic cleavage site. At five or six residues "upstream" from the labeled amino acid, a Met or a basic amino acid, respectively, must therefore have been located. The sequence Lys(Arg)-Met, fortunately, occurs only once in the AChR primary structure; this allowed the unambiguous localization of the label to position δ-Ser 262. By the same experimental procedure, the photolabel was localized to Ser 248 in the α chain and Ser 254 in the β chain. Labeling of γ subunit was not sufficient for localization.

Acknowledgments

Work from the author's lab described in this chapter was supported by the Deutsche Forschungsgemeinschaft (SfB 312) and the Fonds der Chemischen Industrie.

References

1. Papazian, D. M., Schwarz, R. L., Tempel, B. L., Jan, Y. N., and Jan, L. Y. (1987) Cloning of genomic and complementary DNA from Shaker, a putative potassium channel gene from Drosophila. *Science* **237,** 749–753.
2. Pongs, O., Kecskemethy, N., Müller, R., Krah-Jentgens, I., Baumann, A., Kiltz, H. H., Canal, I., Llamazares, S., and Ferrus, A. (1988) Shaker encodes a family of putative potassium channel proteins in the nervous system of Drosophila. *EMBO J.* **7,** 1087–1096.
3. Noda, M., Shimizu, S., Tanabe, T., Takai, T., Rayanao, T., Ikeda, T., Takahashi, H., Nakayama, H., Ranoka, Y., Miniamino, N., Rangawa, R., Natsuo, H., Raftery, M. A., Hirose, T., Inayama, S., Hayashida, H., Miyata, T., and Numa, S. (1984) Primary structure of Electrophorus electricus sodium channel deduced from cDNA sequence. *Nature* **312,** 121–127.
4. Tanabe, T., Takeshima, H., Mikami, A., Flockerzi, V., Takahashi, H., Rangawa, K., Rohima, M., Matsuo, H., Hirose, T., and Numa, S. (1987) Primary structure of the receptor for calcium channel blockers from skeletal muscle. *Nature* **328,** 313–318.
5. Claudio, T., Ballivet, M., Patrick, J., and Heinemann, S. (1983) Nucleotide and deduced amino acid sequences of *Torpedo californica* acetylcholine receptor subunit. *Proc. Natl. Acad. Sci. USA* **80,** 1111–1115.
6. Schofield, P. R., Darlison, M. G., Fujita, N., Burt, D. R., Stephenson, F. A., Rodriques, H., Rhee, L. M., Ramachandran, J., Reale, V., Glencourse, T. A.,

Seeburg, P. H., and Barnard, E. A. (1987) Sequence and functional expression of the GABA_A receptor shows a ligand-gated receptor super-family. *Nature* **328**, 221–227.

7. Grenningloh, G., Rienitz, A., Schmitt, B., Methfessel, C., Zensen, M., Beyreuther, L., Gundelfinger, E. D., and Betz, H., (1987) The strychnine-binding subunit of the glycine receptor shows homology with nicotinic acetylcholine receptor. *Nature* **328**, 215–220.

8. Hucho, F. and Hilgenfeld, R. (1989) The selectivity filter of a ligand-gated ion channel. *FEBS Lett.* **257**, 17-23.

9. Changeux, J.-P., Devillers-Thièry, A., and Chemouilli, P. (1984) Acetylcholine receptor: an allosteric protein. *Science* **225**, 1335–1345.

10. Maelicke, A. (1988) Structure and function of the nicotinic acetylcholine receptor, in *Handbook of Experimental Pharmacology*, vol. 86 (V. P. Whittaker, ed.), Springer-Verlag, Berlin/Heidelberg, pp. 267–313.

11. Hucho, F. (1986) The nicotinic acetylcholine receptor and its ion channel. *Eur. J. Biochem.* **158**, 211–226.

12. Toyoshima, C. and Unwin, N. (1988) Ion channel of acetylcholine receptor reconstructed from images of postsynaptic membranes. *Nature* **336**, 247–250.

13. Kunath, W., Giersig, M., and Hucho, F. (1989) The electron microscopy of the nicotinic acetylcholine receptor. *Electron Microsc. Rev.* **2**, 349–466.

14. Hille, B. (1984) *Ionic Channels of Excitable Membranes.* Sinauer, Sunderland.

15. Levitzki, A. (1985) Reconstitution of membrane receptor systems. *Biochim. Biophys. Acta* **822**, 127–153.

16. Guy, H. R. and Hucho, F. (1987) The ion channel of the nicotinic acetylcholine receptor. *TINS* **10**, 8, 318–321.

17. Maelicke, A., ed. (1988) Molecular biology of neurotransmitter receptors and ion channels (NATO ASI Series H), *Cell Biology*, vol. 32. Springer-Verlag, Berlin, Heidelberg.

18. Noda, M., Takahashi, H., Tanabe, T., Toyosato, M., Rikyotani, S., Furutani, Y., Hirose, T., Takashima, H., Inayama, S., Miyata, T., and Numa, S. (1983) Structural homology of *Torpedo californica* acetylcholine receptor subunits. *Nature* **302**, S28–S32.

19. Imoto, K., Methfessel, C., Sakmann, B., Mishina, M., Mori, Y., Ronno, T., Fukuda, K., Kurasaki, M., Bujo, H., Fujita, Y., and Numa, S. (1986) Location of a δ-subunit region determining ion transport through the acetylcholine receptor channel. *Nature* **324**, 670–674.

20. Hucho, F., Oberthür, W., and Lottspeich, F. (1986) The ion channel of the nicotinic acetylcholine receptor is formed by the homologous helices M II of the receptor subunits. *FEBS Lett.* **205**, 137–142.

21. Imoto, K., Busch, C., Sakmann, B., Mishina, M., Konno, T., Nakai, J., Bujo, H., Mori, Y., Fukuda, K., and Numa, S. (1988) Rings of negatively-charged amino acids determine the acetylcholine receptor channel conductance. *Nature* **335**, 645–648.

22. Dani, J. A. (1986) Ion-channel entrances influence permeation—net charge, size, shape and binding considerations. *Biophys. J.* **49**, 607–618.

23. Oberthür, W., Muhn, P., Baumann, H., Lottspeich, F., Wittmann-Liebold, B., and Hucho, F. (1986) The reaction site of a noncompetitive antagonist in the δ-subunit of the nicotinic acetylcholine receptor. *EMBO J.* **5,** 8, 1815–1819.

24. Hucho, F. and Oberthür, W. (1988) Phottoaffinity labelling and localization by microsequencing of an ion channel protein, in *Modern Methods in Protein Chemistry,* vol. 3 (Tschesche, H., ed.), Walter der Gruyter, Berlin.

24a.Kiefer, H., Lindstrom, J., Lemmox, E. S., and Singer S. J. (1970) *Proc. Natl. Acad. Sci. USA* **67,** 1688–1694.

25. Hucho, F., Layer, P., Kiefer, H. R., and Bandini, G. (1976) Photoaffinity labeling and quaternary structure of the acetylcholine receptor from *Torpedo californica. Proc. Natl. Acad. Sci. USA* **73,** 2624–2628.

26. Hucho, F. (1979) Photoaffinity derivatives of α-bungarotoxin and α-Naja naja siamensis toxin. *FEBS Lett.* **103,** 27–32.

27. Witzemann, V., Muchmore, D., and Raftery, M. A. (1979) Affinity-directed cross-linking of membrane-bound acetylcholine recetor polypeptides with photolabile α-bungarotoxin derivatives. *Biochemistry* **24,** 5511–5518.

28. Tsetlin, V., Pluzhnikov, K., Karelin, A., and Ivanov, V. (1983) Acetylcholine receptor interaction with the neurotoxin II photoactivatable derivatives, in *Toxins as Tools in Neurochemistry* (Hucho, F. and Ovchinnikov, Yu, eds.), Walter de Gruyter, Berlin.

29. Dennis, M., Giraudat, J., Kotzyba-Hibert, F., Doeldner, M., Hirth, C., Chang, J.-Y., Lazure, C., Chrétien, A., and Changeux, J.-P. (1988) Amino acids of the *Torpedo marmorata* acetylcholine binding site. *Biochemistry* **27,** 2346–2357.

30. Changeux, J.-P. (1981) The acetylcholine receptor: an allosteric membrane protein. *Harvey Lect.* **75,** 85–254.

31. Lauffer, L. and Hucho, F. (1982) Triphenylmethylphosphonium is an ion channel ligand of the nicotinic acetylcholine receptor. *Proc. Natl. Acad. Sci. USA* **79,** 2406–240.

32. Spivak, C. E. and Albuquerque, E. X. (1985) Triphenylmethylphosphonium blocks the nicotinic acetylcholine receptor noncompetitively. *Mol. Pharmacology* **27,** 246–255.

33. Muhn, P. and Hucho, F. (1973) Covalent labeling of the acetylcholine receptor from Torpedo electric tissue with the channel blocker [^{3}H]-triphenylmethylphosphonium by ultraviolet irradiation. *Biochemistry* **22,** 421–425.

34. Oberthür, W. and Hucho, F. (1988) Photoaffinity labeling of functional states of the nicotinic acetylcholine receptor. *J. Prot. Chem.* **7,** 141–150.

35. Fahr, A. and Hucho, F. (1986) A stopped-flow apparatus for photoaffinity labeling studies in the milliseconds time range. Application in investigations of the nicotinic acetylcholine receptor. *J. Neurosci. Meth.* **16,** 29–38.

36. Muhn, P., Fahr, A., and Hucho, F. (1984) Rapid laser flash photoaffinity labeling of binding sites for a noncompetitive inhibitor of the acetylcholine receptor. *Biochemistry* **23,** 2725–2730.

37. Cox, R. N., Kaldany, R.-R., Brandt, P.W., Ferren, B., Hudson, R. A., and Karlin, A. (1984) A continuous-flow, rapid-mixing, photolabeling technique applied to the acetylcholine receptor. *Anal. Biochem.* 136, 476–486.

38. Cox, R. N., Kaldany, R.-R., DiPaola, M., and Karlin, A. (1985) Time-resolved photolabeling by quinacrine azide of a noncompetitive inhibitor site of the nicotinic acetylcholine receptor in a transient, agonist-induced state. *J. Biol. Chem.* **260,** 7186–7193.

39. Heidmann, T. and Changeux, J.-P. (1984) Time-resolved photolabeling by the noncompetitive blocker chlorpromazine of the acetylcholine receptor in its transiently open and closed ion channel conformations. *Proc. Natl. Acad. Sci USA* **81,** 1897–1901.

40. Hertling-Jaweed, S., Bandini, G., and Hucho, F. (1990) Purification of nicotinic acetylcholine receptors, in *Receptor Biochemistry: A practical approach,* (E. Hulme, ed.), Oxford University Press, Oxford, UK.

41. Laemmli, U. K. (1970) Cleavage of structural proteins during the assembly of the head of bacteriophage T4. *Nature* **227,** 680–685.

42. Gross, E. and Winthrop, B. (1962) Nonenzymatic cleavage of peptide bonds: the methionine residues in bovine pacreatic ribonuclease. *J. Biol. Chem.* **237,** 1856–1860.

43. Fahr, A., Lauffer, L., Schmidt, D., Heyn, M. P., and Hucho, F. (1985) Covalent labeling of functional states of the acetylcholine receptor. *Eur. J. Biochem.* **147,** 483–487.

CHAPTER 19

Voltage-Gated Ion Channels

Electrophysiological Approaches

Nicholas B. Standen and Peter R. Stanfield

1. Introduction

1.1. Voltage-Gated Channels

Ion channels are membrane-spanning protein molecules. They form pores or channels, through which ions flow down their electrochemical gradients. Channels are characterized by two properties, selectivity and gating. Thus, the pore selects for one or a few ion species, allowing only these to permeate, and the pore can open and close in response to changes in the membrane voltage field, or to the binding of chemical transmitters. Because movement of ions through the pore can be measured by the electrical current they carry, it has been possible to study the functioning of these proteins with high resolution. Channel openings and closings, reflecting changes in protein conformation, can be detected on the time scale of a few tens of microseconds in some cases, and over the past 10 years, methods allowing study of events in a single-channel protein molecule have become widely used.

In this chapter, we consider only those channels gated by membrane voltage, often taking as an example the "delayed rectifier" potassium selective channel involved in the action potentials that form the electrical signals transmitted in nerve and muscle cells. We summarize the techniques used for electrical recording from ion channels, both as populations and individually. We then describe the ways

From: *Methods in Molecular Biology, Vol. 13: Protocols in Molecular Neurobiology*
Edited by: A. Longstaff and P. Revest Copyright © 1992 The Humana Press, Totowa, NJ

in which a kinetic picture describing channel function may be built up. Such pictures form the basis for interpreting the functional effects of alterations in channel structure produced, for example, by site-directed mutagenesis, and the functional state of channels reincorporated into lipid bilayers or expressed in native cell membranes.

1.2. Voltage Clamp

Much of what we know about ion channel properties and function comes from the use of variants of one basic technique, developed 40 years ago—the voltage clamp *(1–3)*. Here we explain what voltage clamp is, and describe some of the basic ways in which it may be implemented and used, both in the study of populations of many channels and of single channels.

Voltage-clamp techniques provide means for controlling or clamping the electrical potential experienced by channel proteins, normally by using electronic circuits to control the voltage across the membrane, native or artificial, in which the channels are situated. Why should this seemingly esoteric technique prove so powerful for studying channels? There are a number of advantages. First, the current measured from cellular membranes consists of the sum of a capacitative current, flowing when the voltage across the membrane lipid bilayer changes, and an ionic or resistive current, most of which flows through ion channels. Although components of capacity current related to conformational changes in channel proteins (gating currents) can be detected, most studies of channel behavior use ionic current to investigate permeation and gating of channels. The capacitative component of membrane current is proportional to the rate of change of voltage, dV_m/dt, and so is zero when voltage is clamped at a constant value, enabling ionic current alone to be recorded. Secondly, the electrochemical gradient for permeant ions is usually fixed, giving a constant ion flux through an open channel, so that changes in current represent changes in channel opening. Finally, when membrane voltage can change freely, there is a complex interdependence between it and the opening and closing of voltage-gated channels. Thus, current flowing through channels affects membrane voltage, which in turn affects channel gating and therefore the current. By imposing external control on membrane voltage, the voltage clamp breaks this interdependence, making it much easier to interpret the way in which voltage controls channel gating.

Channels are often studied using the voltage clamp to impose a step change in voltage, which is shifted very rapidly from one steady level to another. In simple kinetic terms, a channel that has just two states, closed (C) and open (O) can be described by

$$C \underset{\beta}{\overset{\alpha}{\rightleftharpoons}} O \tag{1}$$

so that the probability that the channel is open (P_{open}) is determined by the rate constants α and β, being $\alpha/(\alpha + \beta)$. For voltage-gated channels, these rate constants reflect the response of a charged part of the protein molecule, the voltage sensor, to the electrical potential, and they will be exponential functions of voltage (e.g., ref. *4*). An imposed voltage step produces an effectively instantaneous change in the value of the rate constants, with a corresponding new equilibrium P_{open}, and the relaxation to this new value (which for scheme [1] will occur with a time constant $1/[\alpha + \beta]$) can be studied. Although the kinetic scheme for most channels is more complicated (*see Section 3.*), the basic argument and methodology are the same.

1.2.1. Macroscopic and Unitary Currents

Voltage-clamp methods are available for measuring the ionic currents both from individual channel proteins (unitary currents) and through large populations of channels. The latter are called macroscopic currents, and here we consider the relation between these and the underlying single-channel currents, together with the advantages and disadvantages of the two types of measurement.

At the level of individual proteins and at any given voltage, the current flowing through most voltage-gated ion channels changes predominantly between only two levels, zero, corresponding to a closed pore, and the single-channel current that flows when the pore is open (Fig. 1). The size of this current and the slope of its dependence on voltage, the single channel conductance ($=\Delta i/\Delta V$), are basic parameters measured in single-channel recording, giving information under suitable experimental conditions both on pore size and on selectivity among ions. The open and closed current levels may each correspond to several different underlying states of the channel, distinguishable as described later. The other basic parameters measured from single-channel currents are the times spent at the closed and open current levels (dwell times). Because random thermal vibrations of the pro-

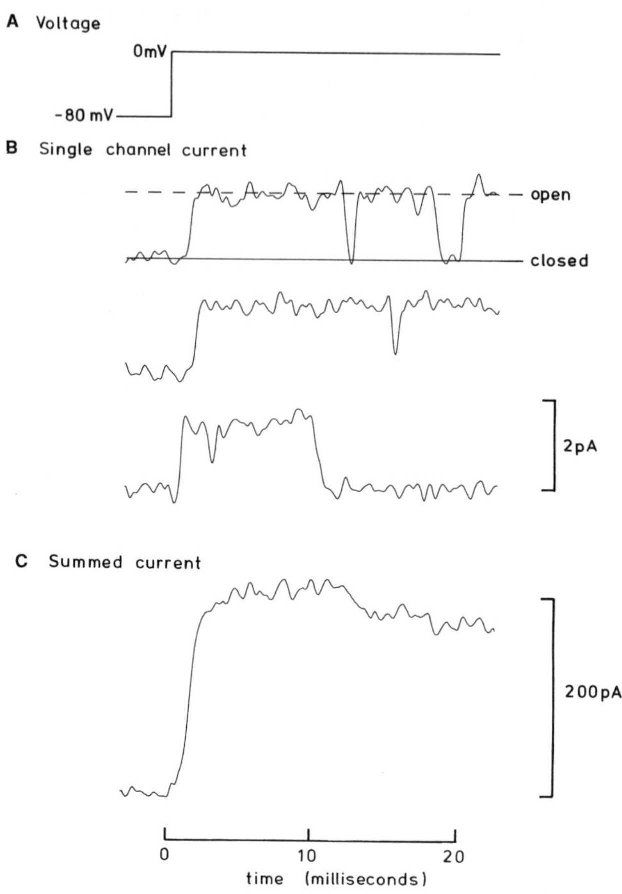

Fig. 1. Recording from single ion channels. The records shown are from delayed rectifier K[+] channels of frog muscle, recorded in an inside-out patch of membrane (*see* Fig. 3). **A.** The membrane voltage was changed from −80 mV to 0 mV. **B.** Three examples of the currents recorded in response. Channel opening leads to an abrupt change from the closed level, where no current flows through the channel, to the open level, where the single-channel current is about 1.3 pA. The records are noisy because of the high gain of the recording, and the apparent slope of the transitions between levels is caused by the 1.2 kHz low-pass filtering of the records. **C.** Summed current obtained by adding 115 records like those of **B** (some of which contained openings of more than one channel). The summed current is proportional to the P_{open} of the channels, and is equivalent to the macroscopic current, which may be recorded from a large area of membrane containing many channels. Note that the summed current rises with a sigmoidal time-course.

tein are what lead to the occasional changes of state, these dwell times will themselves be random variables, with their mean values determined by the rate constants of the underlying processes. This randomness can be seen in Fig. 1B, and gating at the single channel level may be described as a Markov process, in which the probability of a channel undergoing a state transition within a given time depends only on the rate constants for leaving its current state, and not on its previous history, and may be analyzed in these terms *(5–7)*. In the following, we show how a kinetic picture or state diagram may be built up from single-channel information in this way.

Single-channel recording, then, gives directly information about the rate at which ions pass through ion channels, together with the detailed dwell time information needed for a kinetic picture. Because of the randomness of individual dwell times, it is necessary to record many events to resolve their average behavior, requiring stability both of channel properties and recording conditions. The currents that flow through individual channels are small, in the order of one to a few picoamperes, and measuring such currents also imposes constraints. In order to clearly discriminate open and closed levels, it is usually necessary to low-pass filter data at a few kilohertz, therefore losing resolution in time.

Macroscopic currents are summed currents through many channels recorded from quite large areas of membrane or from whole cells. When only one type of channel is active, the macroscopic current at any moment will be proportional to the open-state probability of the channels, though actual calculation of P_{open} would require knowledge both of the single-channel current and the total number of channels in the population. Figure 1C shows the way in which macroscopic current is generated by the summed behavior of many individual channels. Of course, macroscopic currents are larger than unitary currents, making them easier to record, and the averaging has already been done because of the large number of channels, so that many repetitions of the same voltage step are not needed. Although single-channel dwell times cannot be measured directly, it is still possible to obtain substantial kinetic information from macroscopic currents. Perhaps the major difficulty with macroscopic currents occurs when several different types of channel contribute to the current flow in the area of membrane being investigated. The currents will simply sum, so that the result no longer represents the P_{open} of any one channel type. It may be possible to extract the component of current through one type of channel by various current separation methods, for example by using selective channel blocking drugs or by choosing a voltage range in which only one channel type is activated (*see,* e.g., refs. *4,8*). If such separation can be achieved, macroscopic currents often give quickly and easily of information that must be extracted much more laboriously from single-channel recordings. Techniques also exist for obtaining some information about underlying unitary events from macroscopic currents by measuring the fluctuations (or noise) in their amplitude *(4,9,10)*.

2. Methods and Materials

2.1. Recording Macroscopic Currents

Many different voltage-clamp techniques have been developed for application to cells or areas of cell membrane differing in size, geometry, and so on. Further information can be found, for example, in refs. *4* and *11*. Here we describe briefly the methods applicable to systems used to express channels: frog oocytes, and mammalian cell lines or primary cultures that may be transfected with DNA coding for channel proteins.

2.1.1. Microelectrode Clamps

Two-microelectrode voltage-clamp methods are applicable to relatively large cells that are either approximately spherical or are short cylinders. Thus, they are well suited to frog oocytes. They also serve as a good illustration of the general characteristics of voltage-clamp techniques. The basic layout is shown in Fig. 2A. The electrical potential inside the cell is measured by penetrating the membrane with a very fine glass micropipet electrode, filled with a conducting salt solution. The voltage across the cell membrane is measured as the difference between this intracellular potential and that recorded by an electrode in the bathing solution outside the cell. A negative feedback amplifier (A2 in Fig. 2A) compares this voltage to that of an applied command signal and passes current into the cell through a second micropipet in order to clamp the cell membrane to the required voltage. The current flowing across the membrane is measured, using an electrode in the bathing solution, by amplifier A3, a current-to-voltage converter. Amplifiers for this type of circuit are commercially available (e.g., Axoclamp IIA) or can be built quite inexpensively in the laboratory (e.g., ref. *11*). Many more practical details can be found in the original literature (*see* refs. in *4,11*).

A. Two-microelectrode voltage clamp

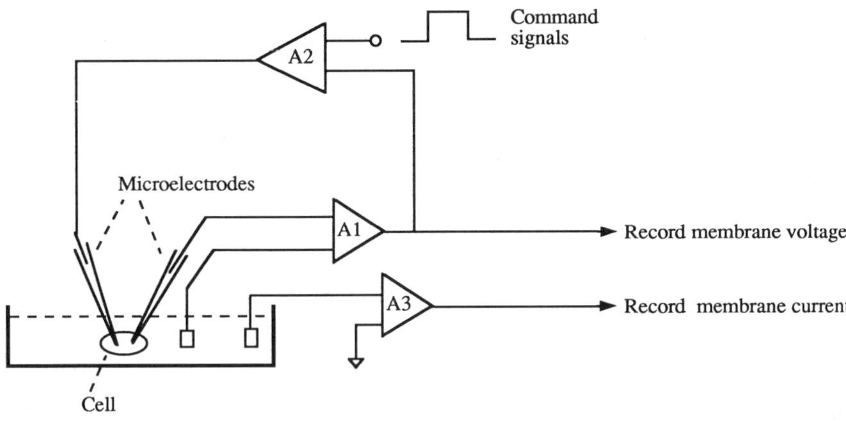

B. Whole-cell patch clamp

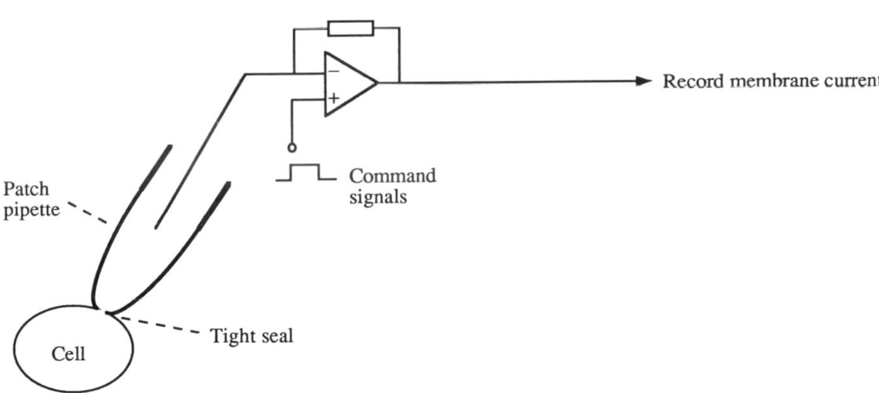

Fig. 2. Voltage-clamp methods for recording macroscopic currents. **A.** Two-microelectrode clamp, described in the text. A1 is a high-input impedance preamplifier, which is necessary to record the voltage accurately at the tip of a microelectrode that typically has a resistance of 5–100 MΩ. **B.** Patch clamp in the whole-cell mode. The amplifier controls the voltage in the patch pipet, and so across the cell membrane, by passing current through the feedback resistor. The method works best for small cells that generate comspondingly small currents and is thus well suited to a wide variety of mammalian cell types.

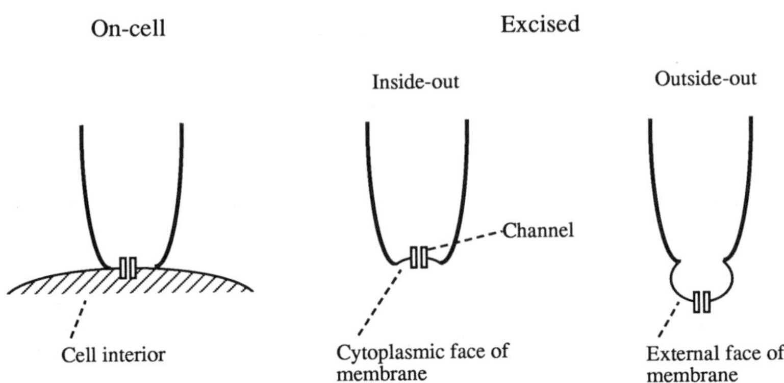

Fig. 3. Single-channel recording. The different configurations for single-channel recording from native cell membranes using patch clamp. The excised configurations allow either the inside or outside face of the membrane to be exposed to the bathing solution. The patch pipet is connected to the same basic circuit as in the whole-cell clamp of Fig. 2B, though a larger feedback resistor may be used to give higher gain and lower noise.

For smaller cells that do not tolerate two microelectrodes well, a single-microelectrode technique has been developed. In this, one intracellular electrode is used both to measure membrane voltage and to pass current, switching rapidly betweeen the two modes at a frequency of a few kilohertz. Amplifiers, such as the Axoclamp IIA, can be used in this mode, and further details and references may be found ref. *11*.

2.1.2. Patch Clamp

The patch-clamp technique, developed by Neher and Sakmann *(12)*, and refined by Hamill et al. *(13)* has revolutionized voltage clamp recording from cells too small or fragile for methods that use penetrating microelectrodes. It is the method of choice for most mammalian cells, and allows recording of both macroscopic and single-channel currents.

The basic patch clamp uses a glass microelectrode, but with a fire-polished tip that is pressed against the cell membrane, where it does not penetrate, but forms a tight seal between the pipet rim and the membrane. This electrically isolates a small patch of membrane (Fig. 3), which may be used to record single-channel currents as described in the next section. For recording macroscopic currents, however, the membrane patch in the pipet tip is broken by suction or by voltage pulses, so that the cell interior becomes connected to the solution inside the pipet. The membrane voltage may then be controlled, and current recorded, with a patch-clamp amplifier circuit of which the basics are shown in Fig. 2B. Commercial amplifiers are most widely used for this purpose; examples are those made by List-electronic (Darmstadt/Eberstadt, Germany), Axon Instruments (Foster City, CA), and Dagan (Minneapolis, MN), all of which may be used for recording either macroscopic or unitary currents. Since the solution in the patch pipet perfuses the cell interior, substances may be delivered to the inside of the cell by this route. In some situations, the loss of soluble cytoplasmic constituents has adverse effects on channel function and may be largely avoided by a recent method that uses the pore-forming antibiotic nystatin to make electrical contact through the membrane patch in the pipet *(14)*. Many further details of patch-clamp recording methods, both for macroscopic and unitary currents, may be found in refs. *15* and *16*.

2.2. Single-Channel Recording

Two basic methods have been used to record currents from individual ion channels, the patch clamp described earlier and reincorporation methods in which channel proteins, either in membrane fragments or purified, are inserted into lipid bilayers. The bilayer is formed over a small hole in a partition separating two solutions and may be voltage clamped using electrodes in these solutions. A hybrid method has also been used in which a bilayer is formed on the tip of a patch electrode. Details of bilayer recording may be found in ref. *17*; here we give a brief account of patch methods.

Patch-clamp, single-channel recording uses one of three configurations (Fig. 3). Currents can be recorded through single channels while the membrane patch stays cell-attached. In this condition, the internal face of the membrane remains exposed to cytoplasm, but it is necessary to allow for the resting membrane potential of the cell when estimating the voltages achieved by the voltage clamp. Alternatively, patches may be excised from the cell so as to expose either their cytoplasmic face (inside-out) or their extracellular face (outside-out) to the external solution. In these configurations, the solutions on either side of the membrane patch and the channels it contains are under experimental control, as is the voltage across the patch. The electrical resistance of the tight seal between pipet and membrane is very high, in the order of 10^{10} Ω, so that the noise level is correspondingly low, and it is this that allows the current through individual channels to be resolved.

3. Theoretical Interpretation

3.1. Kinetic Behavior at the Macroscopic Level

The earliest pictures of the kinetic behavior of ion channels were built up from measurements by Hodgkin and Huxley of membrane currents in the squid axon *(18)*. Such macroscopic currents change with time in a complex way as channel P_{open} itself changes. For a delayed rectifier potassium current, for example, a depolarizing change in membrane potential results in an increase in potassium current along an S-shaped time-course (Fig. 1C), whereas return of the membrane potential to a negative level may result in an apparently simple first-order decline in current. Such behavior itself tells the experimenter that the channels are more complex kinetically than scheme (1) and makes it likely that, before opening, the channel runs through a series of closed states under depolarization. Other channels, such as sodium channels and some calcium and potassium channels, are more complicated still, opening and subsequently closing again (inactivating) under depolarization.

If a channel exists in a number of states between which voltage-dependent transitions occur, channel opening under depolarization will reflect possible occupancy of the various states. As a result, the relationship between channel P_{open} and time will be the sum of a number of exponentials. It is perhaps easiest to illustrate this with a relatively simple case.

If a channel runs through transitions between two closed states and then from closed to open, as follows:

$$C2 \underset{k_{-2}}{\overset{k_2}{\rightleftharpoons}} C1 \underset{k_{-1}}{\overset{k_1}{\rightleftharpoons}} O \tag{2}$$

P_{open} will change with time as the sum of two exponentials according to

$$P_{open}(t) = P_{open}(\infty) + A_1 \exp(-\mu_1 t) + A_2 \exp(-\mu_2 t) \tag{3}$$

and A_1, A_2, μ_1, and μ_2 will be determined in a complex way by the transition rates (k_1, k_{-1}, and so on) between the states. P_{open} will now be $[1 + k_{-1}/k_1(1 + k_{-2}/k_2)]^{-1}$ in the steady state. In general, the number of exponential arguments will reflect the number of states and will be one less than the number of states. It is independent of the way the states are linked together and of the consequent number of possible transitions. Details of this treatment are given in ref. *19*, developed for a different situation, whereas a more general discussion is given in ref. *4*.

3.1.1. The Hodgkin–Huxley Scheme

Although the fitting of changes of permeability or channel P_{open} with exponentials (as above) is the more general solution, many electrophysiologists still prefer to use the description of kinetic behavior developed by Hodgkin and Huxley *(18)*. This description, which predates knowledge that ionic permeability is associated with ion channels, sought to account for the complexity of voltage-gated permeability changes by making the simplifying assumption that gating is a first-order process, with each gate moving from closed to open according to

$$closed \underset{\beta_n}{\overset{\alpha_n}{\rightleftharpoons}} open \tag{4}$$

with n being the probability of a gate being in the open state. Channel opening may then be explained if there are several gates for each channel, all of which must be open for ions to permeate. For the delayed rectifier potassium channel, the best fits to the experimental results were obtained by supposing that each channel has four gates, so that the probability of a channel being open is n^4; n rises and falls exponentially with voltage with a time constant equal to $1/(\alpha_n + \beta_n)$; n^4 rises along an S-shaped time-course and falls exponentially just as channel P_{open} does. Figure 4 summarizes the Hodgkin–Huxley fitting of delayed rectifier potassium channels, using values for α_n and β_n developed to fit potassium currents in skeletal muscle. Although it cannot account fully for channel kinetics, such treatment is often used descriptively. For example, it has been used to describe the consequences of changes of channel structure by site-directed mutagenesis *(20)*. Probably, it remains the method of choice to describe macroscopic currents.

3.2. Kinetic Behavior at the Microscopic, Single-Channel Level

The technique of single-channel recording allows a more complete analysis of the kinetic behavior of ion channels than any other method, within a time resolution of around 10 kHz. A source of experimental frustration lies in the need to isolate (and be sure that one has isolated) one, and not more than one, channel. The certainty that a single channel has been isolated is less at lower values of P_{open}, owing to the reduced likelihood of seeing openings of two or more channels superimposed. A further source of frustration lies in a lack of certainty as to how different states of the channel are linked together.

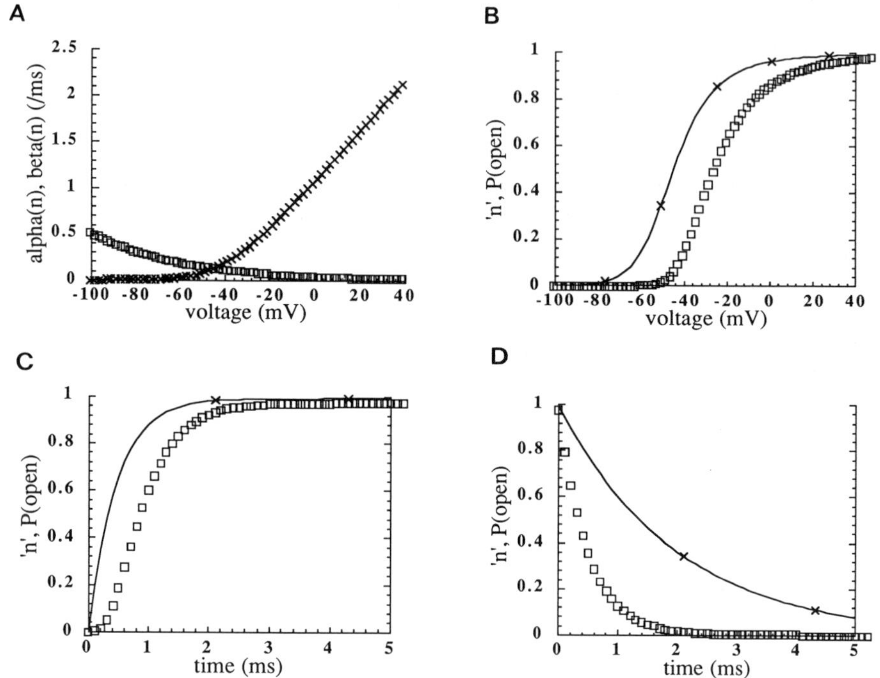

Fig. 4. The Hodgkin–Huxley potassium channel. **A.** The relationships between the rate constants (ms^{-1}; ordinate) for opening (α_n, x) and for closing (β_n, ☐) of the proposed "*n*" gate and membrane potential (mV; abscissa). The relationships are given by:

$$\alpha_n = [0.0269(V+40)/1 - \exp - (V+40/7)]$$

$$\beta_n = 0.113 \bullet \exp - (V+40/40)$$

where *V* (mV) is the membrane potential (*see*, for example, ref. *25)*. **B.** The continuous line gives the relationship between the steady-state probability (*n*; ordinate) of the gate being open and voltage (mV; abscissa). The open squares give the values for channel P_{open} (= n^4). **C, D.** The lines give the relationship between *n* and time under a depolarization from –100 mV to + 40 mV starting at 0 ms (**C**) and under repolarization (**D**), again starting at 0 ms. In each case, the open squares show how channel P_{open}, changes with time: in **C**, under depolarization, and in **D**, under repolarization.

The kinetic picture of the channel is built up from measurements of the P_{open}, of the distribution of dwell times in open and closed states, and of the distribution of times taken for the channel to open for the first time under a step change in membrane potential (the first latency). Additional information comes from the way channel openings may be linked together in bursts, the way bursts of openings are linked together as clusters, and the way correlations may exist between different events. The picture that has to be developed is of the number of states occupied by the channel, the way these states are linked together, and the values of the transition rates among states.

For the simple channel illustrated in scheme (1), a channel that has only two states, the value of P_{open} is $\alpha/(\alpha + \beta)$, as already described. Again as already explained, the dwell times within a given state are determined by the transition rates for leaving that state, and the distribution of open times, closed times, and of first latencies will be described by probability density functions as follows:

$$f(t)_{open} = \beta \cdot \exp(-\beta t)$$

$$f(t)_{closed} = \alpha \cdot \exp(-\alpha t)$$

$$f(t)_{first\ latency} = \alpha \cdot \exp(-\alpha t) \tag{5}$$

Hence, the transition rates will be equal to the reciprocal of the time constants of the argument of each exponential, time constants that will in turn give the mean of each distribution. Thus, the mean dwell time in the open state (the mean open time) will be $1/\beta$, whereas the mean closed time will be $1/\alpha$.

As already indicated, no channel type actually behaves as simply as the model outlined above. All channels have more than one closed state, and many have more than one open state. The method allows one some insight into how many states are occupied by the channel. When there are multiple open or closed states, the probability density functions become the sum of a number of exponentials. As with the fitting of exponentials to macroscopic currents, the number of exponentials reflects the number of states, rather than the number of ways those states are linked together. Though it is often difficult to decide how many exponentials are present, a closed time distribution that has two exponentials indicates that there are at least two closed states, and so on.

In the section on macroscopic currents, the following state diagram was briefly considered:

$$C2 \xrightleftharpoons[k_{-2}]{k_2} C1 \xrightleftharpoons[k_{-1}]{k_1} O \tag{6}$$

In this case, the mean open time is $1/k_{-1}$, but the closed times will be distributed according to two exponentials. The dwell times in the closed states will be $1/(k_1 + k_{-2})$ and $1/(k_2)$, but the probability density function for the closed states can be shown to be (*see* ref. *21*):

$$f(t)_{closed} = (k_1/\lambda_1 - \lambda_2) \, [(k_2 - \lambda_2) \exp (-\lambda_2 t) + (\lambda_1 - k_2) \exp (-\lambda_1 t)] \tag{7}$$

where

$$\lambda_1 = \{k_1 + k_2 + k_{-2} + [(k_1 + k_2 + k_{-2})^2 - 4k_1 k_2]^{1/2}\}/2$$

$$\lambda_2 = \{k_1 + k_2 + k_{-2} - [(k_1 + k_2 + k_{-2})^2 - 4k_1 k_2]^{1/2}\}/2 \tag{8}$$

The time constants here do not give the dwell times in each of the closed states, though sometimes they may have numerical values that are little different from those dwell times. The first latency distribution will be as follows.

$$f(t)_{first\ latency} = (\lambda_1 \lambda_2/\lambda_1 - \lambda_2) \, [\exp (-\lambda_2 t) - \exp (-\lambda_1 t)] \tag{9}$$

Note that the arguments of the exponentials are identical for the distributions of closed times and of first latencies. The arguments will not be identical with those with which P_{open} rises: There the arguments are quantities influenced by the transition rate for leaving the open state. The channel will tend to open in bursts, particularly if $C1$ is short lived compared with $C2$, and the number of openings per burst will be given by $1 + k_1/k_{-2}$.

How do we actually measure the times a channel spends in open and closed states? The simplest way of achieving such measurements is to set a cursor midway between the closed and open levels of a digitized single-channel record and log events as the current record crosses the cursor. Provided the signal:noise ratio is high, the chance of recording false events will be low, but the achievement of low noise requires filtering, which will result in brief events not reaching the cursor. Corrections can be made for missed events, and resolution can be improved by deconvolving the response to the filter from the current record (22).

Once measured, events have to be put into histograms of open and closed times and of first latencies, and these then have to be fitted with exponentials to find out the minimum number of open and closed states that occur. "Log binning" of the distribution of open and closed times (plotting time on a log scale) often helps the identification of the number of exponentials in the probability density function *(23)*. Bursts and clusters of bursts are identified by finding an appropriate "critical time" that delineates short and long closures of the channel *(22)*. A hypothesis must then be erected as to how the various states are linked together. Finally, the hypothesis of the state diagram is tested by showing that it can account for all aspects of channel behavior: P_{open}, open time distributions, closed time distributions, first latencies, and bursts. Often, because of the difficulties in measuring multiple exponentials accurately and because of the difficulty in solving the state diagram, channels have been constrained in research papers to three-state models of the type described earlier. The kinetic scheme then becomes a way of summarizing aspects of the channel behavior, rather than an accurate description of the way the channel actually behaves. More complex cases can, however, be dealt with, and they are generally solved using matrix algebra *(see* refs. *5–7,24)*. The Hodgkin–Huxley potassium channel is such a complex case.

3.2.1. Microscopic Kinetics and the Hodgkin–Huxley Potassium Channel

We have considered already the delayed rectifier potassium channel as a representative voltage-gated ion channel. We now consider further the Hodgkin–Huxley description of this channel. In single-channel terms, the channel will have four closed states and a single open state, linked together as follows:

$$C4 \underset{\beta_n}{\overset{4\alpha_n}{\rightleftharpoons}} C3 \underset{2\beta_n}{\overset{3\alpha_n}{\rightleftharpoons}} C2 \underset{3\beta_n}{\overset{2\alpha_n}{\rightleftharpoons}} C1 \underset{4\beta_n}{\overset{\alpha_n}{\rightleftharpoons}} O \tag{10}$$

The mean open time will be determined by the transition rate for leaving the open state (it will be $1/4\beta_n$ in terms of the Hodgkin–Huxley model), but the closed time distribution will be much more complicated. Figure 5 shows probability density functions for the open, closed, and first latency distributions for a Hodgkin–Huxley potassium channel computed from scheme *(4)* with the rate constants used to fit results obtained in skeletal muscle *(25)*.

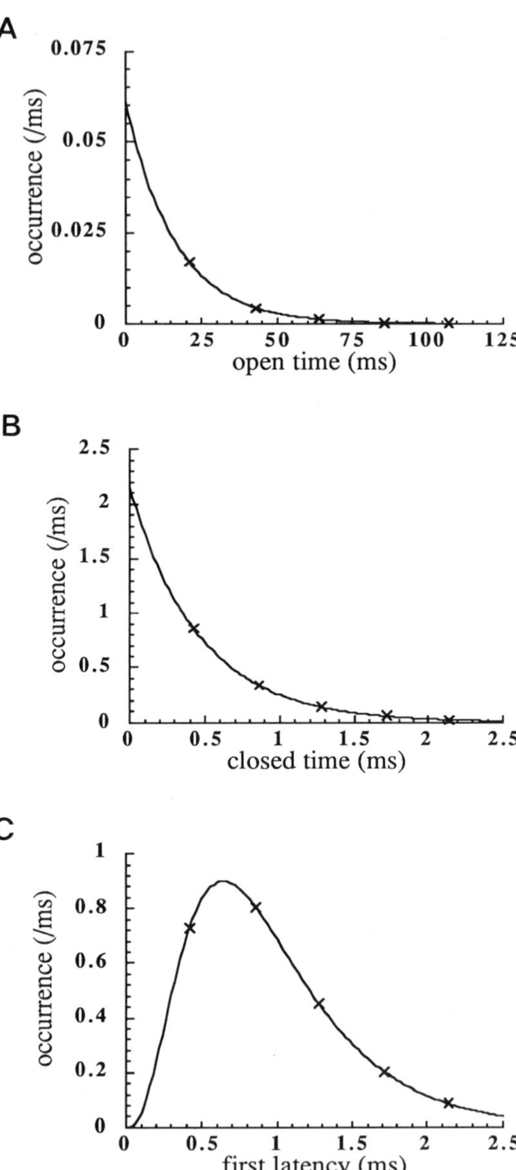

Fig. 5. Open and closed times for the Hodgkin–Huxley potassium channel. The state diagram given in the text (scheme 4) has been solved using the matrix methods of Colquhoun and Hawkes (5,6) using a program developed by K. Magleby and colleagues, and the appropriate values for α_n, and β_n, for a depolarization from −100 mV to 40 mV. The probability density functions given are for open (**A**) and closed times (**B**), and for the latency for the first opening after depolarization (**C**).

The probability density functions are as follows:

Open time:
$$f(t) = 0.061 \cdot \exp(-t/16.35)$$
Closed time:
$$f(t) = 3.0 \times 10^{-6} \cdot \exp(-t/0.115) + 9.4 \times 10^{-4} \cdot \exp(-t/0.154) + 0.086 \cdot$$
$$\exp(-t/0.231) + 2.065 \cdot \exp(-t/0.475)$$
First latency:
$$f(t) = -8.328 \cdot \exp(-t/0.115) + 24.877 \cdot \exp(-t/0.154) - 24.564 \cdot$$
$$\exp(-t/0.231) + 8.015 \cdot \exp(-t/0.475)$$

where t is time (ms).

In the steady state at $+40$ mV, the dwell times in the various states, and the occupancies will be:

State	Mean lifetime (ms)	Occupancy
C4	0.116	<<0.001
C3	0.155	<<0.001
C2	0.231	<0.001
C1	0.455	0.028
O	16.347	0.972

It can be seen that it is impossible to dissect out the expected four exponentials in the closed time distribution, though the distribution is the sum of four exponentials. This difficulty is a consequence of the left-hand closed states (C4, C3) being shorter lived than the right-hand closed states (C2, C1) at positive membrane potentials, and to their being visited very infrequently. A further consequence is that the openings will not readily be resolvable into bursts. The first latency distribution does, however, show clearly the presence of multiple exponentials.

In fact, as may be seen from Fig. 1, delayed rectifier potassium channels do have their closed times distributed in such a way that they open in readily resolvable bursts, and this and certain other information suggest that the Hodgkin–Huxley model, though it will fit macroscopic currents, cannot account for the single-channel events that underlie the macroscopic behavior. In consequence of this, new models must be developed (e.g., *26,27*), an on-going process likely to be helped by increasing knowledge of channel structure.

3.3. Functional Consequences
of Modifications of Channel Structure

The development of a state diagram for an ion channel and its solution undoubtedly depends on factors other than the results of single-channel recording. Models of channels gated by ligand binding (nicotinic acetylcholine receptors and calcium-activated potassium channels, for example) generally link transitions to binding of the ligand, and the number of transitions will be linked to the number of molecules or ions bound by each channel *(22,28,29)*. For a voltage-gated channel, the transitions will obviously be linked to voltage, but the number of transitions is less easy to guess at. As more is understood about the structure of ion channels, more sensible guesses will come to be made, with the number of states linked to the number of voltage-sensing sequences in the ion channel.

The first voltage-gated ion channel to be sequenced was the sodium channel *(30)*, and subsequent work has shown that calcium channels and potassium channels are built up in a similar fashion *(31,32)*. In particular, the channels possess similar membrane-spanning, charged sequences of amino acids (with a lysine or an arginine as each third amino acid), which probably act as voltage sensors. The polypeptide that forms sodium channels and that which forms dihydropyridine-sensitive calcium channels are large and contain four such voltage-sensing elements. The peptides forming potassium channels contain one such sequence, and it is likely that potassium channels are assemblies of four peptide elements.

There is an increasing amount of work examining channels expressed in oocytes or in mammalian cells in culture (e.g., *20,33,34*). Walter Stühmer and his colleagues have succeeded in identifying in rat brain five homologous potassium channel-forming proteins that, when expressed in frog oocytes, assemble as potassium channels with different physiological and pharmacological properties. The channels vary in their unitary conductance, their kinetic properties, and their sensitivity to potassium channel blockers, such as the aminopyridines, tetraethylammonium ions, dendrotoxin, and so on. Such work on channels is only in its infancy, but coupled with modifications of channel structure through site-directed mutagenesis, single-channel recording offers the hope of a complete understanding of how the structure of a membrane protein is linked to its function. Indeed, work on sodium channels has already shown how changes of the voltage-sensing element of the protein lead to changes in voltage dependence of the channel *(20)* and to the identification of the part of the channel sequence that forms the receptor for the channel blocker tetrodotoxin *(35)*.

References

1. Cole, K. S. (1949) Dynamic electrical characteristics of the squid axon membrane. *Arch. Sci. Physiol.* **3,** 253–258.
2. Marmont, G. (1949) Studies on the axon membrane; I. A new method. *J. Cell. Comp. Physiol.* **34,** 351–382.
3. Hodgkin, A. L., Huxley, A. F., and Katz, B. (1952) Measurement of current-voltage relations in the membrane of the giant axon of *Loligo. J. Physiol.* **116,** 424–448.
4. Hille, B. (1984) *Ionic Channels of Excitable Membranes,* Sinauer, Sunderland.
5. Colquhoun, D. and Hawkes, A. G. (1977) Relaxation and fluctuations of membrane currents that flow through drug-operated channels. *Proc. Royal Soc. London B* **199,** 231–262.
6. Colquhoun, D. and Hawkes, A. G. (1982) On the stochastic properties of bursts of single ion channel openings and of clusters of bursts. *Philos. Trans. Royal Soc. B* **300,** 1–59.
7. Colquhoun, D. and Hawkes, A. L. (1983) The principles of the stochastic interpretation of ion channel mechanisms, in *Single Channel Recording* (Sakmann B. and Neher E., eds.), Plenum, New York, pp. 135–175.
8. Standen, N. B. (1987) Separation and analysis of ionic currents, in *Microelectrode Techniques. The Plymouth Workshop Handbook* (Standen, N. B., Gray, P. T. A., and Whitaker, M. J., eds.), Company of Biologists, Cambridge, pp. 29–40.
9. Neher, E. and Stevens, C. F. (1977) Conductance fluctuations and ionic pores in membranes. *Ann. Rev. Biophys. Bioengineering* **6,** 345–381.
10. Cull-Candy, S. G. (1984) New ways of looking at synaptic channels: noise analysis and patch-clamp recording. *Recent Adv. Physiol.* **10,** 1–28.
11. Halliwell, J. V., Plant, T. D., and Standen, N. B. (1987) Voltage clamp techniques, in *Microelectrode Techniques. The Plymouth Workshop Handbook* (Standen, N. B., Gray, P. T. A., and Whitaker, M. J., eds.), Company of Biologists, Cambridge, pp. 13–28.
12. Neher, E. and Sakmann, B. (1976) Single channel currents recorded from membrane of denervated frog muscle fibres. *Nature* **260,** 799–802.
13. Hamill, O. P., Marty, A., Neher, E., Sakmann, B., and Sigworth, F. J. (1981) Improved patch clamp techniques for high-resolution current recording from cells and cell-free membrane patches. *Pflügers Arch.* **391,** 85–100.
14. Horn, R. and Marty, A. (1988) Muscarinic activation of ionic currents measured by a new whole-cell recording method. *J. Gen. Physiol.* **92,** 145–159.
15. Sakmann, B. and Neher, E., eds. (1983) *Single-Channel Recording.* Plenum, New York.
16. Ogden, D. C. and Stanfield, P. R. (1987) Introduction to single channel recording, in *Microelectrode Techniques, The Plymouth Workshop Handbook* (Standen, N. B., Gray, P. T. A., and Whitaker, M. J., eds.), Company of Biologists, Cambridge, pp. 63–81.
17. Miller, C. (ed.) (1986) *Ion Channel Reconstitution.* Plenum, New York.

18. Hodgkin, A. L. and Huxley, A. F. (1952) A quantitative description of membrane current and its application to conduction and excitation in nerve. *J. Physiol.* **117,** 500–544.

19. Chiu, S. Y. (1977) Inactivation of sodium channels: second order kinetics in myelinated nerve. *J. Physiol.* **273,** 573–596.

20. Stühmer, W., Conti, F., Suzuki, H., Wang, X., Noda, M., Yahagi, N., Kubo, H., and Numa, S. (1989) Structural parts involved in activation and inactivation of the sodium channel. *Nature* **339,** 597–603.

21. Hagiwara, S. and Ohmori, H. (1983) Studies of single calcium channel currents in rat clonal pituitary cells. *J. Physiol.* **336,** 649–661.

22. Colquhoun, D. and Sakmann, B. (1985) Fast events in single-channel currents activated by acetylcholine and its analogues at the frog muscle endplate. *J. Physiol.* **369,** 501–557.

23. McManus, O. B., Blatz, A. L., and Magleby, K. L. (1987) Sampling, log binning, fitting, and plotting durations of open and shut intervals from single channels and the effects of noise. *Pflügers Arch.* **410,** 530–553.

24. Colquhoun, D. (1987) The interpretation of single channel recordings, in *Microelectrode Techniques. The Plymouth Workshop Handbook* (Standen, N. B., Gray, P. T. A., and Whitaker, M. J., eds.), Company of Biologists, Cambridge, pp. 105–135.

25. Standen, N. B., Stanfield, P. R., Ward, T. A., and Wilson, S. W. (1984) A new preparation for recording single-channel currents from skeletal muscle. *Proc. Royal Soc. B* **217,** 1–10.

26. Spruce, A. E., Standen, N. B., and Stanfield, P. R. (1989) Rubidium ions and the gating of delayed rectifier potassium channels of frog skeletal muscle. *J. Physiol.* **411,** 597–610.

27. Zagotta, W. N. and Aldrich, R. W. (1990) Voltage-dependent gating of *Shaker* A-type potassium channels in *Drosophila* muscle. *J. Gen. Physiol.* **95,** 29–60.

28. Magleby, K. L. and Pallotta, B. S. (1983) Calcium dependence of open and shut interval distributions from calcium-activated potassium channels in cultured rat muscle. *J. Physiol.* **344,** 585–604.

29. Magleby, K. L. and Pallotta, B. S. (1983) Burst kinetics of single calcium-activated potassium channel in cultured rat muscle. *J. Physiol.* **344,** 605–623.

30. Noda, M., Shimizu, S., Tanabe, T., Takai, T., Kayano, T., Ikeda, T., Takahashi, H., Nakayama, H., Kanaoka, Y., Minamino, N., Kangawa, K., Matsuo, H., Raftery, M. A., Hirose, T., Inayama, S., Hayashida, H., Miyata, T., and Numa, S. (1984) Primary structure of *Electrophorus electricus* sodium channel deduced from cDNA sequence. *Nature* **312,** 121–127.

31. Tanabe, T., Takeshima, H., Mikami, A., Flockerzi, V., Takahashi, H., Kangawa, K., Kojima, M., Matsuo, H., Hirose, T., and Numa, S. (1987) Primary structure of the receptor for calcium channel blockers from skeletal muscle. *Nature* **328,** 313–318.

32. Schwartz, T. L., Tempel, B. L., Papazian, D. M., Jan, Y. N., and Jan, L. Y. (1988) Multiple potassium-channel components are produced by alternative splicing at the *Shaker* locus in *Drosphila*. *Nature* **331,** 137–142.

33. Stühmer, W., Ruppersberg, J. P., Schroter, K. H., Sakmann, B., Stocker, M., Giese, K. P., Perschke, A., Baumann, A., and Pongs, O. (1989) Molecular basis of functional diversity of voltage-gated potassium channels in mammalian brain. *EMBO J.* **8,** 3235–3244.
34. Leonard, R. J., Karschin, A., Jayashree-Aiyar, S., Davidson, N., Tanouye, M. A., Thomas, L., Thomas, G., and Lester, H. A. (1989) Expression of *Drosophila* shaker potassium channels in mammalian cells infected with recombinant vaccinia virus. *Proc. Nat. Acad. Sci. USA* **86,** 7629–7633.
35. Noda, M., Suzuki, H., Numa, S., and Stühmer, W. (1989) A single point mutation confers tetrodotoxin and saxitoxin insensitivity on the sodium channel II. *FEBS Lett.* **259,** 213–216.

CHAPTER 20

An Electrophysiological Approach to the Regulation of Neuronal Voltage-Activated Calcium Channels by Guanine Nucleotide Binding Proteins

Roderick H. Scott and Annette C. Dolphin

1. Introduction

Neuronal voltage-activated Ca^{2+} channels have received considerable attention from scientists because of the central role played by Ca^{2+} in the transduction of electrical activity to chemical signals. The importance of neuronal electrical activity resulting in a rise in intracellular Ca^{2+} is highlighted at the presynaptic nerve terminal. Depolarization of the presynaptic nerve terminal results in influx of Ca^{2+} through voltage-activated Ca^{2+} channels. This Ca^{2+} influx contributes to the depolarization, but more importantly, the increase in intracellular Ca^{2+} activates biochemical events culminating in vesicular release of neurotransmitter. Electrophysiological studies on presynaptic nerve terminals have proved difficult because of access problems and the small size of most terminals. A number of preparations, however, have been amenable to studying presynaptic Ca^{2+} channel activity. These include:

1. The squid giant presynaptic terminal *(1)*;
2. Peptidergic nerve terminals of the rat neurohypophysis *(2)*; and
3. Frog motor endings *(3)*.

From: *Methods in Molecular Biology, Vol. 13: Protocols in Molecular Neurobiology*
Edited by: A. Longstaff and P. Revest Copyright © 1992 The Humana Press, Totowa, NJ

347

Recordings have also been made from a "neurosecretosome" preparation of rat pituitary terminals *(4)*. Model systems using cells in culture have also been useful. Patch-clamp and Ca^{2+} imaging techniques have revealed the distribution of Ca^{2+} channels on cell bodies, neurites, and growth cones of frog sympathetic neurons *(5)*. Dissociation of dorsal root ganglia (DRG) and culturing the neurons result in a model system with several types of voltage-activated Ca^{2+} channels. Furthermore, receptors for neurotransmitters normally found only on the presynaptic terminal are expressed on DRG neuron cell bodies. These receptors are functionally linked to Ca^{2+} channels and, in vivo, are believed to mediate presynaptic inhibition or autoregulation of neurotransmitter release. Depolarization of DRG neurons activates Ca^{2+} channels. However, certain neurotransmitters and neuromodulators, including norepinephrine, GABA, dopamine, and adenosine, inhibit the resultant whole-cell Ca^{2+} current *(6–9)*. So although the Ca^{2+} channels are gated by voltage, they are also functionally coupled to receptors, which can regulate their activity. These findings raise the question: How are the receptors linked to the voltage-gated Ca^{2+} channels?

Biochemical studies have shown that guanosine triphosphate (GTP) reduces the affinity of a number of neurotransmitter receptors for agonist ligands. These data suggest that guanine nucleotide binding proteins (G-proteins) might be involved in coupling receptors with voltage-activated Ca^{2+} channels. Most G-proteins are heterotrimers composed of α, β, and γ subunits, and are associated with the cytoplasmic face of cell membrane. G-proteins were first discovered coupling receptors with stimulation (G_s) and inhibition (G_i) of adenylyl cyclase *(10)*. It has now been established that G-proteins are involved in receptor-mediated changes in other second-messenger systems, and in the activation and inhibition of ion channels *(11)*. In the inactive state, guanosine diphosphate (GDP) is bound to the α subunit of the G-protein. The binding of an agonist to receptors induces G-protein activation as GTP replaces GDP on the α subunit. The α subunit may dissociate from the $\beta\gamma$ complex and may interact with the effector (enzyme or ion channel). The α subunit with bound GTP has a relatively short life-span of a few seconds before the inherent GTP-ase activity of the α subunit converts GTP to GDP. Recently, work from a number of laboratories has indicated that the $\beta\gamma$ complex may also play a role in signaling *(12)*, and there appears to be potential for the $\beta\gamma$ complex to bind an α subunit from other types of G-proteins. A number of G-proteins have been described, but not all of them are heterotrimers. Examples include G_s, G_i, G_o, G_k, G_t (transducin), p21 *ras* protein, and tubulin. The different G-proteins have distinctive α subunits; the $\beta\gamma$ subunits are also heterogeneous, but less so. Different α, β, and γ subunits arise at least in part as a result of transcription of different genes.

The study of G-proteins in the regulation of Ca^{2+} channels presents a problem in that the G-protein is on the inner face of the membrane. Therefore, to manipulate the G-proteins, active compounds need to be introduced into the cell while whole-cell Ca^{2+} currents or Ca^{2+} channel activity is recorded. This chapter describes techniques that enable the electrophysiologist to investigate G-protein regulation of voltage-activated Ca^{2+} channels.

2. Materials

2.1. Whole-Cell Recording

1. Extracellular recording solution: 130 mM NaCl , 3 mM KCl, 0.6 mM $MgCl_2$, 2.5 mM $BaCl_2$, 10 mM Hepes, 4 mM glucose, 25 mM tetraethylammonium bromide, and 0.0025 mM tetrodotoxin, pH is adjusted to 7.4 with NaOH, and osmolarity to 320 mosM with sucrose.
2. Intracellular patch solution: 140 mM cesium acetate (glu, asp, or Cl⁻), 0.1 mM $CaCl_2$, 1.1 mM EGTA, 2 mM ATP, 2 mM $MgCl_2$ (pCa 8.0), 10 mM Hepes, pH adjusted to 7.2 with Tris-HCl, and the osmolarity to 310 mosM with sucrose.

To record whole-cell Ca^{2+} channel currents, other currents (Na^+ and K^+) activated over the same voltage range that mask Ca^{2+} currents need to be inhibited. Tetrodotoxin inhibits the voltage-activated Na^+ channels and the intracellular Cs^+, extracellular Ba^{2+} and tetraethylammonium reduce the K^+ conductance. Ca^{2+}, Sr^{2+}, or Ba^{2+} can be used as charge carriers passing through the Ca^{2+} channels. Although Ca^{2+} is the physiologically relevant ion, using Ba^{2+} as the charge carrier has a number of advantages. First, Ba^{2+} contributes to the inhibition of K^+ channels; second, it gives rise to a larger high-threshold Ca^{2+} channel current. Third, Ba^{2+} does not activate the wide variety of enzymes, intracellular processes, and ion channels that Ca^{2+} does. Finally, Ba^{2+} does not support Ca^{2+}-dependent inactivation of Ca^{2+} channels; however, it may induce release of Ca^{2+} from intracellular stores, so intracellular Ca^{2+} buffering is still important *(13)*. Voltage-activated Ba^{2+} current will be referred to in this chapter as Ca^{2+} channel currents. The concentration of extracellular divalent cation may need to be varied (1–20 m*M*) depending on the number of channels contributing to the whole-cell current. Data from whole-cell, low-threshold currents indicate that changing extracellular Ca^{2+} from 1 to 95 m*M* increases the current amplitude by a factor of 5–10 *(14)*. When large numbers of channels are activated, it may be necessary to reduce extracellular divalent cation concentration to maintain voltage control. Cs^+ in the intracellular solution contributes to the block of K^+ conductance, but is not ideal because under certain circumstances Cs^+ can permeate through ion channels. At present, Cs^+ may still be the best option because larger organic monovalent ions, such as *N*-methyl-D-glucamine, have been found to modify the kinetics and voltage-dependence of calcium currents *(15)*. G protein–GTP interaction is Mg^{2+}-dependent, so the inclusion of $MgCl_2$ in the patch solution is important.

2.2. Single-Channel Recording

1. Bathing solution: 140 m*M* potassium aspartate, 10 m*M* EGTA, 1 m*M* $MgCl_2$, 10 m*M* Hepes, adjusted to pH 7.4 with KOH.
2. Patch solution: 110 m*M* $BaCl_2$, 10 m*M* Hepes, and pH adjusted to 7.4 with $Ba(OH)_2$. It is important to balance the osmolarity of the bathing and patch solutions. The potassium aspartate bathing solution zeros the potential difference across the cell membrane and allows the cell-attached membrane patch potential to be set at a known level *(16)*. Single voltage-activated Ca^{2+} channel activity has also been recorded using more conventional bathing solutions (replacing Na^+ with choline chloride or blocking Na^+ channels with tetrodotoxin), and patch solutions with much lower concentrations of divalent cation (5–60 m*M*), which gives better stability *(17–19)*.

3. Methods

3.1. Patch-Clamp Techniques

Patch clamp techniques *(20)* are particularly useful for investigating the actions of G-proteins on Ca^{2+} channels, because they enable access to the cytoplasmic face of the cell membrane. These techniques initially involve placing the tip of a patch pipet with a resistance between 1 and 5 MΩ for whole-cell recording and up to 20 MΩ for single-channel recording on the cell. Once the patch pipet is touching the cell, negative pressure (suction) is applied to the inside of the patch pipet. Gradually, the resistance at the tip of the patch pipet increases as the cell membrane seals onto it. The rate of formation of a high-resistance (GΩ) seal depends on patch pipet glass composition

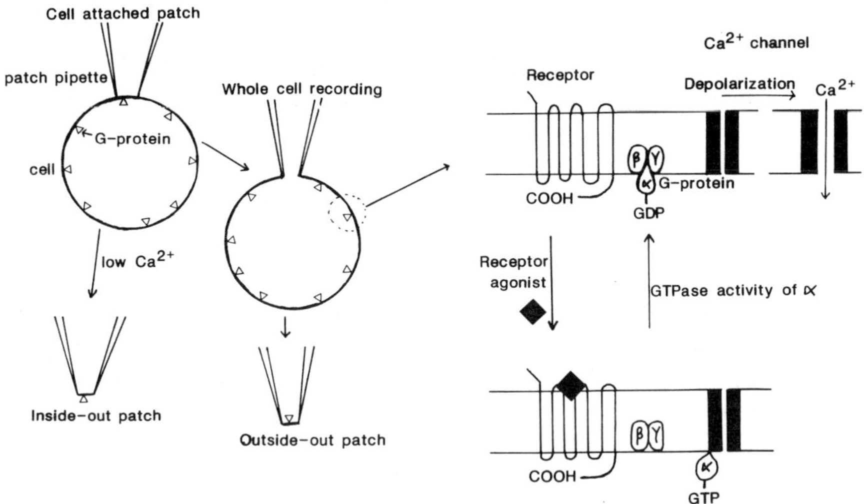

Fig. 1. Schematic diagram showing cell-attached patch, whole-cell recording outside-out patch, and inside-out patch configurations. The cytoplasmic face of the membrane is labeled with a Δ to represent a G-protein. The enlargement on the right-hand side shows the possible direct interaction between activated G-protein and Ca^{2+} channel.

(borosilicate glass seals slowly), what is present in the patch solution (antibodies hinder the formation of good seals), as well as temperature and state of cells. Once a tight seal (10 GΩ) between the patch pipet and cell is formed, single-channel activity can be recorded. This configuration is called the cell-attached patch. Additional manipulations can result in other patch pipet-membrane configurations *(20)* (Fig. 1).

3.1.1. Whole-Cell Recording

The suction can be increased until the patch of membrane breaks. The patch pipet is then in continuum with the inside of the cell. This is the whole-cell recording configuration and allows whole-cell currents to be recorded as well as the introduction of large molecules into the inside of the cell *(21)*. The whole-cell recording variant of the patch-clamp technique can be used to activate different Ca^{2+} channel currents. Three example currents are illustrated Fig. 2. First, step depolarization from V_H (holding potential) –90 mV to V_c (clamp potential) –30 mV can activate a low threshold T-type current. Second, a larger depolarization to between V_c 0 and 10 mV activates the maxi-

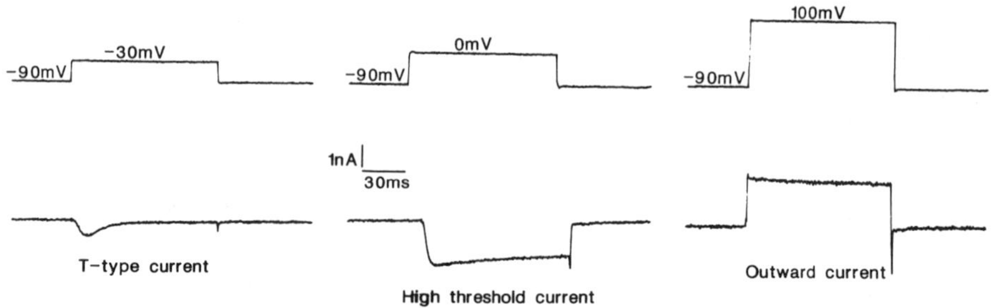

Fig. 2. Traces showing Ca^{2+} channel currents activated by different voltage step commands. On the left, a low-threshold T-type current activated from V_H (holding potential –90 mV) by a 100 ms voltage step command to –30 mV. In the middle, a high-threshold Ca^{2+} channel current activated at 0 mV and composed of T-, L-, and N-type current. Both low- and high-threshold currents illustrated were carried by Ba^{2+}. These currents were activated at low frequency one step every 30 s to reduce run-down. The third trace shows an outward Ca^{2+} channel current carried by Cs^+ and K^+, and activated at +100 mV.

mum inward high-threshold current (some T-type current may also be present). Both low- and-high threshold currents are observed as inward currents as divalent cations enter the cell down a steep electrochemical gradient. Outward currents can also be recorded at very depolarized potentials (V_c + 100 mV) *(22,23)*. Under these conditions, K^+ and Cs^+ leave the cell through Ca^{2+} channels, which are also slightly permeable to these monovalent cations. Net currents are calculated by subtracting the scaled linear leakage and capacitance currents.

3.1.2. Outside-Out Patch

Following the formation of the whole-cell recording configuration, the patch pipet can be lifted up off the cell. With luck it will take a patch of membrane with it. The patch of membrane will be orientated with the cytoplasmic face of the membrane toward the patch solution and the extracellular face toward the bath solution, hence the term outside-out patch.

3.1.3. Inside-Out Patch

Alternatively, from the cell-attached patch configuration, the patch pipet can be lifted up to form an inside-out membrane patch. This procedure may also result in the formation of a membrane vesicle.

Vesicle formation can be prevented by low Ca^{2+} concentration in the bathing solution (no Ca^{2+} added), or by substituting Cl$^-$ for F$^-$ (35 mM) *(24)*. Vesicles can also be ruptured by removing the patch pipet from the bathing solution and then replacing it in the bathing solution again *(20)*.

Recording single Ca^{2+} channel activity presents a number of problems; the channel currents are small and often are observed as short, unresolved openings. Ca^{2+} channels are often concentrated in hot spots, so isolating a single channel can be difficult. Another unfortunate feature is that some types of Ca^{2+} channels run down in a highly variable fashion in isolated membrane patches (outside-out and inside-out patches). Run down is believed to be due to Ca^{2+} channel activity dependence on phosphorylation of the channel *(25)*. Several measures can be taken to optimize the recording conditions. High divalent cation concentration in the patch solution, a very high resistance seal (10 GΩ or more), and using fire-polished and Sylgard-coated patch pipets *(20)* enable even small T-type channel activity to be observed. The channel open times of 1,4-dihydropyridine-sensitive channels can be increased by agonists (light sensitive: (–)–Bay K8644, and (+) –202-791, concentrations between 0.1–5 µM). Caution is required, however, because calcium-channel ligands, including 1,4-dihydropyridines, may alter the interaction between activated G-proteins and Ca^{2+} channels *(26)*. Reducing the tip diameter of the patch pipet reduces the number of channels isolated in a patch, which makes interpretation easier.

3.2. Actions of GTP and GDP Analogs on Whole-Cell Currents

GTP analogs, including GTP-γ-S (guanosine 5'-0-(3-thiotriphosphate) and GMP-PNP (5'-guanylylimidodiphosphate) (Sigma [St. Louis, MO] or Boehringer-Mannheim [Indianapolis, IN]), have proven useful because unlike GTP they are relatively hydrolysis resistant. Including GTP-γ-S (50–500 µM) in the patch pipet solution enables this compound to be introduced into the inside of the cell using the whole-cell recording technique. Inside the cell, GTP-γ-S activates G-proteins essentially irreversibly. This results in a reduction in the amplitude of Ca^{2+} channel current and gives rise to a noninactivating current (over 100 ms) and an apparent slowing in the activation of the current (Fig. 3A) *(27,28)*. GMP-PNP induces responses, but is less effective than GTP-γ-S presumably because the G-proteins have a lower affinity for GMP-PNP than for GTP-γ-S. GTP (1 mM) itself can be used, but is

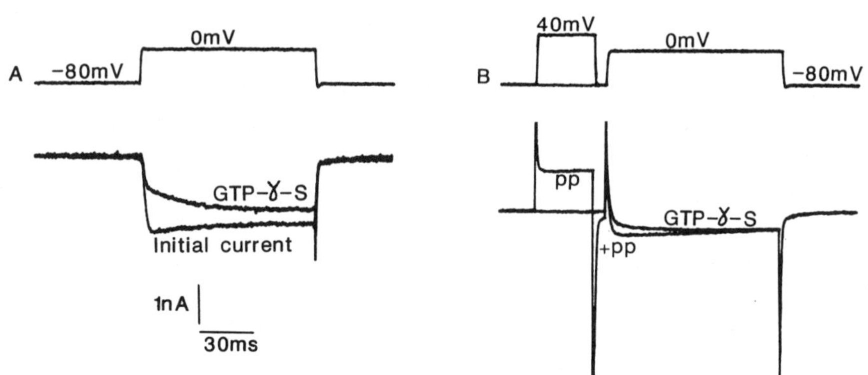

Fig. 3. **A.** Ca^{2+} channel currents recorded using a patch pipet containing GTP-γ-S (500 μ*M*) in the patch solution. Two currents are illustrated, one recorded immediately after entering the whole-cell recording configuration (initial current). The second current was recorded 5 min later after GTP-γ-S had activated G-proteins (GTP-γ-S). Note that both the amplitude and kinetics of the high-threshold Ca^{2+} channel current are altered. **B.** Ca^{2+} channel currents recorded using a patch pipet containing patch solution with GTP-γ-S. Test pulses from V_H −80 mV to 0 mV were given either preceded by a prepulse (+pp) or in the absence of a prepulse (GTP-γ-S). When only the test pulse was given, the current appeared to develop slowly because of G-protein activation by GTP-γ-S and subsequent regulation of the Ca^{2+} channels underlying the current. When the test pulse was preceded by a large prepulse, the Ca^{2+} channel current was activated at a faster rate, suggesting that the effect of G-protein on whole-cell Ca^{2+} channel current kinetics was voltage-dependent. Depolarization is suggested to result in G-protein dissociation from the Ca^{2+} channels, which then become available to open.

much less effective because of rapid hydrolysis and subsequent buildup of GDP. Intracellular GTP-γ-S also enhances the inhibitory actions of the $GABA_B$ agonist (−)−baclofen and other neurotransmitters and neuromodulators that reduce Ca^{2+} channel currents. Bacterial toxins, including cholera toxin, botulinum C_3 toxin, and pertussis toxin are useful tools in determining possible types of G-protein involved in responses. To be effective, the cells must be preincubated in the toxin for some time prior to carrying out an experiment.

Suitable treatments include:

1. Pertussis toxin (PHLS-CAMR Porton Down, Salisbury, Wilts, UK) 500 ng mL^{-1} at 37°C for 3–5 h.
2. Cholera toxin (Sigma) 500 ng mL^{-1} at 37°C for 16 h.

Pretreatment of DRG neurons with pertussis toxin has proved effective at preventing both GTP-γ-S and neurotransmitter-mediated inhibition of Ca^{2+} channel currents. These findings provide evidence that the GTP-γ-S and the neurotransmitter act via a G-protein, rather than produce their effects by direct action on the Ca^{2+} channels. However, the α subunits for G_o, G_i, G_t, and the G-protein associated with the activation of phospholipase C (in some cells only) are all ADP-ribosylated by pertussis toxin. Therefore, positive results using bacterial toxins only reduce the number of possible candidate G-proteins that may mediate a process. Bacterial toxins can also fail to be active even if the appropriate G-protein substrate is present, because the toxin may be unable to bind and internalize, owing to a lack of receptor sites on the cell surface. Negative results with bacterial toxins must also be interpreted with care.

AlF_4^- is another useful tool for studying G-protein-mediated processes, because it can activate G-proteins without prior dissociation of GDP. In a patch solution, AlF_4^- can be formed by including NaF (5–10 mM) and $AlCl_3$ (5–10 μM).

Recently, double-pulse protocols have been used to show that activated G-proteins interact with Ca^{2+} channels in a voltage-dependent manner *(29)*. In the presence of intracellular GTP-γ-S or extracellular GABA, a single test pulse produces a slowly activating Ca^{2+} channel current. If, however, the test pulse is preceded by a large depolarizing prepulse, then the test pulse produces a rapidly activating Ca^{2+} channel current (Fig. 3B). The effect is dependent on the amplitude and duration of the prepulse, as well as the interval between the prepulse and the test pulse. The optimal prepulse protocol is a 20–40 ms prepulse to a very positive potential (V_c +40 to +120 mV) to activate a net outward current and reduce influx of Ca^{2+}, which would tend to inactivate the Ca^{2+} channel current subsequently activated by the test pulse. The interval between the prepulse and test pulse is critical and should be minimized (<10 ms) *(29)*.

Another approach is to reduce G-protein activation by including the GDP analog GDP-β-S (guanosine 5'-O-2 thiodiphosphate) (Sigma, Boehringer-Manneheim) in the patch solution. The concentration of GDP-β-S in the patch solution needs to be high (1–5 mM) to see an effect, because GDP-β-S does not interact irreversibly with G-proteins and has to compete with endogenous GTP. Intracellular GDP-β-S does enhance Ca^{2+} channel currents recorded from DRG neurons *(27)* and also reduces the inhibitory actions of (–)-baclofen and norepinephrine *(27,30)*. These data support the hypothesis that G-proteins couple receptors for neurotransmitters to voltage-gated Ca^{2+} channels.

The experiments outlined in this section suffer from the limitation that control, GTP-γ-S, and GDP-β-S data are derived from different cells. Recently, a technique has been developed to control application of GTP-γ-S and GDP-β-s inside cells, enabling control and experimental data to be recorded from the same cell. This technique involves photorelease of active compounds from "caged" photolabile precursors and is outlined in the next section.

3.3. "Caged" Compounds as Tools to Study G-Protein Ca^{2+} Channel Interactions

When illuminated, "caged" compounds are cleaved or isomerized and release free substrate. This technique can start or stop responses very rapidly (<1 ms) and avoids problems resulting from diffusional delays, contaminants (particularly GTP-γ-S purity), and substrate instability. Different caged compounds, such as GTP-γ-S, GMP-PNP, inositol trisphosphate, and Ca^{2+}, have been used to investigate modulation of ion channels and intracellular processes *(31–34)*. Synthesis of caged compounds (1[2-nitrophenyl]ethyl esters) involves esterification of the substrate (GTP-γ-S, and so on) with 1(2-nitrophenyl) diazoethane *(35)*. Caged compounds are also commercially available from Molecular Probes Inc. (PO Box 22010, 4849 Pitchford Avenue, Eugene, OR 97402) and Calbiochem Corporation (PO Box 12087, San Diego, CA 92112-4180).

Photoactivation of GTP-γ-S or GDP-β-S can be effected by a low-cost, high-energy xenon flash lamp, model JML (G. Rapp, Goethestrasse 66, D6915 Dossenheim, West Germany) *(36)*. This compact system has the advantage that it can be mounted directly above the cells and does not require the use of a liquid light guide, which is a source of energy loss. The JML system has certain limitations:

1. The time resolution is >1 ms;
2. The system cannot be triggered at a high repetition rate (maximum 2 flashes min^{-1}); and
3. The system cannot be used to photorelease active compounds within a localized region of a cell.

If any of the above are required, then a more complex and expensive laser system is needed. The energy output of the lamp can be measured using a Scientech energy meter (Scientech, 5649 Arapaheo Avenue,

Boulder, CO 80303) or in the UK from AG Electro-optics Ltd. (Tarporley, Cheshire, UK). The light pulses should be filtered to prevent UV light damaging the cells. We have previously used a Corning 754 band pass filter (300–380-nm optimum wavelengths for photoactivation of caged compounds). A single light flash (charge 200 V) after filtering yields 63 mJ of energy, focused onto a 20 mm^2 area of a dish containing cultured cells. The efficiency of light-induced hydrolysis can be estimated by placing 50 μL droplets of 100 μM caged compound on cling film in place of the cells. Each 63 mJ flash photoreleases approx 6–8% of the free substrate from caged GTP-γ-S or caged GDP-β-S as measured by reverse-phase HPLC *(31)*. This estimation of photoreleased compound probably represents an overestimation of the amounts of GTP-γ-S or GDP-β-S liberated inside DRG neurons. Caged GTP-γ-S or GDP-β-S can be included in the patch solution at concentrations between 100 μM and 5 mM, and several flashes can be given to determine the dose-dependent action of the photoreleased compound. Experiments should be conducted under conditions of low illumination, and safety glasses should be worn to protect eyes from the light flash. The actions of photoreleased GTP-γ-S are slow ($t_{1/2}$ 1.5 min) compared to the rapid actions of neurotransmitters on Ca^{2+} channel currents. This is believed to be owing to the high affinity of the G-protein for GDP when the associated receptor is unoccupied. It therefore takes some time for GTP-γ-S to displace GDP. In the presence of an agonist for the receptor, photorelease of GTP-γ-S results in much faster (few seconds) action of activated G-protein on Ca^{2+} channel currents. Although data on the kinetics of G-protein action have been difficult to obtain because GDP dissociation is the rate-limiting step, caged GTP-γ-S has been a very useful tool for confirming original observations made with GTP-γ-S and studying the actions of different concentrations of GTP-γ-S. Recent studies have revealed that several different G-proteins may modulate Ca^{2+} channel currents *(37)*, and low (6 μM) and high (20 μM) concentrations of GTP-γ-S have enhancing and inhibiting actions, respectively *(38)*.

Four different types of controls should be carried out with a photorelease study.

1. The effects of a light flash from the xenon flash lamp on current amplitude and kinetics need to be examined.
2. The effect of the unflashed "caged" compound should be studied.
3. The byproduct of photocleavage of caged GTP-γ-S, GDP-β-S, and ATP is 2-nitrosoacetophenone. This compound may also alter the Ca^{2+}

channel current. Intracellular photorelease of ATP and 2-nitroso-acetophenone from their caged precursor should be carried out, and effects, if any, studied. Dithiothreitol (DTT, 10 mM) can be included in the patch solution to react with and eliminate deleterious effects of 2-nitrosoacetophenone. DTT and other thiol agents have recently been shown to alter activation of G-proteins, so caution is necessary *(39)*. Our studies have shown that neither 2-nitrosoacetophenone nor DTT alters Ca^{2+} channel current in DRG neurons *(31)*.

4. Pretreating cells with pertussis toxin should prevent the effects of photoreleased GTP-γ-S if G_o or G_i is involved. This is another good check *(31)*.

3.4. The Use of Purified Subunits of G-Proteins to Elucidate Which Type of G-Protein Modulates Ca^{2+} Channel Currents

Intracellular application of highly purified α-subunits *(40)* can be used to restore G-protein-mediated responses following ADP ribosylation of endogenous G-proteins by pertussis toxin treatment. The inhibitory action of neuropeptide Y on neuronal Ca^{2+} current can be restored by including GTP (1 mM) and α_o (0.01–0.1 μM) in the patch solution *(41)*. However, it appears that α subunits of both G_i and G_o can restore opioid D-Ala and D-Leu enkephalin responses. Although α_o is ten times more potent than G_i *(42)*, in this study, α subunits were infused into differentiated N × G cells for 15 min (concentrations 3.8–0.4 nM). The $\beta\gamma$ complex (44 nM) did not influence the action of α_o and failed to restore opioid responses even at 93 nM following pertussis toxin treatment. Controls for these studies using purified α subunits include examining the effects of G_t (195 nM ineffective) and G-protein denatured by heating at 95°C for 10 min *(42)*.

3.5. Actions of Antibodies Raised Against G-Proteins

Another approach has been developed to clarify which pertussis toxin-sensitive G-protein couples receptors to Ca^{2+} channels. This approach involves using antibodies raised against purified G-proteins *(43)*. Dopamine-induced inhibition of voltage-activated Ca^{2+} currents was greatly attenuated by intracellular injection of anti-α_o antibodies (fivefold dilution to 165 μg/mL in Hepes buffered saline). Antibodies inactivated by heat treatment (65°C for 15 min) did not reduce the dopamine response compared to controls *(43)*. Recently, antiserum against G_i and G_o has been injected into N × G 108-15 cells, which have α-adrenergic receptors coupled to voltage-activated Ca^{2+} channels. The cells injected with antibodies were left for at least 1 h to allow the antibodies to bind to G-proteins. The binding was confirmed by immunohistochemical techniques. Antibodies raised against G_o greatly attenuated noradrenergic inhibition of Ca^{2+} currents, whereas control antiserum and antisera against G_i had little effect *(44)*.

3.6. G-Proteins and Single Ca²⁺ Channels

The study of G-protein regulation of single neuronal Ca^{2+} channels has been limited by problems with Ca^{2+} channel run-down in isolated patches and difficulties in getting active compounds into cells while recording from cell-attached patches. One approach has been to record single Ca^{2+} channel activity from a cell-attached patch and apply to the rest of the cell a neurotransmitter that inhibits whole-cell Ca^{2+} currents via a G-protein. Work from two groups using norepinephrine and the $GABA_B$ receptor agonist baclofen has shown that the application of these ligands externally to the patch pipet failed to modulate Ca^{2+} channel activity in the cell-attached patch. Although these were negative results, they suggested that activated G-proteins have a limited sphere of influence and probably do not mediate their effects on Ca^{2+} channels by formation of readily diffusible second messengers *(45,46)*.

Low-threshold T-type channels do not run-down in isolated membrane patches, but are modulated by G-proteins. Future work using isolated patches may investigate the action of GTP in the hope of observing recovery and ruling out the possibility of run-down or nonspecific effects of GTP-γ-S.

Studies on the direct regulation of cardiac Ca^{2+} channels by G-proteins have been more productive. This work has involved recording Ca^{2+} channel activity from cardiac sarcolemmal vesicles incorporated into planar lipid bilayers and isolated membrane patches from cardiac myocytes. Problems with run-down were reduced by GTP-γ-S (100 μM), and Ca^{2+} channel activity was promoted by purified G_s activated by GTP-γ-S (20–100 pM) as well as the α subunit of G_s (20–100 pM). Neither G_s nor $α_s$ is active when applied to the extracellular face of the membrane, and the effects appear specific to G_s because G_k (100–400 pM) had no effects on Ca^{2+} channel activity *(47)*.

4. Notes

Perhaps the most potentially difficult problems faced by electrophysiologists studying the actions of G-proteins on voltage-activated Ca^{2+} channels are:

1. Identification of the different types of Ca^{2+} channel activity.
2. Identification of which single channel activity underlies particular components of the whole-cell Ca^{2+} current.
3. Identification of the physiological roles of different Ca^{2+} channel types. The functions of Ca^{2+} channels may vary in different cell types, and activity of more than one channel type may contribute to a particular physiological event.

At least three types of Ca^{2+} channel have been identified (N, L, and T) in sensory neurons *(48,49)*. There appears to be considerable overlap in the voltage dependence of activation and inactivation of the different types of Ca^{2+} channel, making different current components difficult to isolate by voltage protocols.

Pharmacological tools offer some assistance in identifying different types of currents. However, overlap in sensitivity to different agents is apparent.

1. 1,4-dihydropyridines enhance or inhibit L-type current *(48)*. Uncontaminated L-type current can be measured using the slowly deactivating tail current (10 ms after the test pulse), recorded during application of a 1,4-dihydropyridine agonist *(50)*. However, 1,4-dihydropyridines have recently been shown to interact with low-threshold T-type current *(51)*.
2. 1-octanol (1–20 μM) *(52)* and amiloride (10–100 μM) *(53)* in some neurons selectively inhibit T-type current.
3. $CdCl_2$ has been found to be a more potent blocker of high threshold currents and $NiCl_2$ a more potent blocker of T-type current, but neither inorganic blocker is selective *(48)*.
4. The venom from the marine snail *Conus geographus*, ω-conotoxin (1–10 μM) irreversibly inhibits N-type current in one study *(50)*, but inhibits both N- and L-type current in another *(54)*. Furthermore, at a concentration of 10 μM, ω-conotoxin also reversibly inhibits T-type currents *(54)*. It has therefore become necessary for biophysical and pharmacological studies at both the whole-cell and single channel levels to be carried out before an experimenter can identify types of Ca^{2+} channel activity with any degree of certainty.

It has been suggested that, in some systems, the presynaptic Ca^{2+} channels may have a completely different pharmacological profile compared to their counterparts on the neuronal cell body *(55)*. However, G-protein regulation of presynaptic Ca^{2+} channels has been identified. Experiments have shown that, following pertussis toxin treatment, adenosine-induced inhibition of glutamate release is prevented *(56)*. The action of adenosine is believed to involve inhibition of voltage-activated Ca^{2+} currents *(9)*.

The other question that needs to be addressed is whether the activated G-protein interacts directly with the Ca^{2+} channel, or whether it activates a series of biochemical events culminating in altered Ca^{2+} channel activity. The answers to this problem remain controversial *(41,57,58)* and may differ in different systems and depend on which channel is regulated by which G-protein. It is worth noting that G-protein-mediated inhibition of Ca^{2+} channel currents and modification of the apparent rate of Ca^{2+} channel current activation may be two distinct phenomena. In N × G108-15 cells, norepinephrine and enkephalin inhibited voltage-activated Ca^{2+} currents without altering the rate of current activation. The inhibitory response is mediated by a pertussis toxin-sensitive G-protein, yet intracellular GTP-γ-S does not inhibit or alter the kinetics of voltage-activated Ca^{2+} currents recorded from these cells *(59)*. What is clear is that G-proteins activated by neurotransmitters can activate a variety of enzymes as well as interact with Ca^{2+} and K^+ channels. Specificity of G-protein action may be achieved by neurotransmitter receptor subtypes, as well as receptor G-protein coupling. To come to grips with this question, more data from studies using lipid bilayers and isolated membrane patches are needed. Future work using purified receptors, G-proteins, and ion channels reconstituted into lipid bilayers may well enable us to determine how G-proteins couple receptors to ion channels.

Although this chapter has centered on the electrophysical techniques used to study G-protein regulation of voltage-activated Ca^{2+} channels, the basic approaches may be used to investigate other systems. Collaboration between biochemists, molecular biologists, and electrophysiologists is proving fruitful, and is likely to grow in importance in the future.

References

1. Augustine, G.T. and Charlton, M. P. (1986) Calcium dependence of presynaptic calcium current and postsynaptic response at the squid giant synapse. *J. Physiol.* **381**, 619–640.
2. Mason, W. T. and Dyball, R. E. J. (1986) Single ion channel activity in peptidergic nerve terminals of the isolated rat neurohypophysis related to stimulation of neural stalk axons. *Brain Res.* **383**, 279–286.

3. Mallart, A. (1984) Presynaptic currents in frog motor endings. *Pflügers Arch.* **400,** 8–13.

4. Lemos, J. R. and Nowycky, M. C. (1987) One type of calcium channel in nerve terminals of the rat neurohypophysis is sensitive to dihydropyridines. *Soc. Neurosci. Abstr.* **13,** 222.6.

5. Lipscombe, D., Madison, D. V., Poenie, M., Reuter, H., Tsien, R. Y., and Tsien, R. W. (1988) Spatial distribution of calcium channels and cytosolic calcium transients in growth cones and cell bodies of sympathetic neurons. *Proc. Natl. Acad. Sci. USA* **85,** 2398–2402.

6. Dunlap, K. and Fischbach, G. D. (1981) Neurotransmitters decrease the calcium conductance activated by depolarization of embryonic chick sensory neurones. *J. Physiol.* **317,** 519–535.

7. Deisz, R. A. and Lux, H. D. (1985) γ-aminobutyric acid-induced depression of calcium currents of chick sensory neurons. *Neurosci. Lett.* **56,** 205–210 .

8. Marchetti, C., Carbone, E., and Lux, H.D. (1986) Effects of dopamine and noradrenaline on Ca channels of cultured sensory and sympathetic neurones of chick. *Pflügers Arch.* **46,** 104–111.

9. Dolphin, A. C., Forda, S. R., and Scott, R. H. (1986) Calcium-dependent currents in cultured rat dorsal root ganglion neurones are inhibited by an adenosine analogue. *J. Physiol.* **373,** 47–61.

10. Gilman, A. G. (1987) Guanine nucleotide binding proteins: transducers of receptor-generated signals. *Ann. Rev. Biochem.* **56,** 615–619.

11. Dolphin, A. C. (1987) Nucleotide binding proteins in signal transduction and disease. *TINS* **10,** 53–57.

12. Neer, E. J. and Clapham, D. E. (1988) Roles of G-protein subunits in transmembrane signalling. *Nature* **333,** 129–134.

13. Scott, R. H., McGuirk, S. M., and Dolphin, A. C. (1988) Modulation of divalent cation-activated chloride ion currents. *Br. J. Pharmacol.* **94,** 653–662.

14. Carbone, E. and Lux, H. D. (1987) Kinetics and selectivity of a low-voltage-activated calcium current in chick and rat sensory neurones. *J. Physiol.* **386,** 547–570.

15. Malecot, C. O., Feindt, P., and Trautwein, W. (1988) Intracellular *N*-methyl-D-glucamine modifies the kinetics and voltage-dependence of the calcium current in guinea pig ventricular heart cells. *Pflügers Arch.* **411,** 235–242.

16. Hess, P., Lansman, J. B., and Tsien, R. W. (1986) Calcium channel selectivity for divalent and monovalent cations. *J. Gen. Physiol.* **88,** 293–319.

17. Kostyuk, P. G., Shuba, Ya. M., and Sauchenko, A. N. (1988) Three types of calcium channels in the membrane of mouse sensory neurons. *Pflügers Arch.* **411,** 661–669.

18. Carbone, E. and Lux, H. D. (1987) Single low voltage-activated calcium channels in chick and rat sensory neurones. *J. Physiol.* **386,** 571–601.

19. Carbone, E. and Lux, H. D. (1984) A low voltage-activated fully inactivated Ca channel in vertebrate sensory neurones. *Nature* **310,** 501–502.

20. Hamill, O. P., Marty, A., Neher, E., Sakmann, B., and Sigworth, F. J. (1981) Improved patch-clamp techniques for high-resolution current recording from cells and cell free membrane patches. *Pflügers Arch.* **391,** 85–100.

21. Pusch, M. and Neher, E. (1988) Rates of diffusional exchange between small cells and a measuring patch pipet. *Pflügers Arch.* **411**, 204–211.

22. Bean, B. P. (1989) Neurotransmitter inhibition of neuronal calcium currents by changes in channel voltage dependence. *Nature* **340**, 153–156.

23. Dolphin, A. C. and Scott, R. H. (1990) Activation of calcium channel currents in rat sensory neurones by large depolarisations: effects of guanine nucleotides and (–)-baclofen. *Eur. J. Neurosci.* **2**, 104–108 .

24. Horn, R. and Patlak, J. B. (1980) Single channel currents from excised patches of muscle membrane. *Proc. Acad. Sci. USA* **77**, 6930–6934.

25. Armstrong, D. and Eckert, R. (1987) Voltage-activated calcium channels that must be phosphorylated to respond to membrane depolarization. *Proc. Natl . Acad. Sci. USA* **84**, 2518–2522.

26. Dolphin, A. C. and Scott, R. H. (1989) Interaction between calcium channel ligands and guanine nucleotides in cultured rat sensory and sympathetic neurones. *J. Physiol.* **413**, 271–288.

27. Dolphin, A. C. and Scott, R. H. (1987) Calcium channel currents and their inhibition by (–)-baclofen in rat sensory neurones: modulation by guanine nucleotide. *J. Physiol.* **386**, 1–17.

28. Wanke, E., Ferroni, A., Malgaroli, A., Ambrosini, A., Pozzan, Y., and Meldolesi, J. (1987) Activation of a muscarinic receptor selectively inhibits a rapidly inactivated Ca^{2+} current in rat sympathetic neurons. *Proc. Natl. Acad. Sci. USA* **84**, 4313–4317.

29. Grassi, F. and Lux, H. D. (1989) Voltage-dependent GABA-induced modulation of calcium in chick sensory neurons. *Neurosci. Lett.* **105**, 113–117.

30. Holz, G. G., Rane, S. G., and Dunlap, K. (1986) GTP-binding proteins mediate transmitter inhibition of voltage-dependent calcium channels. *Nature* **319**, 670–672.

31. Dolphin, A. C., Wootton, J. F., Scott, R. H., and Trentham, D. R. (1988) Photoactivation of intracellular guanosine triphosphate analogues reduces the amplitude and slows the kinetics of voltage activated calcium channel currents in sensory neurones. *Pflügers Arch.* **411**, 628–636.

32. Gray, P. T. A., Ogden, D. G., Trentham, D. R., and Walker, J. W. (1989) Elevation of cytosolic free calcium [Ca^{2+}]; by rapid photolysis of "caged" inositol trisphosphate in single rat parotid acinar cells. *J. Physiol.* **410**, 90P.

33. Gurney, A. M., Tsien, R. Y., and Lester, H. A. (1987) Activation of potassium current by rapid photochemically generated steps of intracellular calcium in rat sympathetic neurons. *Proc. Natl. Acad. Sci. USA* **84**, 3496–3500.

34. Morad, M., Davies, N. W., Kaplan, J. H., and Lux, H. D. (1988) Inactivation and block of calcium channels by photo-released Ca^{2+} in dorsal root ganglion neurons. *Science* **241**, 842–844.

35. Walker, J. W., Reid, G. P., McCray, J. A., and Trentham, D. R. (1988) Photolabile 1-(2-nitrophenyl)ethyl phosphate esters of adenine nucleotide analogues. Synthesis and mechanism of photolysis. *J. Am. Chem. Soc.* **110**, 7170–7177.

36. Rapp, G. and Guth, K. (1988) A low cost high intensity flash device for photolysis experiments. *Pflügers Arch.* **411**, 200–203.

37. Rosenthal, W., Hescheler, J., Hinsch, K-D., Spicher, K., Trautwein, W., and Schultz, G. (1988) Cyclic AMP-independent, dual regulation of voltage-dependent Ca^{2+} currents by LHRH and somatostatin in a pituitary cell line. *EMBO J.* **7**, 1627–1633.

38. Dolphin, A. C., Scott, R. H., and Wootton, J. F. (1989) Photorelease of GTP-γ-S inhibits the low threshold calcium channel current in cultured rat dorsal root ganglion (DRG) neurones. *J. Physiol.* **410**, 16P.

39. Pedersen, S. E. and Ross, E. M. (1985) Functional activation of β-adrenergic receptors by thiols in the presence or absence of agonists. *J. Biol. Chem.* **260**, 14,150–14,157.

40. Sternweis, P. C. and Robishaw, J. D. (1984) Isolation of two proteins with high affinity for guanine nucleotides from membranes of bovine brain. *J. Biol. Chem.* **259**, 13,806–13,813.

41. Ewald, D. A., Sternweis, P. C., and Miller, R. J. (1988) Guanine nucleotide-binding protein G_o-induced coupling of neuropeptide Y receptors to Ca^{2+} channels in sensory neurones. *Proc. Natl. Acad. Sci. USA* **85**, 3633–3637.

42. Hescheler, J., Rosenthal, W., Trautwein, W., and Schultz, G. (1987) The GTP-binding protein G_o regulates neuronal calcium channels. *Nature* **325**, 445–447.

43. Harris-Warrick, R. M., Hammond, C., Paupardin-Tritsch, D., Homburger, V., Rouot, B., Bockaert, J., and Gerschenfeld, H. M. (1988) An α_{40} subunit of a GTP-binding protein immunologically related to G_o mediates a dopamine-induced decrease of Ca^{2+} current in snail neurons. *Neuron* **1**, 27–32.

44. McFadzean, I., Mullaney, I., Brown, D. A., and Milligan, G. (1989) Antibodies to GTP binding protein G_o, antagonize noradrenaline-induced calcium current inhibition in NG108-15 hybrid cells. *Neuron* **3**, 177–182.

45. Forscher, P., Oxford, G. S., and Schultz, D. (1986) Noradrenaline modulates calcium channels through tight receptor-channel coupling. *J. Physiol.* **379**, 131–144.

46. Green, K. A. and Cottrell, G. A. (1988) Actions of baclofen on components of the Ca-current in rat and mouse DRG neurones in culture. *Br. J. Pharmacol.* **94**, 235–245.

47. Yatani, A., Codina, J., Imoto, Y., Reeves, J. P., Birnbaumer, L., and Brown, A. M. (1987) A G-protein directly regulates mammalian cardiac calcium channels. *Science* **238**, 1288–1292.

48. Fox, A. P., Nowycky, M. C., and Tsien, R. W. (1987) Kinetic and pharmacological properties distinguish three types of calcium currents in chick sensory neurones. *J. Physiol.* **394**, 149–172.

49. Fox, A. P., Nowycky, M. C., and Tsien, R. W. (1987) Single-channel recordings of three types of calcium channels in chick sensory neurones. *J. Physiol.* **394**, 173–200.

50. Plummer, M. R., Logothetis, D. E., and Hess, P. (1989) Elementary properties and pharmacological sensitivities of calcium channels in mammalian peripheral neurons. *Neuron* **2**, 1453–1463.

51. Akaike, N., Kostyuk, P. G., and Osipchuk, Y. V. (1989) Dihydropyridine-sensitive low-threshold calcium channels in isolated rat hypothalamic neurones. *J. Physiol.* **412,** 181–195.
52. Llinas, R. (1988) The intrinsic electrophysiological properties of mammalian neurons: insights into central nervous system function. *Science* **242,** 1654–1664.
53. Tang, C-M., Presser, F., and Morad, M. (1988) Amiloride selectively blocks the low threshold (T) calcium channel. *Science* **240,** 213–215 .
54. McCleskey, E. W., Fox, A. P., Feldman, D. H., Cruz, L. J., Olivera, B. M., Tsien, R. W., and Yoshikami, D. (1987) ω-conotoxin: direct and persistent blockade of specific types of calcium channels in neurons but not muscle. *Proc. Natl. Acad. Sci. USA* **84,** 4327–4331.
55. Seabrook, G. R. and Adams, D. J. (1989) Inhibition of neurally-evoked transmitter release by calcium channel antagonists in rat parasympathetic ganglia. *Br. J. Pharmacol.* **97,** 1125–1136.
56. Dolphin, A. C. and Prestwich, S. A. (1985) Pertussis toxin reverses adenosine inhibition of neuronal glutamate release. *Nature* **316,** 148–150.
57. Dolphin, A. C., McGuirk, S. M., and Scott, R. H. (1989) An investigation into the mechanism of inhibition of calcium channel currents in cultured rat sensory neurones by guanine nucleotide analogues and (–)-baclofen. *Br. J. Pharmacol.* **97,** 263–273.
58. Rane, S. A. and Dunlap, K. (1986) Kinase C activator 1,2-oleoylacetylglycerol attenuates voltage-dependent calcium current in sensory neurones. *Proc. Natl. Acad. Sci. USA* 184–188.
59. McFadzean, I. and Docherty, R. J. (1989) Noradrenaline- and enkephalin-induced inhibition of voltage-sensitive calcium currents in NG 108-15 hybrid cells. *Eur. J. Neurosci.* **1,** 141–147.

Index

A

Acetylcholine receptor,
    nicotinic, 308
    noncompetitive antagonists, 312, 313
    pharmacology, 312
    structure, 308–311
Adenosine receptor, 261, 262
    purification, 266
Affinity crosslinking, *see* Receptors,
    affinity crosslinking
Alkaline phosphatase,
    endogenous activity, 173
    reaction product, 173
Alkaline phosphatase linked probe, 168
    hybridization, 171, 172
    preparation, 170, 172
    structure, 171
Amino acid sequencing, automatic
    microsequencing, 318
Annealing temperature, *see*
    Oligonucleotide, melting
    temperature
Antibody production, 97–99, 105,106
Antigenic proteins, 111
Antisense DNA or RNA, 207
Antisera,
    isoform specific, 95, 96
    isoform specificity, 106–108
Autophosphorylation, *see* Receptor
    autophosphorylation
Autoradiography, identification and
    localization of hybridization
    signal, 163, 164

B

Bacterial toxins, 354
Binding studies, 250–252, 265–267
    assay of membrane bound
        receptors, 235, 265–267
    assay of solubilized receptor,
        236, 267
    dihydropyridine binding site,
        280–283
    membrane preparation, 234, 235
    phenylalkamine binding site,
        281, 283

C

"Caged" compounds, 356–358, *see
    also* Photoactivation
Calcium channel blockers (CaCB),
    273
    binding site, 273, 274
        irreversible binding, 282, 283
        reversible binding, 280–282
Calcium channels, 347–349
    anti G-protein antibodies, 358, 359
    blockers, 360
    cardiac, 359
    G-protein regulation, 349
    modulation by G-protein
        subunits, 358
    run down in isolated patches, 353
    single channel currents, 359
    subtypes, 273, 360
    T-type, 359

voltage-dependent, 273
whole cell currents, 354
Calcium release channel complex,
        287, 288
    characterization, 293–295
    isolation and purification, 291–293
    reconstitution, 295–297
    ryanodine binding, 294, 295
cDNA, *see* Complementary
        deoxyribonucleic acid
Cell culture, 231–233, 249, 250
    bacterial culture plate
            preparation, 83
        dissociation of cells, 231–233
        glial cells, 231–233
        neurons, 231–233
        precoating culture dishes, 231
Cell-specific complementary
        deoxyribonucleic acid, 79
Cell-specific gene expression, 115,
        155, 167, 181
Chemical amplification, 168, 169
Chromatography,
    affinity, 261
        gel regeneration, 269
    DEAE ion-exchange, 278, 279
    reversed-phase (HPLC), 318
    WGA-lectin, 278
Chromosome mapping, *see*
        Restriction mapping
Clone characterization, 72
Collagenase purity, 223
Complementary deoxyribonucleic
        acid (cDNA),
    ^{32}P labeling of probes, 88, 89
    cloning, 60, 69–71
    first strand, 73
    libraries, 79, 85
Complementary ribonucleic acid, 156
    probe hydrolysis, 162, 164
    probe synthesis, 160–162
    stability, 207
    synthesis, 207

cRNA, *see* Complementary
        ribonucleic acid
Deoxyribonucleic acid (DNA),
    degradation, 22
    genomic, 73
        from isolated nuclei, 18, 19
        hot phenol extraction, 19, 20
        invertebrate, 15
    human cell, 8, 9
    linearized, 216
    methylation, 125
    microinjection, 193–196
    mitochondrial (mtDNA), 25
        deletions, 25, 41, 43, 48
    mutations, 25
    plasmid vectors, 157
    quality and yield, 20–22
    template, 216
    unsheared, preparation of, 7
    yeast chromosome, 7, 8

D

Deoxyribonucleic acid (DNA)
        sequencing, 34, 35
    dideoxy method, 72, 73
    fluorescence-based, 31–33, 36, 47
Desalting of samples, 267, 268
Differential colony hybridization, 80–82
Dihydropyridines, 273, 360
DNA, *see* Deoxyribonucleic acid
cDNA, *see* Complementary
        deoxyribonucleic acid
DNA sequencing, *see* Deoxyribo-
        nucleic acid sequencing
DNA-protein interactions, 139
DNase I purity, 222

E

Electrophoresis, *see also* Gel
        electrophoresis,
    thin-layer, 247
End-labeling, 145

DNA, 122–125
primer, 131
Enhanced chemiluminescence
(ECL), 37, 38
Enzyme histochemistry
β-galactosidase, 197, 198

F

Fluorometric enzyme assay, 198, 199
Fusion genes, 182, 185
Fusion proteins, 96, 109–112
production, 102–104
structure, 100
immunoreactivity, 104, 105

G

G-proteins, 348, 349
antibodies, 358, 359
different types, 349
direct interaction with calcium
channels, 361
purified subunits, 358
subunits, 348, 349
G-protein activation,
aluminium fluoride, 355
bacterial toxins, 354
GTP analogs, 353
reduction by GDP analogs, 355, 356
Gel electrophoresis, 38, 39
agarose, 35
high resolution denaturing, 148, 149
polyacrylamide, 116, 124, 127,
140, 293, 294, 299, 300
adenosine receptors, 270
running conditions, 10
sequencing gel preparation, 128–130
Gel retardation assay, 140, 145–151
Gel-blotting, 11, 12
Gene expression,
quantitation, 198, 199
Genome size, 15, 16, 181

Glial cell culture, *see* Cell culture
GTP, *see* Guanosine triphosphate
Guanine nucleotide binding
proteins, *see* G-proteins
Guanosine diphosphate (GDP), 348
Guanosine triphosphate(GTP), 348
analogs, 353, 354

H

Hybrid arrest, 207, 208, 218, 219
Hybridization histochemistry, *see In
situ* hybridization

I

Immunoblotting, 104, 300, 301
Immunocytochemistry, *see*
Immunohistochemistry
Immunofluorescence,
dual labeling, 198
Immunohistochemistry, 156
β-galactosidase, 198
stability, 201
dual localization, 198
SV40 T antigen localization,
199, 200
In situ hybridization, 155, 156, 167
control experiments, 174–176
frozen sections, 163
paraffin sections, 162, 163
tissue preparation, 170, 171
Insulin and insulin-like growth
factor receptors (IGF-I and
IGF-II), 227
Ion channels, 273, 307
gating, 326
ligand-gated, 307, 308
modeling, 318, 319
modification, functional
consequences, 341, 342
voltage-gated, 307, 325–342, 347
Isoforms, 94

L

Ligand-gated ion channels, *see* Ion
     channels, ligand-gated
Lipid bilayers, 295–297, 333, 359

M

Macroscopic currents, 327, 329
     Hodgkin-Huxley scheme, 335
     kinetic behavior, 334, 335
Maxam-Gilbert "G>A" methylation,
     see Deoxyribonucleic acid,
     methylation
Melting temperature, *see*
     Oligonucleotide, melting
     temperature
Membrane binding studies, *see*
     Binding studies
Membrane proteins, 206
Messenger ribonucleic acid (mRNA),
     block of expression, *see* hybrid
          arrest
     cell-specific, 155
     contamination of, 206, 207
     localization, 167
     poly(A), 206, 207, 214–216
Methylated G cleavage, 148
Methylation interference
          footprinting, 140, 141, 146–149,
          152, 153
Microinjection,
     ribonucleic acid, 205, 221
Microinjection pipets, 194, 200
Microsomal membrane preparation, 277
Minigenes, 183, 184
mRNA, *see* Messenger ribonucleic acid
Multigene families, 57, 62, 93

N

Na/K-ATPase, 93–96
     isoform expression, 108–110
     isoforms, 94–96
     subunits, 94
Neuronal cell culture, *see* Cell culture
Nicotinic acetylcholine receptor, *see*
     Acetylcholine receptor
Northern blotting, 183
Nuclear extract preparation, 144, 145,
     149

O

Oligo-dT cellulose columns, 214–216
Oligodeoxynucleotides, *see*
     Oligonucleotides
Oligonucleotides,
     alkaline phosphatase-labeled, 168
     best-guess, 56, 58–61
     concentration determination, 51
     consensus, 62, 63
     degenerate, 56–59
          inosine substitution, 64
     design, 55
     end-labeling, 131
     enzyme labeling, 170
     labeling, 168, 169
     melting temperature (T_m), 63, 64,
          74, 75, 126, 131, 132
     partial methylation, 146, 147
Oocytes, 205, 206, 330, 342
     disadvantages, 206
     incubation, 222
     patch clamping, 222
     preparation, 220, 221
     RNA injection, 221

P

Patch clamp, 332, 333, 350–353
     inside-out patch, 352, 353
     outside-out patch, 352
     single channel, 301, 302, 332, 333,
          350
     whole cell, 331, 349–352

bacterial toxins, 354
GTP analogs, 353
Phenol preparation, 85
Phosphoamino acid analysis, 245–
248, 256, 257
Phosphopeptide mapping, 248
Phosphorylation,
autophosphorylation, 239–241
cell-free, 239–242, 254, 255
exogenous substrates, 241–243
in intact cells, 242-245, 255, 256
serine/threonine kinases, 245
tyrosine kinase, 239
WGA-purified preparations, 243,
244
whole cell extraction, 242, 243
Photoactivation of "caged"
compounds,
control experiments, 357, 358
efficiency, 357
laser, 356, 357
xenon flash lamp, 356
Photoaffinity labeling, 311, 312
apparatus, 314, 315
impure peptides, 319, 320
membrane preparation, 316
methods, 314–316
Photoreceptor specific complementary DNA, 81, 82
Plasmid construction, 99–102
Plasmid deoxyribonucleic acid
denaturation and immobilization,
86
hybridization and washing, 86,
87
preparation, 192, 193
Poly(A) RNA, *see* Messenger
ribonucleic acid, poly(A)
Polyacrylamide gel electrophoresis
(PAGE), *see* Gel
electrophoresis
Polymerase chain reaction (PCR),
26, 56, 57, 67

asymmetric, 34
cDNA amplification, 69
contamination of, 74
control reactions, 74
primer-shift method, 40–42, 47–50
primers, 27–29, 56, 60–62
S1 method, 43–46, 50, 51
symmetric, 32–34
Potassium channels, 325
delayed rectifier, 328, 335, 336
Hodgkin-Huxley, 339–341
Primer concentration, 74
Primer extension, 116, 117
Primer extension assay, 130–133,
135–137
Primers,
concentration, 51
nomenclature, 51
Probes, *see* Oligonucleotides
Protein determination,
Amido-schwarz method, 268, 269
Protein-DNA binding factors, 182
Proteinase K digestion, 164
Pulsed-field gel electrophoresis
(PFGE), 1, 2
apparatus, 4–6

R

Receptors, 206, 308
affinity crosslinking, 237, 238
antagonists, 312, 313
autophosphorylation, 239–241
glycoprotein, enzymatic
modification, 238, 239
labeling, 277, 278
labeling of different receptor
states, 318
radiolabeling with TPMP+,
316, 317
structural studies, 236, 237, 253,
254
subunits, 273, 274, 283

Receptor binding assay, *see* Binding
 studies
Receptor purification, 235, 252,
 253, 261, 264, 265, 276–280,
 288, 289, 291–293, 317
 contaminants, 283
Receptor solubilization, 235, 252,
 253, 277, 278
Receptor superfamily, 62, 67
Receptor-rich membranes,
 preparation of, 316
Replica plating, 86
Reporter genes, 183–185
Restriction endonucleases, *see*
 Restriction enzymes
Restriction enzyme digestion, 9, 10
Restriction enzymes, 3, 22, 23, 101, 102
Restriction mapping, 2
Retina-specific clones, 85
Reverse transcriptase, 130
 problems with, 135–137
RNA, *see* Ribonucleic acid
cRNA, *see* Complementary
 ribonucleic acid
Ribonucleic acid (RNA),
 capped, 207
 concentration determination, 213
Ribonucleic acid extraction, 133, 134
 Chomczynski-Sacchi method, 214
 guanidinium thiocyanate-phenol-
  chloroform method, 120–122
 guanine hydrochloride method, 212
 LiCl-urea method, 211
 phenol-chloroform method,
  212–214
Ribonucleic acid transcription, 115
 in vitro, 161, 207
  capped RNA, 216–218
 initiation site, 115, 116
RNA-DNA hybridization, 126
RNase contamination, 133
RNases, 206, 207
RNase H, 207, 208

Ryanodine, 287
Ryanodine binding, *see* Calcium
 release channel complex,
 ryanodine binding
Ryanodine receptor, *see* Calcium
 release channel complex

S

S1 nuclease assay, 116, 117, 122–130,
 134, 135
Saccharomyces cerevisiae, 7
Sanger sequencing, *see*
 Deoxyribonucleic acid
 sequencing, dideoxy method
Single channel currents, 327, 359
 first latencies, 337–339
 kinetic behavior, 335–339
 open and closed times, 337–339
Single channel recording, *see* Patch
 clamp, single channel
Skeletal muscle
 sarcoplasmic reticulum, 287
  preparation of vesicles, 290, 291
 transverse tubules, 273
Sodium-potassium ATPase, *see* Na/
 K-ATPase
Southern blotting, 11, 22
 nonisotopic, 37
 peroxidase labelling of probes, 39,40
 primer pairs, 39
Stringency, 59
Sucrose buffer osmolarity, 21
Sucrose density gradients, 279, 280

T

T4 kinase end-labeling, 145, 146
Taq polymerase, 67
Thin-layer electrophoresis, 247
Tissue processing
 frozen unfixed tissue, 160
 parrafin embedding of
  fixed tissue, 159, 160

T_m, *see* Melting temperature
Torpedo californica, 308
Trans-acting factors, 115, 139, 140
Transfected cells, 181
Transgenic mice, 182–185
    identification of, 196, 197
    production of, 190–197
Triphenylmethylphosphonium
    (TPMP$^+$), 312
Tyrosine kinase activity, *see*
    Phosphorylation, tyrosine kinase

V

Voltage clamp, 326
    single-microelectrode, 332
    two-microelectrode, 330, 331

Voltage-gated ion channels, *see* Ion
    channels, voltage-gated

W

Western blotting, 108
Whole cell currents, *see* Patch
    clamp, whole cell *and*
    Macroscopic currents
Whole cell recording, *see* Patch
    clamp, whole cell

X

Xenopus laevis, 205
    maintenance and surgery, 219, 220